Luminescent Thermometers

Luminescent Thermometers deals with all aspects of the subject from principles of methods to their applications in different areas. This book familiarizes the readers with the fundamentals of luminescence thermometry, materials used for the development of different luminescence thermometers, viz. metal-organic frameworks (MOFs) including lanthanide-doped MOFs referred as LOFs, quantum dots (QDs), rare earth-doped phosphors, and upconversion phosphors. Further, some advanced and next generation approaches for luminescent thermometers such as carbon-based materials, nanocomposites, double perovskites, and garnet systems are assimilated. The applications of luminescent thermometers in temperature sensing of biological cells and tumors, thermal imaging of biological cells, flexible temperature sensors, health monitoring with wearable thermometers, and environmental monitoring are the key features of this volume. It is a valuable contribution to the literature for material scientists and engineers in academia and R&D as well as researchers working in biology and environment science.

Key Features:

- Covers entire range of luminescent thermometers from fundamentals to applications;
- Describes state-of-the-art materials and next generation approaches for luminescence-based nanothermometery;
- Discusses the high-end utilities of luminescent thermometers in different aspects of human life.

Luminescent Thermometers

Fundamentals, Materials and Applications

Edited by

Pragati Kumar, Nupur Saxena,
Sanjay J. Dhoble and Hendrik C. Swart

CRC Press
Taylor & Francis Group
Boca Raton London New York

CRC Press is an imprint of the
Taylor & Francis Group, an **informa** business

Front cover image: jijomathaidesigners/Shutterstock

First edition published 2025
by CRC Press
2385 NW Executive Center Drive, Suite 320, Boca Raton FL 33431

and by CRC Press
4 Park Square, Milton Park, Abingdon, Oxon, OX14 4RN

CRC Press is an imprint of Taylor & Francis Group, LLC

© 2025 selection and editorial matter, Pragati Kumar, Nupur Saxena, Sanjay J. Dhoble and Hendrik C. Swart; individual chapters, the contributors

ISBN: 978-1-032-62808-0 (hbk)
ISBN: 978-1-032-66150-6 (pbk)
ISBN: 978-1-032-66153-7 (ebk)

DOI: 10.1201/9781032661537

Typeset in Sabon
by codeMantra

Contents

About the editors

Pragati Kumar has been a senior Assistant Professor in the Department of Nanoscience and Materials, Central University of Jammu since 2016. Dr. Kumar has more than 15 years of research and teaching experience. His research includes nanostructures, thin films, composites, 2D materials, ion beam techniques, optoelectronic devices, sensors, and photocatalysis. Dr. Kumar has published 37 research papers, 14 chapters, and edited two books with international publishers.

Nupur Saxena is a DST-Women Scientist in the Department of Physics, IIT Jammu. Previously, she has been awarded a number of PDFs and other fellowships including CSIR-Pool Scientist, UGC-DSK, IUAC-RA, CSIR-SRF, IUAC-JRF, etc. She has more than 18 years of research experience in the synthesis of nanostructures, thin films, composites, 2D materials, ion beam techniques, optoelectronic devices, sensors, and photocatalysis. Dr. Saxena has published 30 research papers and 9 chapters as well as presenting more than 40 papers in conferences. She is a lifetime member of many scientific societies and reviewed for almost all international publishers.

Sanjay J. Dhoble is a senior Professor in the Department of Physics, RTM Nagpur, University Nagpur. He has about 31 years of teaching and research experience. He has guided 76 students, published 62 patents (South Korean, South African, Australian, and Indian patents), 921 articles, edited/authored 30 international books, and delivered 177 plenary/keynote/invited talks and guest lectures. His research is well acknowledged globally and cited more than 15,000 times. He has secured a place among the top 2% of scientists in the world in the last three years.

Hendrik C. Swart is an internationally acclaimed researcher and currently a senior Professor in the Department of Physics at the University of the Free State. He brought luminescence materials to South Africa in the beginning of 1996 since then he has led research in the degradation of phosphors for field emission displays as well as developing materials for nano solid-state lighting. He has more than 900 publications, edited/authored 70 books/ book chapters with more than 24,486 citations, H-index of 68, and i10 index of 587. He is an editor in critical reviews in Solid State and Materials Sciences.

Contributors

B. Amrithakrishnan
Department of Physics
University of Kerala
Thiruvananthapuram, Kerala,
India

P. Antonyananiya
Department of Physics
KGiSL Institute of Technology
Coimbatore, Tamil Nadu, India

Nikhilesh Bajaj
Physics Department
Toshniwal Arts, Commerce and
Science College
Shengaon, India

Manju Bala
Department of Physics
Hindu College, University of Delhi
Delhi, India

Vivek Chopra
Department of Botany
Hindu College, University of Delhi
Delhi, India

S.J. Dhoble
Department of Physics, RTM
Nagpur University
Nagpur, India

Madeshwari Ezhilan
Department of Biomedical
Engineering
Vel Tech Rangarajan
Dr. Sagunthala R & D Institute
of Science and Technology
Tamil Nadu, India

Abhilasha Jain
Department of Physics
R.T.M. Nagpur University
Nagpur, India

I. N. Jawahar
Department of Physics
University of Kerala
Thiruvananthapuram, Kerala,
India

Daljit Kaur
Department of Physics
DAV University
Jalandhar, Punjab, India

Gurpreet Kaur
Department of Physics
Indian Institute of Technology
Roorkee
Roorkee, Uttarakhand, India

x Contributors

G. Kiruthiga
Department of Physics
Avinashilingam Institute for Home
Science and Higher Education
for Women
Coimbatore, India

Richa Kothari
Department of Environmental
Sciences
Central University of Jammu
Bagla, Rahya-Suchani, India

Arockia Jayalatha Kulandaisamy
School of Electrical & Electronics
Engineering
SASTRA Deemed University
Thanjavur, Tamil Nadu, India

Deepak Kumar
Department of Physics
Punjab University
Chandigarh, India

Gulshan Kumar
University School of Basic and
Applied Sciences
Guru Gobind Singh Indraprastha
University
New Delhi, India

T. Pradeepkumar
Department of Mathematics
KGiSL Institute of Technology
Coimbatore, India

Sariga C. Lal
Department of Physics
University of Kerala
Kerala, India

Saurav Kumar Maity
University School of Basic and
Applied Sciences

Guru Gobind Singh Indraprastha
University
New Delhi, India

Girdega Muruganandam
School of Chemical & Biotechnology
SASTRA Deemed University
Thanjavur, Tamil Nadu, India

Noel Nesakumar
School of Chemical &
Biotechnology
SASTRA Deemed University
Thanjavur, Tamil Nadu, India

Rudrashish Panda
School of Physical Sciences
National Institute of Science
Education and Research (NISER)
Bhubaneswar, An OCC of Homi
Bhabha National Institute
Jatni, Odisha, India

Yatish R. Parauha
Department of Physics
Shri Ramdeobaba College of
Engineering and Management
Nagpur, India

Ashvini Pusdekar
Department of Physics
Anand Niketan College
Warora, India

Vaishali Raikwar
Physics Department
Ramniranjan Jhunjhunwala College
Mumbai, India

Shubham Raina
Department of Environmental
Sciences
Central University of Jammu
Bagla, Rahya-Suchani, India

Alka Rani
Nanomaterials and Sensors
  Research Laboratory,
  Department of Physics
Babasaheb Bhimrao Ambedkar
  University
Lucknow, U.P., India

John Bosco Balaguru Rayappan
School of Electrical & Electronics
  Engineering
SASTRA Deemed University
Thanjavur, Tamil Nadu, India

S. Satyanarayana Reddy
Department of Physics
RV Institute of Technology and
  Management
Bengaluru, India

Pratap Kumar Sahoo
School of Physical Sciences, National
  Institute of Science Education and
  Research (NISER), Bhubaneswar,
  An OCC of Homi Bhabha
  National Institute
Jatni, Odisha, India

Nupur Saxena
Department of Physics, Indian
  Institute of Technology
Nagrota, Jagti, India

Vinita Sharma
Department of Environment
  Science & Engineering
Guru Jambheshwar University of
  Science and Technology
Haryana, India

Soorya Srinivasan
PG & Research Department of
  Physics
T.B.M.L College
Poraiyar, Tamil Nadu, India

Subodh Ganesanpotti
Department of Physics
University of Kerala
Thiruvananthapuram, Kerala,
  India

Uplabdhi Tyagi
University School of Chemical
  Technology
Guru Gobind Singh Indraprastha
  University
New Delhi, India

Nilesh Ugemuge
Department of Physics
Anand Niketan College
Anandwan, Warora, India

Arpit Verma
Nanomaterials and Sensors
  Research Laboratory,
  Department of Physics
Babasaheb Bhimrao Ambedkar
  University
Lucknow, U.P., India

Manoj Verma
Department of Physics
Hindu College, University of Delhi
Delhi, India

Bal Chandra Yadav
Nanomaterials and Sensors
  Research Laboratory,
  Department of Physics
Babasaheb Bhimrao Ambedkar
  University
Lucknow, U.P., India

Renuka Bokolia
Department of Applied Physics,
Delhi Technological University,
  Delhi, India

Chapter 1

The predominant role and the need of thermometers

Dr. G. Kiruthiga and Nupur Saxena

1.1 INTRODUCTION

In today's industrial, technological, scientific research, and medical domain, temperature measurement encircles a broad variety of applications and needs [1]. To handle these demands, industries have developed a huge number of devices and sensors. This chapter provides an opportunity to understand the need and predominant role of thermometers. For most scientific researchers, medical professionals, and mechanical engineers, temperature is a very essential and largely measured variable. For controlling or monitoring temperature, various processes have been followed in various industries [2]. This may vary from monitoring the water temperature of an engine or device loaded, which is simple, to something more complicated, such as the temperature of a weld in a laser welding application [1,2]. From a power generating station, the measurement of temperature of chimney, smoke stack gas, blast furnace, or the exhaust gas of a rocket may need to be monitored, which is more challenging. The temperatures of fluids in processes or process support applications are more common. From a power generating station, it's compulsory to monitor the temperature of solid objects such as metal plates and bearings [1,3]. This extraordinary device, which is a universal instrument, is used to measure temperature. The name "thermometer" derives from the following two words: can be followed, i.e., Thermos and Metron. Thermos represents "hot" and Metron represents "measure". Heat and temperature are two different words that very often confuse the people. The first one is a form of energy and the unit of this parameter is Joules. The second one is the measure of that heat. People may raise questions like, "How can the hotness of an object is expressed? Which one is the basis or measure for that hotness? For all these queries, the answer is "Temperature". More than the thermometer, some other essential devices for monitoring heat incorporate the following: thermistors, thermostats, thermocouples, and resistance temperature detector (RTD). It can measure the temperature of solids like food, liquids like water, and gases like air. The three most general units of temperature measurements are Fahrenheit, Celsius, and Kelvin. The Celsius scale is part

DOI: 10.1201/9781032661537-1

of the metric system. Thermometer is a Greek word, and it is the fusion of two words. Thermo means warm and Meter means to measure. The first thermometer was invented in the 1600s [3,4]. The correct year of invention is not known, but the chronicler estimates the year of invention is 1593. Behind a thermometer, the basic principle is the contraction by cold and the expansion of air by heat. Based on different principles, the device measures temperature gradient. The two basic important elements are temperature sensor (e.g., some physical change occurs with temperature in the bulb on a mercury thermometer) and a tool for converting this physical change into measurement value (e.g., on a mercury thermometer, the scale is inserted). Usually, thermometers are classified into two separate groups according to the knowledge level underlying thermodynamic quantities and laws. The classifications are (i) primary thermometers and (ii) secondary thermometers [5]. By using a primary thermometer without any unknown quantities, temperature property is measured. The most widely thermometers are secondary thermometers due to their convenient use. The secondary thermometers are highly sensitive ones than the primary thermometers. To allow direct measurement of temperature, knowledge of the measured property is not adequate for secondary devices. At a number of fixed temperatures, or at least at one temperature, secondary thermometers are calibrated against a primary thermometer. In general, thermometers are broadly classified into two categories: (iI) laboratory thermometer and (ii) clinical thermometer. Usually, the primary one is not utilized for measuring the temperature of the body. It is used to measure the chemical's temperature in laboratories. For Temperature measurement, the following units are used: Celsius (°C), Kelvin (°K), and Fahrenheit (°F).

1.2 TYPES OF THERMOMETER

On our daily basis, the following thermometers are the different types that we are using in different fields clinics, laboratories, industries, etc.

- Galileo thermometer
- Beckmann differential thermometer
- Clinical thermometer/medical thermometer
- Mercury-in-glass thermometer
- Infrared ear thermometer
- Bimetal mechanical thermometer
- Constant volume thermometer
- Platinum resistance thermometer
- Constant-pressure gas thermometer
- Thermocouple thermometer
- Alcohol thermometer

- Digital thermometer
- Laboratory thermometer
- Pyrometer thermometer

1.2.1 Need of Galileo thermometer

This thermometer was not invented by Italian physicist Galileo (1564–1642), despite its name, but it was invented by scientific society called Accademia del Cimento of Florentine after two decades of his death. This scientific society was founded by Grand Duke Ferdinando II de' Medici (former student of Galileo) and his brother Prince Leopoldo in 1657. All over the world, for decoration purposes, Galileo thermometers are now built and sold [6]. In the early 1600s, Galileo did invent a thermoscope (a thermometer without a numeric scale) as illustrated in Figure 1.1a. It did not employ floaters, but it was very rudimentary and was influenced by atmospheric pressure. It is Galileo thermometer constructed of a sealed glass cylinder containing purified water, and the density of a series of objects is such that they fall or rise if a change in temperature occurs. On the principle of buoyancy, Galileo thermometers work, and they determine whether objects sink or float. If there is any change in temperature, the glass balls in the thermometer either float to the top (a fall in temperature) or sink to the bottom (a rise in temperature). In Galileo thermometer, mercury in a glass bulb is not used. It's a conventional instrument.

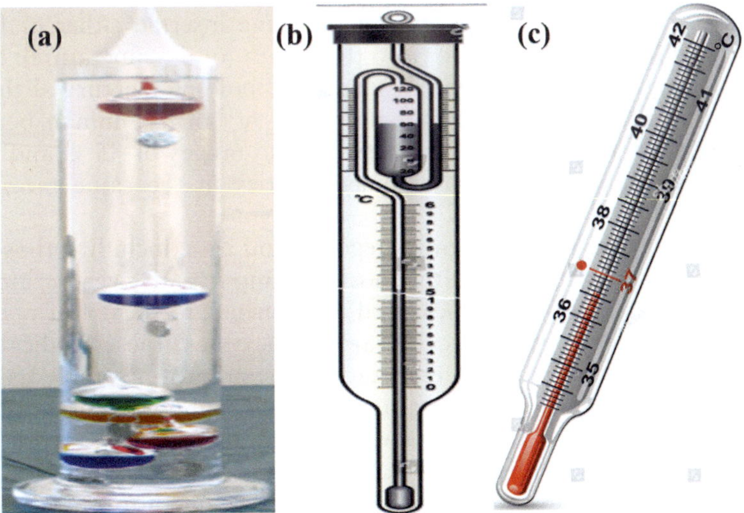

Figure 1.1 (a) Galilean thermometer. (Adapted with permission from ref. [6] © 2012 ACS.) (b) Beckmann differential thermometer, and (c) medical thermometer. ((b and c) Adapted from Shutterstock.)

1.2.2 Need of Beckmann differential thermometer

The small difference in temperature can be measured by a Beckmann thermometer, and the measured values are not absolute values of temperature. A German Scientist, namely Emst Otto Beckmann (1853–1923), invented this device for measuring colligative properties in 1905. The other name of this device is known as differential thermometer. For measuring melting points, boiling points, calorimetry, Beckmann thermometers can be used. This kind of Beckmann differential thermometer was made in Germany in 1940 and is in the collection of the Robert A. Paselk Scientific Instrument Museum at state University of Humboldt. The overall length of the thermometer is 57.5 cm, and the graduations of thermometers vary from 0.1°C to 0.6°C by 1/100° on the main scale, and from −9°C to 144°C by single degrees on the upper setting scale [7]. In contrast to the Galileo thermometer, this type is constructed with mercury in glass thermometer used in calorimetry. It offers an accuracy of up to ±0.001 K, but its range is about only 5 K [7]. Even though this kind of thermometer is not able to measure absolute temperature, this device is used to measure the change in temperature, i.e., in the freezing point of the solution and solvent, and it can perfectly measure the little changes in the temperature.

1.2.3 Need of clinical/medical thermometer

This type of thermometer is used for monitoring the body temperature of human beings, with the thermometer tip being inserted either under the armpit (axillary temperature) or into the mouth (oral temperature). It consists of a lengthy glass tube with a small bulb holding mercury at the bottom of thermometer. The body temperature of normal human beings is about 37°C, and it can oscillate between the ranges of 35°C and 42°C. Consequently, the medical thermometer varies in the same range, i.e., between 35°C and 42°C.

When caring for patients of all categories and ages, including those with health concerns, an accurate diagnosis is, importantly, very critical. In order to deliver precision in measured assessments, medical thermometers facilitate healthcare providers. In various locations, these medical thermometers can measure body temperature for monitoring stable temperatures inclusive of rectal, axillary (armpit), oral, temporal (forehead), and tympanic (ear) regions. Illness or fever in a patient is indicated by the reading of temperature of the body exceeding 38°C commonly. For monitoring fever and assessing health conditions, precise temperature measurement is critical, as is treatment guidance. For a specific type of medical thermometer, it is important to follow the instructions of the manufacturers and also the proper treatment guidelines. Maintenance of proper hygiene is essential to avoid infections spreading, and every cleaning and handling procedure is very critical [8]. Due to the toxicity of the mercury element, these kinds of thermometers are replaced by digital thermometers in the present day. For a clean check-up and safety, this medical thermometer ought to be sterilized.

1.2.3.1 The ways to handle and read clinical thermometer

- With water at a normal temperature, thermometer must be washed initially.
- Few jerks will be given. After the jerk, the mercury level will be down. We ensure that the normal human body temperature falls below the 37°C mark or the 98°F mark.
- To obtain the reading on the thermometer, the device is then kept under the arms or the tongue.
- To observe the correct reading, thermometer is held close to the eye.
- In between the two bigger marks, the difference in temperature indicated is noted. Also, the number of divisions (shown by smaller marks) between these marks is noted. For example, if the bigger mark reads 10 and the number of divisions is 5, then each division reads the value equal to 10/5=2 [9].

In the clinical thermometer, the level of mercury can rise or fall depending on the temperature of the object. Figure 1.2a illustrates the variation in mercury level corresponding to different temperatures.

1.2.4 Need of mercury-in-glass thermometer

A glass tube is filled with mercury in a mercury thermometer, and a typical temperature scale is marked on the tube. The level of mercury expands and contracts with changes in temperature, and the temperature can be read from the scale. To monitor the temperature of vapor

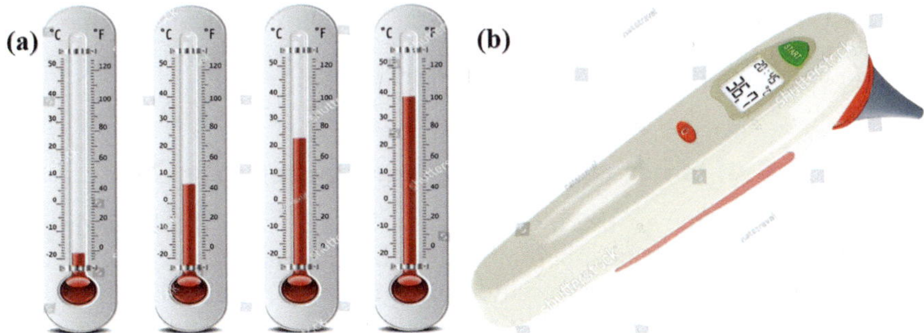

Figure 1.2 (a) The variation of mercury in clinical thermometer and (b) infrared ear and forehead thermometer [10,11]. (Adapted from Shutterstock.)

or liquid, human body mercury thermometers can be used. The better accuracy value attainable with mercury thermometers for measurement of temperature close to 0°C is approximately ±5 mK for field measurements and ±4 mK for laboratory measurements. The working of this thermometer is the same as that of clinical thermometer. The common difference is that a medical or clinical thermometer is used to measure the temperature of a human body, whereas the mercury in glass thermometer is conventional and is used to measure the substance temperature as well. There is a constriction that the clinical thermometer does not permit flowing back of the mercury liquid [12].

1.2.5 Need of infrared ear and forehead thermometer

In clinical settings, the temperature of the body is frequently measured by infrared ear and forehead thermometers. This kind of thermometer measures the infrared energy released from the eardrum of patient in a calibrated length of time [10]. Figure 1.2b represents such a thermometer. In the ear, a protective sleeve is connected to a short tube, and to allow radiation from a tympanic membrane, a shutter is opened to fall on infrared detector for a time period that usually varies from 0.1 to 0.3 seconds. When the collection of data is completed, the device beeps and the measured temperature is read out, which is then displayed on a liquid crystal display [10,11].

1.2.6 Need of bimetal mechanical thermometer

Based on the functional principle, the metals used in bimetal thermometers expand abnormally depending on the temperature change. The two different metal strips in a bimetal thermometer always have a different thermal expansion coefficient. Using bimetallic strip, this thermometer converts the temperature of a media into mechanical displacement as shown in Figure 1.3a. This kind of thermometer is used to protect circuits from excess current in miniature circuit breakers. The bimetallic metals have the following properties: corrosion, mechanical resistance, thermal expansion, and electromagnetic conductivity [13].

1.2.7 Need of constant volume thermometer

The constant-volume air thermometer contains a U-shaped tube with filled mercury open to the atmosphere on the right side and on the left side connected to a glass bulb, as represented in Figure 1.3b. Inside the glass bulb, air is trapped, and inside the bulb, the pressure can be varied by increasing and lowering the right side. In the glass bulb the air pressure is maintained at constant-volume air thermometer is calibrated at different temperatures

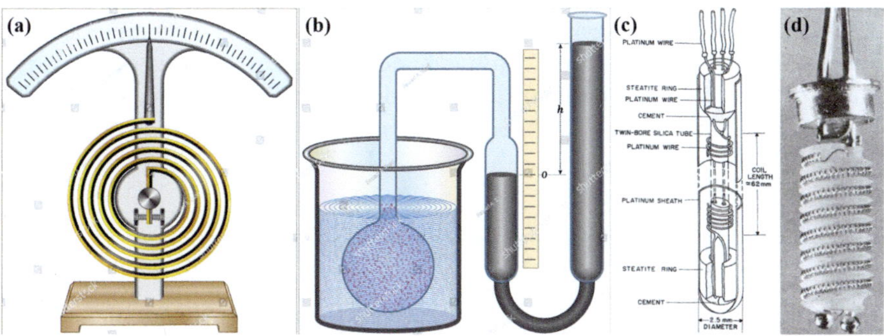

Figure 1.3 (a) Bi-metal mechanical thermometer, (b) constant volume thermometer. ((a and b) Adapted from Shutterstock, and (c and d) Platinum Resistance Thermometer.) (c) Single layer filament around tune bore silica tubing and (d) coiled filament helically grooved ceramic. (Adapted from Ref. [17] Online available.)

from the ice point of water to the steam point of water. The temperature in Celsius degrees, corresponding to absolute, is calculated to be zero from these evaluated data. This kind of thermometer is needed to measure the change in pressure in the industrial chambers [14–16].

Over a large range of temperature, the gas expands regularly and uniformly. For all gases, the expansion coefficient is nearly the same.

1.2.8 Need of platinum resistance thermometer

For the measurement of temperature in the range of 200°C to 1000°C, a versatile instrument, the platinum resistance thermometer, is used. This version of thermometer is usually utilized for higher accuracy for precision measurements and for regular industrial work. For the temperature measurement in this thermometer, a platinum wire is attached to the electrical resistance as depicted in Figure 1.3c and d. It is used to monitor the outside temperatures [17,18]. The rate of working is very slow, but the readings are accurately taken by using this device. While the temperature increases, then the resistance of the material also increases.

1.2.9 Need of constant-pressure gas thermometer

In this type of thermometer related to temperature changes, the changing volume of the gas is monitored. The measurement by this thermometer is very accurate [14]. Recording on two different constant-pressure gas thermometers, there is no difference in temperatures. This is the advantage of this thermometer. The disadvantage of this device is that it is not easily portable because it is a bulky instrument. Based on Charles law, the apparatus is designed in such a way that the rigid vessel is occupied with a gas, generally helium or hydrogen, at very low pressure, and as its temperature is increased, the corresponding volume is measured. During this process, the pressure is maintained constant. At two different fixed points, such as the steam point at one end and the ice point at another end, the device is calibrated [15,16].

1.2.10 Need of thermocouple thermometer

For the very quick temperature measurement, this kind of thermometer can be used. It is needed in many laboratories. This device measures the temperature in the range of 500 and 2300 K. In this device, a voltage is generated by the electrical resistance that varies with the temperature [19,20]. In this device, two dissimilar metal wires are joined in the sensor. One of the wires is joined to a thermocouple at one end, and another wire is connected to a thermometer at the other end. For laboratory purposes, this thermometer can be used. The image of a commercially available thermocouple thermometer is shown in Figure 1.4a.

1.2.11 Need of alcohol thermometer

An alcohol thermometer is the best alternative to the mercury-in-glass thermometer, and the functions of both the devices are same. If this thermometer is subjected to high temperatures, alcohol will evaporate quickly. So, we can use this device only for the measurement of lower temperature. Alcohol is less toxic in comparison with mercury. Where we need a small range of temperature measurements, we can use this thermometer.

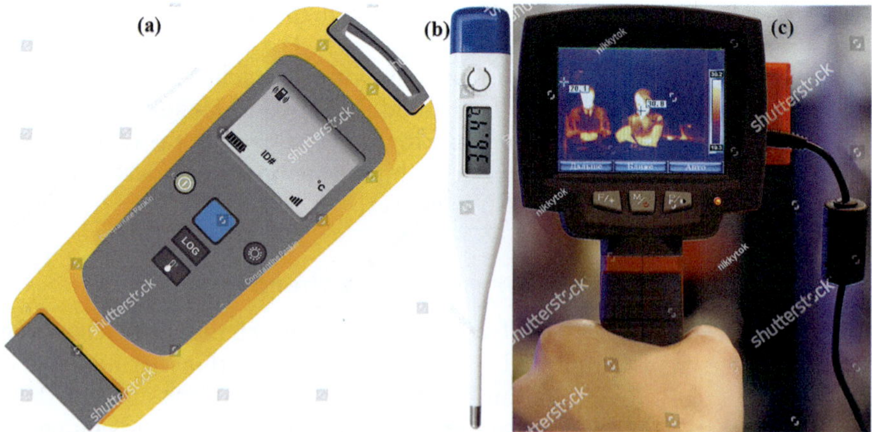

Figure 1.4 (a) Thermocouple thermometer, (b) digital thermometer, and (c) pyrometer thermometer. (Adapted from Shutterstock.)

1.2.12 Need of digital thermometer

To analyze rectal, oral, and armpit temperatures, digital thermometer can be used (Figure 1.4b). For the successful calibration of the smart temperature transmitter, this thermometer is used. Under flowing conditions, this device is also used to verify a smart temperature transmitter. Digital thermometer calculates the dew point, humidity, temperature, air volume, and wet bulb temperature. After testing the device, it is then recalibrated or adjusted, so it can give the proper reading. For that, the thermometer is placed in an ice cold bath or at a boiling point for adjustments using the variation in temperature. Either an oral digital or clinical thermometer can be used to measure oral temperature. The average reading of oral temperature is 98.6°F (37°C). Apart from these digital thermometers, many doctors still consider rectal thermometers to be the best accurate device for children and babies, but forehead and oral thermometers are highly reliable and can be used conveniently.

1.2.13 Need of laboratory thermometer

There are so many varieties of lab thermometers needed, such as mechanical, differential, logging, etc., progressively producing digital display reading that are input capable to computers and software programs for the purpose of logging. Primarily, laboratory thermometers serve for lab purposes like monitoring boiling and freezing points and also checking for other substance temperature. The solvent temperature can also be measured by laboratory thermometer. It consists of many types of thermometers, including total immersion, partial immersion, digital, liquid-filled, and dial thermometers. Other than the temperature of the human body, the laboratory thermometer can be used for measuring other temperatures. For checking freezing and melting points, these types of thermometers are used in many laboratories. High level of precision in temperature measurement for applications such as instrument calibration, maintaining a sterile work environment, experiment monitoring, and materials testing are carried out by laboratory thermometer.

1.2.14 Need of pyrometer thermometer

Based on Stefan's law, Pyrometer thermometer is working. From the radiation of heat radiated by the objects, this device measures the temperature. Without being in contact with the objects, these thermometers can be used. For the huge range of measurement of temperature (up to 2000 K), pyrometer thermometer is used. For non-contact measurement of temperature, pyrometer or infrared thermometer can be used. The measurement of electromagnetic radiation measured from the object is directly proportional to the object temperature. For focusing, EM radiation lens is used in this device. The detector in this device converts the detected radiation into an electrical signal, and after specific mathematical transformation, this value can be displayed in temperature units. The image of commercially available pyrometer thermometer is shown in Figure 1.4c.

1.3 CONCLUSION

In the field of medicine, it is important to measure the temperature of the body. Elevated temperature or fever can be caused by numerous conditions and diseases from the normal cold to COVID-19 or cancer, which indicates the way of body's fighting an infection. Human body temperature can be measured by clinical thermometer. In automobiles for temperature measurement, infrared thermometers are used at a distance. For several firms, in order to meet safety standards, like air leaks check, many of the thermometers are needed. In many food and restaurant industries, the measurement of temperature is carried out by thermometers based on the principle of the expansion of solids and liquids with temperature. Mercury in thermometer begins to rise while the thermometer lamp is dipped in a given substance or solution. In our everyday life, the role of thermometer is essential for healthy survival.

REFERENCES

1. L. Michalski, K. Eckersdorf, J. Kucharski, J. McGhee, (2001) *A Book of Temperature Measurement.* Chichester: John Wiley & Sons, Ltd. doi:10.1002/0470846135, Copyright © 2001.
2. D. Ross Pinnock, P. G. Maropoulos, (2015) Review of industrial temperature measurement technologies and research priorities for the thermal characterisation of the factories of the future. *Proceedings of the Institution of Mechanical Engineers Part B Journal of Engineering Manufacture* 230(5), doi:10.1177/0954405414567929.
3. B. Atakan, D. Roskosch, (2013) Thermographic phosphor thermometry in transient combustion: a theoretical study of heat transfer and accuracy. *Proceedings of the Combustion Institute* 34: 3603–3610.
4. W. E. K. Middleton, (1966) *A History of the Thermometer and Its Use in Meteorology.* Internet Archive. Baltimore: Johns Hopkins Press.
5. J. P. Kauppinen, K. T. Loberg, A. J. Manninen, J. P. Pekola, (1998) Coulomb blockade thermometer: tests and instrumentation. *Review of Scientific Instruments* 69 (12): 4166–4175. doi:10.1063/1.1149265.
6. P. Loyson, (2012) Galilean thermometer not so Galilean. *Journal of Chemical Education* 89 (9): 1095–1096. doi:10.1021/ed200793g.
7. https://www.daviddarling.info/encyclopedia/B/Beckmann_thermometer.html.

8. J. M. S. Pearce, (2002) A brief history of the clinical thermometer, *QJM: An International Journal of Medicine* 95 (4): 251–252. doi:10.1093/qjmed/95.4.251.

9. W. C. Beck, R. Campbell, (1975) Clinical thermometry. *The Guthrie Bulletin* 44: 175–194.

10. I. Pušnik, E. V. D. Ham, J. Drnovšek, (2004) IR ear thermometers: what do they measure and how do they comply with the EU technical regulation? *Physiological Measurement* 25 (3): 699. doi:10.1088/0967-3334/25/3/010.

11. J. Merchant, (2008) Infrared temperature measurement theory and application. *Thermal Processing*: 38–41. https://www.omega.com/en-us/resources/infrared-temperature-measurement-theory-application.

12. B. Littlefield, (2022) Why it is that mercury is chosen for thermometers? UCL. University College London. https://www.ucl.ac.uk/culture-online/case-studies/2022/apr/why-it-mercury-chosen-thermometers.

13. D. Sobel, (1995) *Longitude*. London: Fourth Estate. p. 103. One of the inventions Harrison introduced in H-3... is called... a bi-metallic strip.

14. Y. Zhang, J. Zu, D. Pei, (2012) The miniature internal electronic pressure gauge and dynamic calibration method. *Journal of Computers* 7 (11): 2750–2757.

15. P. Fullick, (1994) *Physics*. New Hampshire: Heinemann Educational. pp. 41–42.

16. P. Tipler, G. Mosca, (2008) 17.2: Gas thermometers and the absolute temperature scale. *Physics for Scientists and Engineers* (6th Ed.). Freeman.

17. J. L. Riddle, G. T. Furukawa, H. H. Plumb, (1973) Platinum Resistance Thermometry, report, April 1973; Washington, DC. https://digital.library.unt.edu/ark:/67531/metadc13167/ accessed March 9, 2024, University of North Texas Libraries, UNT Digital Library, https://digital.library.unt.edu; crediting UNT Libraries Government Documents Department.

18. D. R. White, P. M. C. Rourke, (2020) Standard platinum resistance thermometer interpolations in a revised temperature scale, *Metrologia* 57 (3): 035003. doi:10.1088/1681-7575/ab6b3c.

19. E. Webster, (2021) A critical review of the common thermocouple reference functions, *Metrologia* 58 (2): 025004. doi:10.1088/1681-7575/abdd9a.

20. (1993) *Manual on the Use of Thermocouples in Temperature Measurement* (4th Ed.). ASTM. pp. 48–51.

Chapter 2

Optical thermometers
Importance and advantages over other thermometers

P. Antonyananiya and T. Pradeepkumar

2.1 INTRODUCTION

The thermometer, an indispensable instrument in science, medicine, industry, and daily life, serves the fundamental purpose of measuring temperature. Throughout history, the need to understand and quantify temperature has driven the development and refinement of various thermometric devices. From ANCIENT rudimentary versions to the advanced digital thermometers of today, this three-page introduction explores the evolution, principles, and applications of thermometers, highlighting their significance in various fields and the impact they have on society.

2.1.1 Ancient origins and historical development

The concept of temperature measurement dates back to ancient civilizations where early humans attempted to understand and predict changes in weather patterns. One of the earliest recorded forms of thermometry was the use of the Galilean thermometer in the 17th century, invented by Galileo Galilei, an Italian physicist and astronomer. This simple glass tube containing water and floating spheres demonstrated the principle of density change with temperature. However, it was not until the late 17th century that the mercury-in-glass thermometer, credited to Daniel Gabriel Fahrenheit, brought about a significant advancement in thermometry. Fahrenheit's scale introduced precise temperature readings, and his thermometer quickly gained popularity due to its accuracy.

2.1.2 Principles of thermometry

Thermometers operate based on the principles of thermal expansion and the correlation between temperature and certain physical properties of substances. The common types of thermometers, such as liquid-in-glass thermometers, bimetallic thermometers, and electronic thermocouples, each exploit specific properties for temperature measurement. Liquid-in-glass

DOI: 10.1201/9781032661537-2

thermometers utilize the expansion and contraction of liquids (mercury or alcohol) in a glass tube as temperature changes. Bimetallic thermometers, on the other hand, use the differing thermal expansion rates of two metal strips bonded together to indicate temperature variations. Electronic thermocouples employ the Seebeck effect, which generates a voltage proportional to the temperature difference between two junctions formed by dissimilar metals.

2.1.3 Applications and importance in modern society

The thermometer's impact extends across diverse fields, shaping the way we understand and interact with the world around us. In medicine, thermometers are essential tools for assessing body temperature, diagnosing illnesses, and monitoring patients' health. In weather forecasting, accurate temperature measurements are crucial for predicting atmospheric changes and helping us prepare for extreme weather events. Thermometers play a critical role in various industrial processes, ensuring that temperature-sensitive operations, such as manufacturing, food production, and chemical reactions, are conducted with precision and safety.

In the realm of scientific research, thermometry contributes to countless experiments, ranging from climate studies to material testing. Moreover, advancements in digital technology have revolutionized temperature measurement, enabling remote monitoring, data logging, and real-time analysis. The integration of thermometers into smart devices and Internet of Things (IoT) applications has further expanded their accessibility and utility.

The thermometer's journey from ancient rudimentary tools to cutting-edge digital instruments showcases humanity's relentless pursuit of understanding temperature and its significance in our lives. Whether it is in healthcare, meteorology, industry, or scientific research, thermometers remain pivotal in facilitating progress and enhancing our understanding of the natural world. As technology continues to evolve, we can expect thermometry to become even more sophisticated, empowering us to tackle complex challenges and improve the quality of life for generations to come [1].

2.2 CONDUCTING AND NON-CONDUCTING THERMOMETER

Thermometers can be broadly categorized into two types based on their ability to conduct heat: conducting thermometers and non-conducting thermometers. Each type has its unique characteristics and applications. Let's explore the differences between these two types of thermometers:

2.2.1 Conducting thermometers

Conducting thermometers, also known as contact or mechanical thermometers, rely on the principle of thermal conductivity to measure temperature. These thermometers contain a temperature-sensitive element that directly comes into contact with the substance whose temperature is being measured. The element responds to temperature changes by expanding or contracting, and this movement is converted into a visible or measurable reading on a scale [2–5].

Common types of conducting thermometers include:

a. **Liquid-in-Glass Thermometers:** These traditional thermometers consist of a glass tube filled with a temperature-sensitive liquid, such as mercury or alcohol. As the temperature changes, the liquid expands or contracts, causing it to rise or fall within the calibrated tube, displaying the temperature on a scale [6].
b. **Bimetallic Thermometers:** Bimetallic thermometers consist of two metal strips with different coefficients of thermal expansion, bonded together. As the temperature changes, the two metals expand or contract at different rates, causing the bonded strip to bend, which moves an indicator along a calibrated scale [6].
c. **Liquid Crystal Thermometers:** Liquid crystal thermometers use liquid crystals that change color with temperature variations. These thermometers are often found in applications like forehead strips for fever measurement.

Conducting thermometers are commonly used in various applications, including household thermometers, medical thermometers, weather instruments, and industrial processes where direct contact with the substance being measured is possible [7].

2.2.2 Non-conducting thermometers

Non-conducting thermometers, also known as non-contact or infrared thermometers, measure temperature without direct contact with the substance. Instead, they detect the thermal radiation emitted by the object and use infrared technology to convert the radiation into temperature readings.

Common types of non-conducting thermometers include:

d. **Infrared (IR) Thermometers:** These thermometers use an infrared sensor to detect the infrared radiation emitted by an object. The sensor converts the radiation into an electrical signal, which is then converted into a temperature reading and displayed on a digital screen.
e. **Thermal Imaging Cameras:** Thermal imaging cameras use arrays of infrared sensors to create visual representations of temperature variations. These devices are particularly useful in industrial inspections, building diagnostics, and security applications.

Non-conducting thermometers are preferred in situations where conducting thermometers may not be feasible or safe, such as measuring the temperature of distant or hazardous objects, moving targets, or objects with rapid temperature changes.

In conclusion, both conducting and non-conducting thermometers play essential roles in temperature measurement. Conducting thermometers provide accurate readings when direct contact with the substance is possible, while non-conducting thermometers offer the advantage of non-contact temperature measurement, making them suitable for specific applications and scenarios where direct contact is not practical or safe [8].

2.3 DISTINGUISHING FEATURES OF CONDUCTING AND NON-CONDUCTING THERMOMETERS

Conducting thermometers	*Non-conducting thermometers*
Measurement: Conducting thermometers require direct physical contact with the substance whose temperature is being measured. The temperature-sensitive element (e.g., liquid, bimetallic strip) comes into contact with the object, and its response to temperature changes provides the measurement.	**Measurement:** Non-conducting thermometers measure temperature without requiring physical contact with the substance. They use infrared technology to detect the thermal radiation emitted by the object, converting it into temperature readings.
Temperature Range: Conducting thermometers can typically measure a wide range of temperatures, from very low to high temperatures. For example, liquid-in-glass thermometers can handle a broad range of temperatures, including sub-zero temperatures and high-temperature applications.	**Temperature Range:** Non-conducting thermometers typically have a limited temperature range compared to conducting thermometers. They are well-suited for measuring surface temperatures and are commonly used in moderate temperature ranges.
Response Time: Conducting thermometers generally have a slower response time compared to non-conducting thermometers. The time it takes for the temperature-sensitive element to reach equilibrium with the substance being measured contributes to the response time.	**Response Time:** Non-conducting thermometers have a rapid response time, providing almost instantaneous temperature readings. The lack of physical contact with the object allows for quick measurements.
Calibration: Conducting thermometers often require periodic calibration to ensure accuracy. Environmental factors and wear and tear can affect the calibration of these thermometers over time.	**Calibration:** Non-conducting thermometers often require less frequent calibration compared to conducting thermometers. However, periodic calibration may still be necessary to maintain accuracy.

(Continued)

Conducting thermometers	*Non-conducting thermometers*
(Continued)	
Applications: Conducting thermometers find applications in various fields, including medical, meteorological, industrial, and household settings. They are commonly used when direct contact with the object is feasible and accurate readings are essential.	**Applications:** Non-conducting thermometers are widely used in situations where direct contact with the object is not feasible or safe, such as measuring the temperature of moving objects, distant targets, or hazardous substances. They are commonly employed in industrial, medical, and research applications.

In summary, the key distinguishing features of conducting and non-conducting thermometers lie in their measurement method, temperature range, response time, calibration requirements, and specific applications. Conducting thermometers rely on direct contact for measurement and have a broader temperature range, while non-conducting thermometers offer non-contact measurement with a faster response time and are often used in specialized situations where direct contact is not practical or safe.

2.4 OPTICAL THERMOMETER

Optical pyrometers make it possible to monitor the temperature of items that would otherwise be difficult to measure with contact sensors. Because of its non-contact nature, high spatial resolution, rapid response, and anti-electromagnetic interference, optical thermometry technology based on rare earth (RE) and transition metal (TM) ions doped luminous materials has sparked considerable interest. Recently, temperature was measured using temperature-sensitive optical metrics such as luminous intensity, spectral position, decay duration, bandwidth shift, and fluorescence intensity ratio (FIR). FIR and decay lifetime-based temperature sensing that is independent of spectral losses and external interferences have received a lot of attention [9,10].

2.4.1 Types of temperature measurement using optical methods

Optical thermometers are a class of non-contact temperature measurement devices that rely on various optical principles to determine the temperature of an object without physical contact. There are several types of optical thermometers, each utilizing different optical methods for temperature measurement:

    f. **Infrared Thermometers:** These thermometers use the infrared radiation emitted by an object to determine its temperature. The device contains an infrared sensor that detects the radiation and converts it into temperature readings. Infrared thermometers are commonly used in industrial, medical, and home applications, spot checks of process pipe temperatures, motor winding temperatures, bearing cap temperatures, and other related applications. Its limitation is that the machine's or structure's temperature only represents one point [11].

g. **Fiber Optic Thermometers:** Fiber optic thermometers utilize optical fibers to transmit thermal radiation from the object to a detector. These thermometers are suitable for measuring temperature in hard-to-reach or hazardous environments [12].

h. **Raman Spectroscopy Thermometers:** Raman spectroscopy involves analyzing the scattered light from a material to determine its molecular vibrations, which are temperature-dependent. Raman spectroscopy thermometers are used in precise scientific research and materials analysis [13].

i. **Fluorescence Thermometers:** Fluorescence thermometers exploit temperature-dependent changes in the fluorescence properties of certain materials to measure temperature. They are used in specialized scientific and research applications [14,15].

2.4.2 Principle, construction, working, diagram, advantages, disadvantages, application

Principle: Optical thermometers operate on the principle of detecting and analyzing thermal radiation emitted by an object. Different optical methods, such as infrared sensing, fiber optic transmission, Raman spectroscopy, or fluorescence, are employed to achieve accurate temperature measurements.

Construction: Optical thermometers consist of various components depending on their type. Infrared thermometers typically have an infrared sensor, a lens to focus the radiation, and electronic circuitry to convert the detected radiation into temperature readings. Fiber optic thermometers utilize optical fibers to transmit radiation to a remote detector. Raman and fluorescence thermometers involve specific optical setups for spectroscopic measurements [16].

Working: Optical thermometers work by receiving thermal radiation emitted by the object being measured. The detector or sensor analyses the received radiation and converts it into an electrical signal, which is then calibrated to provide temperature readings. The diagram below depicts an optical pyrometer. It is made up of a lens that directs the heated object's radiated energy toward the electric filament bulb. The filament's intensity is determined by the current that flows through it. Hence, the lamp receives the adjusted current (Figure 2.1a).

In this particular case, because their temperature is lower than that necessary for equal brightness, the filament appears too dark (Figure 2.1b). Once the filament's temperature is higher than the source's, it seems bright

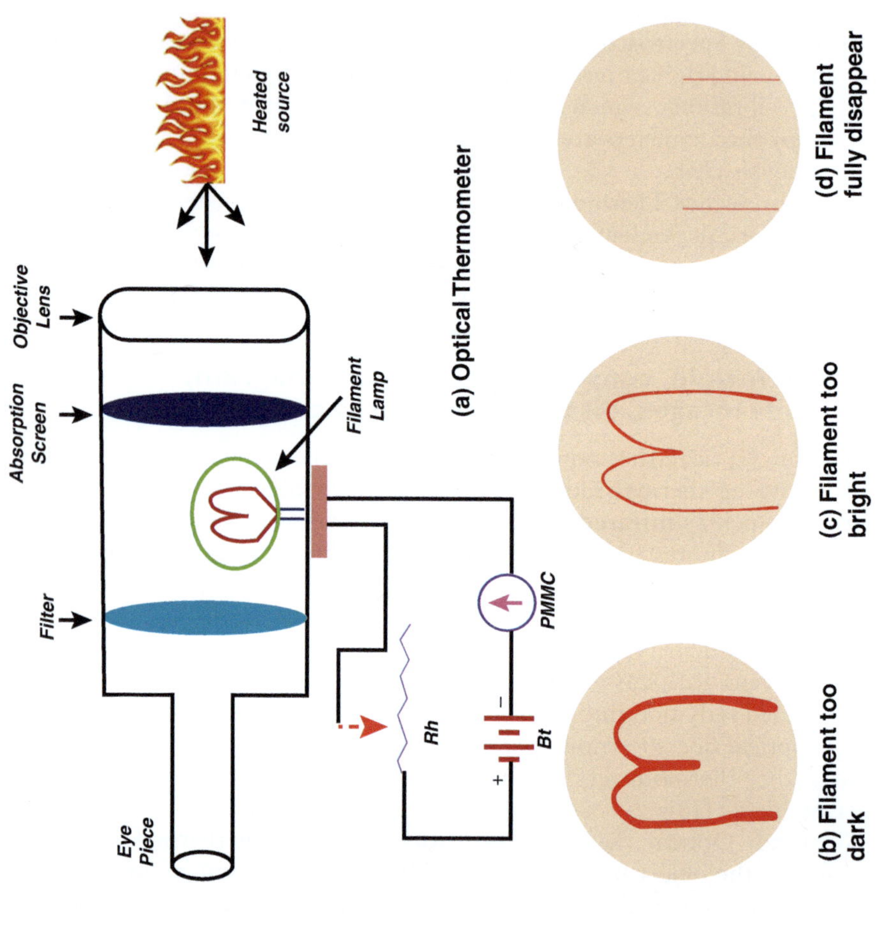

Figure 2.1 (a) Schematic of optical thermometer and (b–d) different conditions of filament. (Adopted from Ref. [17], circuitglobe.com.)

(Figure 2.1c). Until the filament's brightness is comparable to the object's brightness, the current magnitude is changed. The filament's outline totally disappears when the brightness of the filament and the component are equal (Figure 2.1d) [17].

2.4.2.1 Advantages

1. **Non-Invasive and Hygienic:** In medical settings, optical thermometers offer non-invasive temperature measurement, eliminating the need for physical contact with the patient. This feature reduces the risk of cross-contamination and makes them more comfortable for users.
2. **Safety:** Optical thermometers are especially valuable in hazardous environments or high-temperature situations where using contact thermometers could be dangerous for personnel.
3. **Wide Temperature Range:** Some optical thermometers, particularly infrared thermometers, have a wide temperature measurement range, making them suitable for various applications, from low-temperature processes to extremely high-temperature environments.
4. **Remote Sensing:** Optical thermometers allow for remote temperature sensing, enabling measurements from a distance. This capability is advantageous in scenarios where accessing the object directly is challenging or unsafe.
5. **Real-Time Monitoring:** Optical thermometers facilitate real-time temperature monitoring, providing instantaneous feedback on temperature changes. This feature is crucial in industrial processes, quality control, and research applications.
6. **Non-Destructive Testing:** Optical thermometers enable non-destructive testing, making them useful for assessing materials' temperature without altering or damaging the samples.
7. **No Consumables:** Unlike some other thermometers that require consumables, such as mercury or other liquid fillings, optical thermometers do not require any refills or replacements, reducing operational costs and environmental impact.
8. **Precision and Accuracy:** Advanced optical thermometers, such as infrared thermometers, can offer high precision and accuracy, providing reliable temperature data for critical applications.
9. **Non-Contact Measurement:** Optical thermometers offer non-contact temperature measurement, making them suitable for delicate or hazardous environments.
10. **Fast Response Time:** Optical thermometers provide rapid temperature readings, ideal for applications requiring quick data acquisition.
11. **Versatility:** Different optical methods allow optical thermometers to measure a wide range of temperatures and materials.

2.4.2.2 Disadvantages

1. **Surface Measurements:** Optical thermometers measure the surface temperature of an object, which may not represent the internal temperature accurately.
2. **Emissivity Effects:** The accuracy of optical thermometers can be affected by the emissivity of the object's surface.
3. **Calibration:** Some optical thermometers require periodic calibration to maintain accuracy.

2.4.2.3 Applications

1. **Industrial:** Infrared thermometers find applications in industrial processes, maintenance, and quality control.
2. **Medical:** Non-contact infrared thermometers are commonly used for temperature screening and medical diagnostics.
3. **Scientific Research:** Optical thermometers, such as Raman and fluorescence thermometers, are used in research laboratories for specialized measurements.
4. **Environmental Monitoring:** Optical thermometers are employed in environmental monitoring and climate studies [18].

2.4.3 Future of real deployment

The future of optical thermometers holds great promise, driven by ongoing advancements in technology and the need for contactless and precise temperature measurements. As research and development continue, optical thermometers are expected to become even more accurate, faster, and versatile. Improvements in sensor technology, miniaturization, and integration with artificial intelligence will likely expand their applications in various industries, from manufacturing and aerospace to healthcare and environmental monitoring. Additionally, the increasing demand for non-contact and remote temperature measurement in critical situations, such as pandemics or hazardous environments, will further drive the real-world deployment of optical thermometers.

2.5 THE SIGNIFICANCE AND BENEFITS OF VARIOUS FIELDS (THERMOMETERS)

Optical thermometers hold significant importance and offer distinct advantages over other types of thermometers, especially in specific applications where non-contact temperature measurement is essential. Let's explore the importance and benefits of optical thermometers compared to other thermometers:

2.5.1 Importance of optical thermometers

1. **Non-Contact Temperature Measurement:** One of the primary advantages of optical thermometers is their ability to measure temperature without physical contact with the object being measured. This feature is crucial in situations where direct contact is not feasible, such as when dealing with delicate or hazardous substances.
2. **Versatility:** Optical thermometers are versatile and can measure temperature in various environments and surfaces, including moving objects or hard-to-reach locations. They are used in a wide range of applications, from medical and industrial settings to research laboratories and environmental monitoring.
3. **Speed and Efficiency:** Optical thermometers provide rapid temperature readings, making them suitable for applications that require quick data acquisition. The non-contact measurement allows for swift data collection, enabling efficient and continuous monitoring.

Thermometers play a significant role in various fields, providing critical temperature measurement capabilities that have a profound impact on each area's functionality, safety, and efficiency. Let's explore the significance and benefits of thermometers in different fields:

2.5.2 Medical care and medication

Significance: In healthcare, thermometers are essential tools for diagnosing and monitoring patients' health. They help identify fever, a common symptom of various illnesses. Accurate temperature measurements aid in assessing the effectiveness of medical treatments and medications. Thermometers are crucial for detecting temperature changes that may indicate infection or other health conditions.

Benefits: Early detection of fevers and abnormalities allows for timely medical intervention, leading to better patient outcomes. Non-contact infrared thermometers provide a hygienic and convenient way to measure body temperature, reducing the risk of cross-contamination. Medical thermometers are user-friendly, making them accessible for both medical professionals and individuals for home use.

2.5.3 Industrial and manufacturing

Significance: In industrial settings, thermometers are vital for monitoring and controlling various processes that rely on specific temperature ranges. Accurate temperature measurements ensure that manufacturing processes are optimized for quality and efficiency. Thermometers help prevent overheating and ensure the safety of machinery and equipment.

Benefits: Improved process control leads to higher-quality products and reduced wastage in manufacturing. Non-contact infrared thermometers enable temperature measurements in hazardous or hard-to-reach areas without risking personnel safety. Remote temperature monitoring allows for real-time data analysis and preventive maintenance, preventing costly breakdowns.

2.5.4 Meteorology and weather forecasting

Significance: Thermometers are crucial instruments in meteorology for measuring air temperatures and determining weather patterns. Accurate temperature data aids in understanding climate trends and predicting extreme weather events. Monitoring temperature changes helps in studying climate change and its impact on the environment.

Benefits: Better weather forecasts and early warnings enable communities to prepare for and mitigate the effects of severe weather conditions. Long-term temperature data helps scientists and policymakers make informed decisions related to environmental conservation and sustainable practices.

2.5.5 Scientific research and laboratories

Significance: In scientific research, accurate temperature measurements are essential for conducting experiments, especially those involving chemical reactions and biological processes. Thermometers are vital tools in calibration processes, ensuring the accuracy of other instruments.

Benefits: Reliable temperature data contributes to the validity and reproducibility of scientific findings. Laboratory thermometers with high precision and stability provide confidence in research results and data analysis. Non-contact infrared thermometers enable measurements of samples without contaminating or altering them.

2.5.6 Household and everyday use

Significance: Thermometers are used in households for various purposes, such as monitoring room temperature, cooking, and checking body temperature during illnesses.

Benefits: Household thermometers facilitate comfortable living conditions by maintaining indoor temperatures at desirable levels. Cooking thermometers ensure food safety and help achieve desired cooking outcomes. Medical thermometers in households aid in early detection of fever or illness, allowing for timely medical attention.

2.6 CONCLUSION

Optical thermometers are highly valuable instruments with a broad range of applications due to their non-contact nature, versatility, and real-time monitoring capabilities. Their ability to measure temperature remotely and safely in various environments makes them indispensable tools in healthcare, industrial processes, research, and everyday use. The advantages of optical thermometers over other thermometers, particularly in terms of non-invasiveness, safety, and wide temperature range, underscore their significance in modern temperature measurement techniques. Thermometers offer invaluable benefits across various fields, from healthcare and industry to meteorology and scientific research. Their accurate temperature measurements enable critical decision-making, process optimization, and safety in diverse applications, ultimately improving human health, productivity, and overall quality of life.

REFERENCES

1. M. Javaid, A. Haleem, R. P. Singh, S. Rab, R. Suman, Significance of sensors for industry 4.0: roles, capabilities, and application, *Sensors International* 2(2021), 100110 (2021).
2. E. Grodzinsky, Märta Sund Levander, History of the thermometer *Understanding Fever and Body Temperature*, 03, 23–35 (2019) doi:10.1007/978-3-030-21886.
3. J. Early Mod Stud (Bucur), The Weight of the Air: Santorio's Thermometers and the Early History of Medical Quantification Reconsidered, 7(1): 73–103 (2019). PMID: 30854347.
4. U. Grigull, Fahrenheit: a pioneer of exact thermometry, *Proceedings of the 8th International Heat Transfer Conference* 1, 9–18 (1986).
5. W. M. Saslow, A history of thermodynamics: the missing manual, Entropy (Basel) 22(1), 77 (2020). doi:10.3390/e22010077.
6. C. Warren Hurley, J. F. Schooley, Calibration of Temperature Measurement Systems Installed in Buildings, NBS Building Science Series 153, pp. 15–54 (1984).
7. T. D. McGee, *Principles and Methods of Temperature Measurement*. New York: John Wiley & Sons (1988).
8. Y. Zhao, J. H. M. Bergmann, A systematic review non-contact infrared thermometers and thermal scanners for human body temperature monitoring, *Sensors (Basel)* 23(17), 7439 (2023). doi:10.3390/s23177439.
9. X. Zhang, Z. Zhu, Z. Guo, Z. Sun, Y. Chen, A ratiometric optical thermometer with high sensitivity and superior signal discriminability based on Na3Sc2P3O12:Eu2+, Mn2+ thermo chromic phosphor, *Chemical Engineering Journal* 356, 413–422 (2019).
10. J. Rocha, C. D. S. Brites, L. D. Carlos, Lanthanide organic framework luminescent thermometers, *Chemistry a European Journal* 22(42), 14782–14795 (2016).

11. R. Keith Mobley, (2001) 50- *Maintenance Engineering Handbook 2001*, McGraw-Hill, pp. 867, 869–888. doi:10.1016/B978-075067328-0/50052-5.
12. R. Martinek, J. Koziorek, S. Hejduk, J. Vitasek, A. Vanderka, R. Poboril, V. Vasinek, R. Hercik, Temperature measurement using optical fiber methods: overview and evaluation, *Journal of Sensors* 2020, 1–25 (2020). Article ID 8831332. doi:10.1155/2020/8831332.
13. D. Tuschel, Ranam thermometer, *Spectroscopy* 31(12), 8–13 (2016).
14. J. Yang, K. Gu, C. Shi, M. Li, P. Zhao, W.-H. Zhu, Fluorescent thermometer based on a quinolinemalononitrile copolymer with aggregation – induced emission characteristics, *Materials Chemistry Frontiers* 3(8), (2019). doi:10.1039/C9QM00147F.
15. Q. Wang, M. Liao, Q. Lin, M. Xiong, Z. Mu, F. Wu, A review on fluorescence intensity ratio thermometer based on rare-earth and transition metal ions doped inorganic luminescent materials, *Journal of Alloys and Compounds* 850, 156744 (2021).
16. R. W. A. Oliver, A. Stott, An optical thermometer: the design, construction and calibration of a photometric temperature scale for use with a thermometric solution, *Journal of Physics E: Scientific Instruments* 7(4), 275–280 (1974). doi:10.1088/0022-3735/7/4/018.
17. Circuit Globe. Optical Pyrometer. https://circuitglobe.com/optical-pyrometer.html. Accessed 10/04/2024.
18. C. N. Owston, Application of thermistors to optical pyrometry and temperature control, *Journal of Physics E: Scientific Instruments* 1, 1228 (1968). doi:10.1088/0022–3735/1/12/322.

Chapter 3

Luminescent thermometers

Methods, materials, and applications

Rudrashish Panda and Pratap Kumar Sahoo

3.1 INTRODUCTION

Temperature is seen as one of the most significant physical values that can be monitored in daily life. Because of the cooling and heating processes, precise and reliable temperature measurements are crucial for instruments used in biology, chemistry, medicine, and metrology. Temperature sensors make up between 75% and 80% of all sensors sold in the modern world. The ability of materials to react to heat is the primary factor used in conventional temperature sensing measurements. Making such thermometric measurements mostly requires direct contact of the device with the heated objects [1–3]. In recent years, one of the most active areas of research has been non-contact specific thermometry with high resolution for measuring a wide range of temperatures in nanoscale environments [4,5]. Luminescence thermometry has been recognized as the primary non-invasive thermometric approach with high detection sensitivity, spatial resolution, and fast acquisition times [6–8]. Optical thermometry relies on luminescence, which is the light emission that occurs when excited states in materials relax. The emission properties of luminescent materials, such as organic dyes, quantum dots, and phosphors, vary with temperature. The change in photoluminescence characteristics as a function of temperature is the basis for temperature sensors utilizing luminous materials. The emission intensity, peak position, complete width at half maximum of the emission spectrum, and the typical lifetime of the excited state are among the photoluminescence properties [9,10].

Two methods for precisely measuring the thermometric behavior of luminous materials were proposed by Nikolic et al. [11]. The first technique measures the fluorescence intensity ratio (FIR) between the emission from the host and the emission from the doped rare earth ions. The fluorescence intensity of a certain electronic transition is assessed in terms of its time dependence using the second method [12,13]. Luminescent thermometers can be made utilizing a variety of thermal probes, including organic dyes, rare earth ions, and quantum dots (QDs), based on the two approaches mentioned above. A vast amount of research using QD thermometry and organic dyes has been conducted during the last ten years. There have

DOI: 10.1201/9781032661537-3

also been reports of nano-thermometers based on broad band gap rare earth-doped hosts [1–13]. Previous research on luminescence thermometry states that more sophisticated equipment is needed for temperature monitoring via emission lifetime measurements. The FIR methodology is proven to be faster and simpler than the luminescence lifetime method in terms of luminescence thermometric applications [1–13]. The development of thermometers employing broad band gap hosts doped with appropriate rare earth ions via the FIR technique has garnered more attention recently.

Alkaline-earth perovskite-structured $MZrO_3$ (M=Sr, Ba, and Ca) is an appealing option for a variety of optoelectronic applications due to its strong chemical and thermal stability and environmentally favorable characteristics [14–22]. The axial and planar oxygen sites of the $[ZrO_6]$ octahedral [20] may experience some vacancy defects due to the displacement of Zr or M atoms in disordered perovskite $MZrO_3$. Because of these vacancies, which function as luminescence centers, the perovskite $MZrO_3$ exhibits widespread violet-blue emission.

Several investigations on the structural and photoluminescence characteristics of SrZrO3: Eu^{3+} nanocrystals produced by various methods have been conducted in the last few years [23–26], in which it was summarized that the main cause of the high-temperature dependence of fluorescence intensity ratio of the $SrZrO_3$ trap emissions to the emissions of Eu^{3+} ions is the motivation for the use of these materials for luminescence-based optical thermometers. Figure 3.1 shows the schematic energy level diagram for the trap and characteristic emissions of Eu^{3+} ions with temperature excitation.

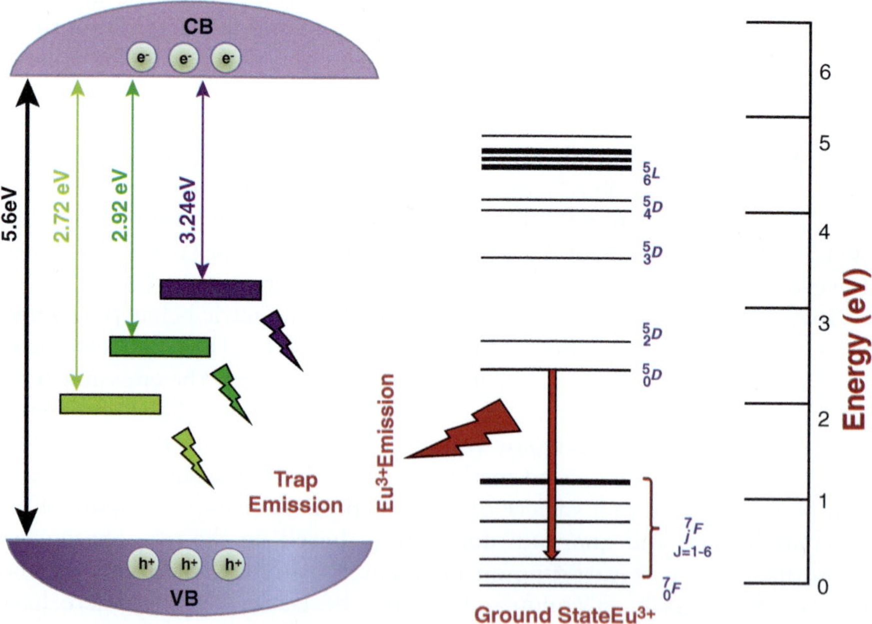

Figure 3.1 The schematic diagram of the energy level of trap emission and Eu^{3+} emission.

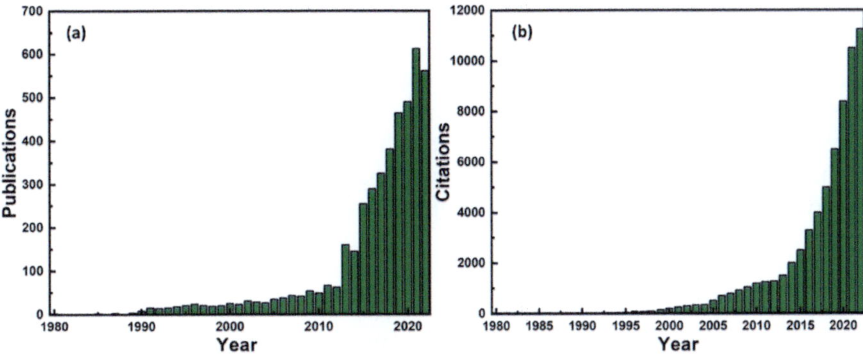

Figure 3.2 The ray diagram representing (a) number of publications and (b) number of citations on this device in the last four decades.

These luminescent thermometers are used by materials scientists to carefully track temperature changes that occur during the synthesis of new materials, phase transitions, and quality control procedures. Luminescence-based optical thermometers provide a non-invasive way to check the temperature of living things, cells, and tissues in the field of biotechnology. Luminescence-based thermometry is used in the semiconductor and aerospace industries to achieve precise process control.

Due to its simplicity and speed, thermometry is a discipline that has seen exponential growth in research efforts in recent years. In particular, over the past ten years, there has been a significant increase in the quantity of publications and associated patents in this field as depicted in Figure 3.2.

This chapter explores the principles, measurement methods, applications, and recent developments of luminescent-based optical thermometers, which are not only fascinating from a scientific standpoint but also have the potential to significantly advance temperature sensing technology and knowledge.

3.2 FUNDAMENTALS AND PRINCIPLES OF LUMINESCENCE-BASED THERMOMETRY

3.2.1 Luminescent materials and mechanisms

Because of their distinctive emission characteristics, luminescent materials are the cornerstone of luminescence-based optical thermometry, serving as the basis for temperature sensing. To fully utilize luminous materials in temperature monitoring applications, it is essential to comprehend the many types of luminescent materials and the fundamental principles that dictate their behavior. The wide variety of luminous materials and the processes underlying their temperature-dependent emission are examined in this section.

3.2.1.1 Fluorescent materials

A substance that absorbs photons and quickly reemits them at a longer wavelength is known as fluorescence. The emission-causing excited state usually has a brief lifetime, measured in nanoseconds. Temperature changes can have an impact on the spectrum features and intensity of fluorescence emission. Temperature variations have an impact on the rate of non-radiative decay processes, which modifies fluorescence emission.

3.2.1.2 Phosphorescent materials

Compared to fluorescence, phosphorescence has an excited state that lasts longer – typically between milliseconds and microseconds. The participation of intersystem crossing between singlet and triplet states is the cause of this. The influence of temperature on the speeds of different relaxation processes inside the excited states is the reason for the temperature sensitivity of phosphorescence. Different temperature-dependent emission characteristics are displayed by phosphorescent materials.

3.2.1.3 Quantum dots

Semiconductor nanocrystals with special electrical characteristics are known as quantum dots. These nanoscale structures produce size-dependent optical characteristics due to the confinement of charge carriers within them. The intensity and emission spectrum of quantum dots can change with temperature. Temperature variations affect the dynamics of charge carriers and the bandgap energy in quantum dots.

3.2.1.4 Organic dyes and molecules

Depending on their molecular makeup, organic dyes and molecules can fluoresce or phosphoresce. Electronic transitions within the molecular orbitals frequently cause the emission. Organic dyes and molecules are temperature-sensitive, just like other luminous materials. Modifications in molecular vibrations and conformational modifications are responsible for variations in luminescence characteristics.

3.2.1.5 Upconversion nanoparticles

Through multiphoton absorption mechanisms, upconversion nanoparticles change low-energy photons into higher-energy ones. Emission at shorter wavelengths than the excitation source is made possible by this. Emission properties can alter with temperature due to the thermal population of excited states and their impact on energy transfer mechanisms within upconversion nanoparticles.

3.2.1.5.1 Mechanism

The relative intensity (FIR) and the time-dependent intensity of a specific transition (lifetime decay curve) are two examples of a luminous material's thermometric behavior [12,13]. Radiative and non-radiative transition mechanisms work together to relax electrons from high excited states of materials toward low excited states or the ground state. As a result, the population density of the luminous species in their excited states is proportional to the emission intensity (I). The following equations [12,13] show how temperature affects emission lifetime:

$$\tau = \frac{1}{w_r + w_{nr}(T)} \tag{3.1}$$

$$\tau = \frac{1}{\tau_0^{-1} + k\,\exp\left(-\dfrac{\Delta E}{K_B T}\right)} \tag{3.2}$$

where w_r is the probability of radiative decay process and w_{nr} is the probability of non-radiative decay process. τ_0 is the radiative decay lifetime at absolute zero temperature, $k=$exponential factor, ΔE is the bandgap energy between the higher excited state and the emitting energy state, and K_B is the Boltzmann constant. Any optical transition's intensity is proportionate to the overall population (number of atoms) in an excited state. The ratio of the two transitions' fluorescence intensity at temperature T can be expressed as [12,13].

$$\frac{I_1}{I_2} = \frac{g_1 A_1 h v_1}{g_2 A_2 h v_2}\exp\left(-\frac{\Delta E_{12}}{K_B T}\right) \tag{3.3}$$

$$\frac{I_1}{I_2} = C \times \exp\left(-\frac{\Delta E_{12}}{K_B T}\right) \tag{3.4}$$

where A is the spontaneous emission rate, h is the Planck constant, v is the frequency, E is the level's energy, and C is a constant. The degeneracy of the corresponding levels is represented by g_1 and g_2 respectively. As a result, temperatures can be determined using the intensity ratio of the transitions. In an ideal scenario, the ratio of the intensity of the temperature-dependent emission (I) to the intensity of the temperature-independent reference emission (I_R) is used to determine the FIR [6,11].

Comprehending the complex processes that regulate luminous materials is essential for the development and enhancement of optical thermometers. These materials' temperature sensitivity, together with developments in material science and nanotechnology, keep expanding the range of applications for luminescence-based optical thermometry.

3.2.2 Temperature-dependent luminescent properties

The fundamental component of luminescence-based optical thermometry is the temperature-dependent luminescent characteristics of materials. These features show themselves as differences in luminous materials'' emission characteristics in response to temperature changes. For the accurate and dependable monitoring of temperature using luminous materials, it is imperative to comprehend the subtleties of these temperature-dependent features. Important features of temperature-dependent luminous characteristics are examined in this section.

Emission Intensity: One essential characteristic that varies with temperature is the luminous emission's intensity. Temperature affects the rates of radiative and non-radiative decay processes in the excited states of luminescent materials, which affects the total emission intensity. This is the basis for its mechanism. Elevated temperatures frequently lead to heightened thermal motion, which impacts the efficacy of radiative processes.

Spectral Shifts: Spectral shifts, as the name implies, are variations in the wavelength or color of luminescent emission caused by changes in temperature. The electronic structure of luminous materials can change with temperature, which can have an impact on the energy levels involved in the emission process. These modifications appear as modifications in the emission spectrum, offering a sensitive gauge of temperature fluctuations.

Decay Lifetimes: When in their excited states, luminescent materials have distinct lifetimes that can vary with temperature. The rate at which different relaxation processes inside the excited states occur is influenced by temperature, and this has an impact on how long it takes for luminous materials to return to their ground state. The luminous characteristics' sensitivity to temperature is partly determined by variations in decay lifetimes.

Thermal Quenching: The term "thermal quenching" describes the decrease in luminescence intensity at high temperatures. Warming up a material can accelerate non-radiative decay mechanisms including energy transfer to impurity sites or transitions aided by phonons. This causes the luminescence intensity to drop, and precise temperature measurements depend on an understanding of thermal quenching.

Cross-Sensitivity: In addition to temperature, luminescent materials can also show cross-sensitivity to other variables. The mechanism of cross-sensitivity is dependent on ambient factors that may cause measurement errors in temperature, including pressure, humidity, and chemical composition. These factors can also affect luminous characteristics. To obtain accurate and trustworthy findings, cross-sensitivity must be understood and mitigated.

Calibration Curves: The temperature-dependent luminous characteristics are related to known temperature values by calibration curves. Systematic measurements of luminous characteristics at various temperatures are necessary for the construction of calibration curves. In real-world applications, these curves are used as a guide when turning luminous signals into precise temperature readings.

The research using the lifetime read-out approach, which uses one or more emission lines to sense temperature changes, has been reviewed in Table 3.1.

Understanding the subtle interplay between temperature and luminous characteristics permits the creation of robust luminescence-based optical thermometers. These characteristics are used by scientists and engineers to create sensors that are suited for certain uses, from basic materials science research to useful applications in biotechnology, environmental monitoring, and industrial operations.

3.2.3 Calibration and temperature sensitivity

To ensure precise and trustworthy temperature readings, luminescence-based thermometry relies heavily on calibration and temperature sensitivity. Temperature sensitivity measures how sensitive a luminous material is to temperature fluctuations, whereas calibration entails establishing a link between the substance's luminescent qualities and the appropriate temperature values. The significance of temperature sensitivity and calibration in luminescence-based thermometry is explored in this section.

3.2.3.1 Calibration

The process of creating a calibration curve or equation involves establishing a correlation between luminous signals and temperature values that are known. Accurate temperature readings from luminous measurements

Table 3.1 An overview of metal oxides with their operating temperature range, sensitivity information, and rise/lifetime read-out method

Probing ion	Co-dopant: host material	Temperature range (K)	Max sensitivity (% K^{-1})	References
Cr^{3+}	Bi^{3+}:$ZnGa_2O_4$	293–473	0.3	[21]
Eu^{3+}	SrY_2O_4	293–473	0.34@473 K	[22]
	BaY_2ZnO_5	330–510	2.2@490 K	[23]
Mn^{2+}	Zn_2GeO_4	250–420	12.2@370 K	[24]
Mn^{3+}	Mn^{4+},Nd^{3+}:YAG	Physiological range	2.69	[25]
Mn^{4+}	β-Li_2TiO_3	10–350	3.21@332 K	[26]
Nd^{3+}	Y_2O_3, $YGdO_3$, $YAlO_3$, YAG, $LiLaP_4O_{12}$, and Gd_2O_3	300–350	4	[27]
Tm^{3+}	Er^{3+},Yb^{3+}:La_2O_3	298–333	0.67[a]	[28]
Yb^{3+}	$Bi_7F_{11}O_5$	293–573	0.24	[29]

depend on this step. Systematic measurements of luminous characteristics at different known temperatures are required for calibration. A mathematical relationship is established using the resultant data, which can subsequently be utilized to take actual temperature readings. Calibration accuracy can be affected by a number of things, such as the type of luminous material used, the surrounding environment, and the equipment used. It could be required to recalibrate on a regular basis to take these factors' changing values into consideration.

3.2.3.2 Temperature sensitivity

The term "temperature sensitivity" describes the luminous material's response to temperature variations. It measures how much luminous characteristics vary for a specific temperature change. To get precise temperature readings, a higher temperature sensitivity is preferred. Sensitivity can be increased by optimizing the luminous material's selection, concentration, and experimental setup. The dynamic range, which denotes the range of temperatures over which the luminous material can produce precise and reliable measurements, is frequently used to describe temperature sensitivity.

3.2.3.3 Influence of environmental factors

In addition to temperature, additional environmental variables that may affect luminescent materials include pressure, humidity, and chemical composition. It is essential to comprehend and reduce cross-sensitivity in order to obtain precise measurements. Certain luminous materials require careful consideration during calibration and application because they may exhibit variable cross-sensitivity with ambient conditions at different temperatures.

3.2.3.4 Material selection and engineering

The selection of luminous material is crucial for both sensitivity and calibration. Materials with particular luminous qualities that meet the needs of the desired application might be chosen or engineered by researchers. Coatings and functionalizations of surfaces can affect how sensitive luminous materials are to temperature. These adjustments are made deliberately to improve sensitivity or deal with particular environmental issues.

Real-time calibration techniques can be used to continuously update the relationship between luminescent signals and temperature values in dynamic environments or applications with changing conditions. The calibration curve might need to take into consideration the non-linear temperature dependence of luminous characteristics in applications that span a large temperature range. The precision and dependability of temperature readings are guaranteed by the calibration and temperature sensitivity, which are essential elements of luminescence-based thermometry. These characteristics need to be carefully taken into account by scientists and engineers when designing, implementing, and optimizing luminescence-based temperature sensors for use in a variety of scientific and industrial applications.

3.3 MEASUREMENT TECHNIQUES/OPERATING PRINCIPLES

3.3.1 Imaging and microscopy

A key component of luminescence-based thermometry is imaging and microscopy techniques, which provide highly sensitive, spatially resolved temperature data. With the use of these techniques, temperature distributions at the micro and nanoscale can be seen, offering important new information on the localized temperature fluctuations within a sample. The concepts, methods, and uses of microscopy and imaging in relation to luminescence-based thermometry are examined in this section.

1. **Confocal Microscopy:**
   Luminescent signals from particular focal planes within a sample can be selectively illuminated and detected using confocal microscopy, which uses a focused laser beam with a pinhole aperture. Detailed temperature maps can be created using confocal microscopy by moving the laser focus over the material. It is especially useful to investigate temperature fluctuations in complex and heterogeneous materials because of the attained spatial resolution.
2. **Fluorescence Lifetime Imaging (FLIM):**
   By measuring the fluorescence signal's decay period, FLIM can determine how long excited states last in luminous materials. Temperature variations can modify the lifespan of fluorescence. Thus, FLIM provides a novel method for luminescence-based thermometry, enabling sub-nanosecond precision in quantitative temperature measurements.
3. **Two-Photon Microscopy:**
   Two photons must absorb simultaneously in order to stimulate luminous materials for two-photon microscopy to work. When compared to conventional one-photon excitation techniques, this technology delivers reduced photodamage and enhanced penetration depth. Two-photon microscopy facilitates three-dimensional temperature mapping in biological tissues and other optically dense samples for luminescence-based thermometry.
4. **Fiber Optic Probes for Luminescence Imaging:**
   Fiber optic probes can be used to expand luminosity-based thermometry to inaccessible or distant locations. These probes enable imaging in difficult situations by delivering excitation light and

gathering luminescent signals. Medical diagnostics, environmental sensing, and industrial process monitoring are just a few of the industries that use fiber optic-based imaging.

5. **Multimodal Imaging:**

A more thorough understanding of the sample can be obtained by combining luminescence-based thermometry with other imaging modalities such as MRI, X-ray, or Raman spectroscopy. This approach provides complementary information. Luminescence-based thermometry is improved by multimodal imaging, which offers a multitude of data on composition, structure, and temperature.

6. **In Vivo Imaging and Biomedical Applications:**

Luminescence-based thermometry makes it possible to monitor temperature changes in living things without causing harm, particularly when combined with cutting-edge imaging methods. Applications include temperature fluctuations in biological tissues for diagnostic purposes and real-time monitoring of temperature during medical operations such as laser surgery.

7. **Time-Resolved Luminescence Imaging:**

Techniques for time-resolved imaging capture the kinetics of luminous processes as they occur. Understanding the kinetics of temperature-induced changes in luminous characteristics is made possible with the help of this temporal information. Time-resolved luminescence imaging is very helpful for examining dynamic processes in materials and biological systems, as well as abrupt temperature changes.

The capabilities of luminescence-based thermometry are further expanded by imaging and microscopy techniques, which provide researchers with previously unheard-of spatial and temporal resolution when examining temperature fluctuations. These techniques are used in a wide range of sectors, including advanced medical diagnostics, industrial monitoring, and fundamental materials science research. The combination of image and luminescence-based thermometry methods offers ongoing innovation in temperature sensing approaches as technology advances.

3.3.2 Fiber optic probes

The use of fiber optic probes is essential for expanding the range of settings and uses for luminescence-based thermometry. By remotely applying excitation light to the luminous material and gathering the ensuing signals, these probes make it possible to assess temperature in difficult-to-reach places. The principles, benefits, and uses of fiber optic probes in luminescence-based thermometry are explored in this section. A bundle of optical fibers makes up a fiber optic probe; one set of fibers provides excitation light to the luminescent material, while another set gathers the signals that are released when the substance glows. Because of their tiny size and versatility, fiber optics can be integrated into a variety of experimental setups. The key advantages and applications of the fiber optic probes are their remote sensing capability in harsh environments, versatility, temperature profiling ability, industrial process monitoring, biomedical applications, and miniaturization for microscale and nanoscale applications. Keeping these advantages in mind, the fiber optic probes will probably become more important as technology develops and contributes to the advancement of luminescence-based thermometry.

3.3.3 Nanoparticle-based sensors

A state-of-the-art method for luminescence-based thermometry, nanoparticle-based sensors show great promise for nanoscale temperature measurements. Specifically, size-dependent optical characteristics of quantum dots can be used for extremely sensitive temperature detection. At the nanoscale, accurate and dynamic temperature measurements are made possible by the temperature-dependent shift in the emission spectra or intensity of these nanoparticles. Because of their small size, these particles can be easily integrated into complex systems, which makes them useful in fields like biotechnology, where they can be used to monitor intracellular temperature. Furthermore, nanoparticle-based sensors show adaptability in a variety of settings, such as semiconductor devices and biological systems, highlighting their potential to improve our knowledge of temperature dynamics at the nanoscale. By combining these sensors with imaging methods, their potential is further expanded, allowing exceptional precision in temperature mapping at the nanoscale regime for spatially resolved mapping. Nanoparticle-based sensors in luminescence-based thermometry have great potential to open new doors to understanding nanoscale temperature changes and their applications in a variety of scientific and technological fields as technology develops.

3.3.4 Multimodal approaches

In order to obtain a more thorough understanding of temperature dynamics, multimodal techniques in luminescence-based thermometry constitute a sophisticated strategy that combines various sensing modalities. The benefits of each modality can be combined with luminescence-based techniques to allow researchers to extract more detailed information from the target system.

Imaging Techniques: Spatial resolution temperature mapping is made possible by combining luminescence-based thermometry with imaging methods like confocal microscopy or two-photon microscopy. This method offers insights regarding the sample's morphology and structure in addition to temperature, making it very useful in complicated biological or material systems [30].

Spectroscopy Integration: By integrating luminescence-based thermometry with spectroscopic techniques such as Raman spectroscopy, it is possible to simultaneously measure both the molecular composition and temperature. By using a multimodal method, researchers can get more detailed information and investigate the complex interaction between chemical fingerprints and temperature fluctuations [31].

Industrial Applications: Multimodal techniques combine luminescence-based thermometry with other sensors, including strain or pressure sensors, in industrial contexts. Strong and adaptable monitoring systems with extensive data on temperature dynamics and other crucial parameters are produced by this synergistic combination [32].

Biomedical Applications: Applications for multimodal methodologies can be found in biomedical research, where non-invasive in vivo temperature monitoring is made possible by combining imaging techniques with luminescence-based thermometry. This gives concurrent information on the structure of the tissue and helps to explain temperature changes in biological tissues [33]. Multimodal techniques offer improved precision, dependability, and a comprehensive understanding of the target system in luminescence-based thermometry. These methods have the potential to be extremely important in solving challenging measurement problems in the scientific, industrial, and biomedical domains as technology develops. Temperature sensing and monitoring will continue to be innovative thanks to the synergistic combination of several sensing modalities.

3.4 FABRICATION PROCESS

Luminescence-based optical thermometers are made by first prepping the luminous materials, then incorporating them into an appropriate matrix or substrate, and then designing the sensing device. An outline of the fabrication process is provided below, along with pertinent citations and references:

1. **Synthesis of Luminescent Materials:**
   Luminescent materials are synthesized utilizing a variety of techniques that are specific to the characteristics of each substance, such as organic dyes, quantum dots, and phosphors [34]. The material selection is based on temperature sensitivity and the intended emission characteristics.

2. **Surface Modification and Functionalization:**
   Strategies for surface modification, such as coating or ligand exchange, are used to improve the biocompatibility, dispersibility, and stability of luminous materials. Further, laser texturing-based surface modifications are helpful to drastically increase the effective surface area of any material surface by multiplying the signal spectrum for the

intended applications [35]. The introduction of particular functional groups for intended applications may be a part of functionalization.

3. **Matrix or Substrate Integration:**

   The sensing layer is formed by depositing luminescent materials onto a substrate or integrating them into a matrix. This may entail embedding within a polymeric matrix, thin-film deposition, or sol-gel techniques [36].

4. **Device Fabrication:**

   The transducer, optical parts, and signal readout systems may all be a part of the overall device construction, which includes the luminous sensing layer. Materials that glow can be precisely placed using methods like drop-casting, spin-coating, or inkjet printing.

5. **Calibration and Characterization:**

   Through the correlation of luminous qualities with known temperatures, the constructed device is calibrated. Evaluating the device's sensitivity, accuracy, and response time in various scenarios is part of its characterization process.

6. **Integration with Optical Systems:**

   Optical thermometers based on luminosity can be combined with optical systems for signal detection and excitation. Temperature readouts are reliable and accurate thanks to this integration. For more complex applications, advanced optical equipment like confocal microscopy or spectroscopy may be used [37].

7. **Application-Specific Modifications:**

   The tool may go through certain changes depending on its intended use, such as encapsulation for environmental protection or biofunctionalization for biological uses [38]. Engineering, chemistry, and materials science are all used in the multidisciplinary process of creating luminescence-based optical thermometers. The goal of ongoing research in this area is to improve manufacturing processes, maximize material properties, and discover novel uses for these temperature sensors.

3.4.1 Performance and reliability

For luminescence-based optical thermometers to be widely used in a variety of applications, their performance is essential. Reliability, accuracy, sensitivity, and response time are examples of key performance indicators. An overview of the dependability and performance of optical thermometers based on luminescence is given below.

1. **Sensitivity and Response Time:**

   An optical thermometer based on luminescence is said to be sensitive if it can pick up on even the smallest temperature variations.

Accurate temperature readings are made possible by increased sensitivity [39]. Also, cross-sensitivity reduction is essential for precise temperature readings [40]. The speed at which a luminescence-based optical thermometer detects a temperature change is known as its reaction time. For dynamic temperature monitoring, faster reaction times are beneficial [41].

2. **Accuracy and Reliability:**

   The degree to which the temperature recorded matches the actual temperature is known as accuracy. Achieving a high degree of accuracy is necessary to get accurate temperature readings. The luminescence-based optical thermometer's stability and consistency over time and in different settings are all included in its reliability. Accurate and consistent readings should be provided using a trustworthy thermometer for which the sample thickness plays a vital role in thermometry in general and Raman thermometry in particular [42].

3. **In Vivo Performance:**

   The performance of the luminescence-based optical thermometer inside living things is crucial for in vivo applications, taking into account elements like signal stability and tissue penetration depth.

4. **Environmental Stability:**

   This device has the potential to demonstrate stability under a variety of environmental circumstances, rendering it appropriate for a multitude of uses ranging from environmental monitoring to industrial processes.

   In order to expand the range of applications for luminescence-based optical thermometers in various scientific and industrial contexts, ongoing research in this subject is concentrated on improving the efficiency and dependability of these devices, investigating novel materials, and tackling particular difficulties.

3.4.2 Failure and stability against harsh conditions

For luminescence-based optical thermometers to be used in real-world scenarios, especially in hard environments such as extreme temperatures, chemical exposure, dampness, and mechanical stress, they must be stable. Elevated temperature might cause material deterioration, which could reduce precision. Similarly, exposure to strong chemicals or moisture can result in chemical and physical alterations that could compromise the luminescence of the material. Devices that operate in challenging conditions must be resistant to radiation and mechanical stress. Improved stability is the goal of techniques like encapsulation and protective coatings, which also serve to shield the luminous materials from outside influences. Stability over time and age resistance are critical for reliable performance over long durations. The goal of research is still to solve these issues and guarantee the dependability of luminescence-based optical thermometers in a variety of harsh environments.

3.5 APPLICATIONS

Because of the crucial advantages like accuracy, non-invasive, non-contact nature, luminescence-based optical thermometers find extensive use in a wide range of fields. These thermometers aid in in vivo temperature monitoring during medical procedures in biomedical imaging, guaranteeing accurate therapeutic control. They aid in temperature mapping at the micro and nanoscale in microelectronics and nanotechnology, improving the thermal performance of electronic devices. These thermometers are useful for environmental monitoring because they make it possible to evaluate temperature differences in industrial and ecological contexts. Luminescence-based optical thermometers are used by industries to monitor and regulate temperatures during operations like chemical reactions and manufacturing. These thermometers are used in the food sector to monitor temperatures at different phases of production, thereby ensuring food safety. Examining and improving the thermal performance of batteries, solar cells, and other energy storage devices is beneficial for energy systems. In order to ensure safe aircraft and spacecraft operation, aerospace and aviation rely on these thermometers to monitor crucial component temperatures. Luminescence-based optical thermometers are used in scientific research to investigate temperature-dependent processes in biological, chemical, and material systems, therefore expanding scientific understanding across disciplinary boundaries.

3.6 RECENT ADVANCEMENTS AND CHALLENGES

The creation of nanomaterials with enhanced sensitivity and precision for nanoscale temperature monitoring is the focus of recent developments, especially with regard to advanced quantum dots and other luminous agents. Combining luminescence-based thermometry with imaging, spectroscopy, and other sensing modalities to provide a more thorough understanding of temperature changes in complex systems, the integration of multimodal techniques marks a significant advancement. The potential for sophisticated and real-time temperature monitoring with the integration of AI and ML algorithms for data analysis and interpretation is promising, as it will augment the capabilities of luminescence-based optical thermometry. Future research endeavors will investigate innovative sensing modalities such as plasmonic and metamaterial-based techniques, with the goal of augmenting the potential and adaptability of luminescence-based thermometry. It is expected that advancements in wireless and remote sensing technologies will make it easier to implement luminescence-based sensors in difficult-to-reach areas, increasing the range of applications for these devices.

Luminescence-based optical thermometry still faces a number of difficulties. Ensuring robustness in challenging environments, such as high temperatures and chemical exposure, is still a top priority, requiring constant research into materials and sensor design to improve endurance. Refinements in calibration processes are necessary to ensure accurate and dependable temperature observations, as calibration accuracy can be challenging, particularly in dynamic contexts. The challenge of cross-sensitivity to environmental influences continues, leading to research into strategies to reduce interference and boost the specificity of luminescence-based thermometry. It is a significant problem to close the gap between basic research and practical application, which calls for multidisciplinary cooperation and methodical validation of these technologies in real-world settings. Furthermore, the absence of established protocols for luminescence-based thermometry impedes its broad implementation, highlighting the necessity of creating uniform protocols and standards to guarantee the dependability and consistency of temperature readings in a range of contexts.

3.7 FUTURE PERSPECTIVES

Luminescence-based optical thermometry has a bright future ahead of it, with a number of possible developments in the works. The continuous development of nanomaterials for temperature sensing with higher sensitivity and specificity is an important direction. It is expected that advances in material design, such as sophisticated quantum dots and other luminous agents, will push the limits of precision at the nanoscale. Another intriguing possibility is the integration of machine learning and artificial intelligence for data analysis and interpretation, which will allow for more accurate and real-time temperature monitoring in intricate systems. Developing new sensing modalities and methodologies (e.g., plasmonic and metamaterial-based approaches) may allow luminescence-based thermometry to achieve even greater capabilities. Furthermore, the deployment of these sensors in difficult-to-reach places may be made easier with the use of wireless and remote sensing technologies. Future study will probably concentrate on addressing the issues that are now being faced, such as cross-sensitivity, calibration accuracy, and sensor resilience. The broad adoption of luminescence-based optical thermometry is anticipated to be driven by the translation of these developments into useful applications across a variety of industries, including industry and healthcare. Ultimately, cutting-edge technology, interdisciplinary cooperation, and more applications are anticipated in the future, cementing luminescence-based optical thermometry's place as a key component of temperature sensing and monitoring.

3.8 CONCLUSIONS

In summary, luminescence-based optical thermometry is evolving into a versatile and efficient tool for non-invasive, non-contact temperature measurement in a broad range of applications. Recent developments in temperature monitoring have greatly improved its accuracy and range, including the creation of nanomaterials and the use of multimodal techniques. Even if there are still problems, such as cross-sensitivity and robustness in challenging environments, more study shows promise in resolving these problems. With predicted advancements in nanomaterial design, the integration of cutting-edge technologies like artificial intelligence, and increased applications across a variety of areas, the future of luminescence-based optical thermometry seems bright. This technology is positioned to play a crucial role in furthering scientific understanding, enhancing industrial processes, and supporting advancements in environmental monitoring and healthcare as it continues to grow and bridge the gap between fundamental study and practical implementation. At the vanguard of temperature sensing, luminosity-based optical thermometry provides a path for future temperature measurements that are more precise, effective, and adaptable.

REFERENCES

1. Van Herwaarden, S. Physical principles of thermal sensors. *Sensor Mater.* 8, 373–387 (1996).
2. Mergny, J. L. & Lacroix, L. Analysis of thermal melting curves. *Oligonucleotides* 13, 515–537 (2003).
3. Narberhaus, F., Waldminghaus, T. & Chowdhury, S. RNA thermometers. *FEMS Microbiol. Rev.* 30, 3–16 (2006).
4. Saxena, N., Kumar, P. & Gupta, V. $CdS:SiO_2$ nanocomposite as a luminescence-based wide range temperature sensor. *RSC Adv.* 5, 73545–73551 (2015).
5. Wang, D. X., Wolfbeis, O. S. & Meier, R. J. Luminescent probes and sensors for temperature. *Chem. Soc. Rev.* 42, 7834–7869 (2013).
6. Brites, C. D. S. et al. Thermometry at the nanoscale. *Nanoscale* 4, 4799–4829 (2012).
7. Liu, H. et al. Intracellular temperature sensing: An ultra-bright luminescent nanothermometer with non-sensitivity to pH and ionic strength. *Sci. Rep.* 5, 14879; doi:10.1038/srep14879 (2015).
8. Fisher, L. H., Harms, G. S. & Wolfbeis, O. S. Upconverting nanoparticles for nanoscale thermometry. *Angew. Chem. Int. Ed.* 50, 4546–4551 (2011).
9. Brites, C. D. S. et al. Ratiometric highly sensitive luminescent nanothermometers working in the room temperature range: Applications to heat propagation in nanofluids. *Nanoscale* 5, 7572–7580 (2013).
10. Brites, C. D. S. et al. Thermometry at the nanoscale using lanthanide-containing organic–inorganic hybrid materials. *J. Lumin.* 133, 230–232 (2013).

11. Nikolic, M. G., Antic, Z., Culubrk, S., Nedeljkovic, J. M. & Dramicanin, M. D. Temperature sensing with Eu^{3+} doped TiO_2 nanoparticles. *Sensor Actuat. B Chem.* 201, 46–50 (2014).

12. Stich, M. I. J., Fischer, L. H. & Wolfbeis, O. S. Multiple fluorescent chemical sensing and imaging. *Chem. Soc. Rev.* 39, 3102–3114 (2010).

13. Heyes, A. L. On the design of phosphors for high-temperature thermometry. *J. Lumin.* 129, 2004–2009 (2009).

14. Gupta, S. K., Ghosh, P. S., Pathak, N., Arya, A. & Natarajan, V. Understanding the local environment of Sm^{3+} in doped $SrZrO_3$ and energy transfer mechanism using time-resolved luminescence: A combined theoretical and experimental approach. *RSC Adv.* 4, 29202–29215 (2014).

15. Zou, Y. et al. Fabricating $BaZrO_3$ hollow microspheres by a simple reflux method. *New J. Chem.* 38, 2548–2553 (2014).

16. Ye, T. et al. Controllable fabrication of perovskite $SrZrO_3$ hollow cuboidal nanoshells. *Cryst. Eng. Comm.* 13, 3842–3847 (2011).

17. Zhang, A. et al. Synthesis, characterization and luminescence of Eu^{3+}-doped $SrZrO_3$ nanocrystals. *J. Alloys Compd.* 22, L17–L20 (2009).

18. Huang, J. et al. Photoluminescence properties of $SrZrO_3:Eu^{3+}$ and $BaZrO_3:Eu^{3+}$ phosphors with perovskite structure. *J. Alloys Compd.* 487, L5–L7 (2009).

19. Gupta, S. K., Mohapatra, M., Natarajan, V. & Godbole, S. V. Site-specific luminescence of Eu^{3+} in gel-combustion-derived strontium zirconate perovskite nanophosphors. *J. Mater. Sci.* 47, 3504–3515 (2012).

20. Arora, S. T., Arora, V. B., Dayawati, R. & Khatkar, S. P. Synthesis, structural and optical properties of $SrZrO_3:Eu^{3+}$ phosphor. *J. Rare Earth* 32, 293–297 (2014).

21. Glais E., Pellerin M., Castaing V., Alloyeau D., Touati N., Viana B. & Chanéac C. Luminescence properties of $ZnGa_2O_4:Cr^{3+}$, Bi^{3+} nanophosphors for thermometry applications, *RSC Adv.* 8(73), 41767–41774 (2018).

22. Lojpur V., Antic Z. & Dramicanin M. D. Temperature sensing from the emission rise times of Eu^{3+} in SrY_2O_4, *Phys. Chem. Chem. Phys.* 16(46), 25636–25641 (2014).

23. Li X., Wei X., Qin Y., Chen Y., Duan C. & Yin M. The emission rise time of $BaY_2 ZnO_5:Eu^{3+}$ for non-contact luminescence thermometry, *J. Alloys Compd.* 657, 353–357 (2016).

24. Chi F., Jiang B., Zhao Z., Chen Y., Wei X., Duan C., Yin M. & Xu W. Multimodal temperature sensing using $Zn_2GeO_4:Mn^{2+}$ phosphor as highly sensitive luminescent thermometer, *Sens. Actuators B* 296, 126640 (2019).

25. Trejgis K. & Marciniak L. The influence of manganese concentration on the sensitivity of bandshape and lifetime luminescent thermometers based on $Y_3Al_5O_{12}:Mn^{3+},Mn^{4+},Nd^{3+}$ nanocrystals, *Phys. Chem. Chem. Phys.* 20(14), 9574–9581 (2018).

26. Dramićanin M. D. et al. $Li_2TiO_3:Mn^{4+}$ deep-red phosphor for the lifetime-based luminescence thermometry, *ChemistrySelect* 4(24), 7067–7075 (2019).

27. Marciniak L., Bednarkiewicz A. & Elzbieciak K. NIR–NIR photon avalanche based luminescent thermometry with Nd^{3+} doped nanoparticles, *J. Mater. Chem. C* 6(28), 7568–7575 (2018).

28. Siaï A., Haro-González P., Horchani-Naifer K. & Férid M. La$_2$O$_3$:Tm,Yb,Er upconverting nano-oxides for sub-tissue lifetime thermal sensing, *Sens. Actuators B* 234, 541–548 (2016).

29. Wang T. et al. Abnormally heat-enhanced Yb excited state lifetimes in Bi$_7$F$_{11}$O$_5$ nanocrystals and the potential applications in lifetime luminescence nanothermometry, *J. Mater. Chem. C* 7(44), 13811–13817 (2019).

30. Wang, Y., Liu, X., Jin, L. & Liu, Y. Multimodal imaging techniques for in situ monitoring of temperature in biological systems. *Adv. Optical Mater.* 9(21), 2100742 (2021).

31. Klimesz, B., Lisiecki, R. & Ryba-Romanowski, W. Sm^{3+}-doped oxyfluorotellurite glasses - spectroscopic, luminescence and temperature sensor properties, *J. Alloys Compd.* 788, 658–665 (2019).

32. Wu, H. & Wang, A. Multimodal sensing technologies for industrial temperature monitoring, *Sensors* 20(22), 6394 (2020).

33. Wang, Y., Liu, X., Jin, L. & Liu, Y. Multimodal imaging techniques for in situ monitoring of temperature in biological systems., *Adv. Optical Mater.* 9(21), 2100742 (2021).

34. Xia, Y. et al. One-pot synthesis of monodisperse colloidal spheres of upconverting NaYF$_4$:Yb^{3+}/Er^{3+} nanocrystals, *ACS Nano* 3(3), 723–729 (2009).

35. Panda, R., Thomas, J. & Joshi, H. C. Exploring the superhydrophilicity of nanosecond laser textured silicon: a Raman analysis, *Applied Optics* 61(23), 6770–6777 (2022).

36. Zhou, J. et al. Water-soluble NaYF$_4$:Yb,Er(Tm)/NaYF$_4$/polymer core–shell–shell nanoparticles with significant enhancement of upconversion fluorescence, *Chem. Mater.* 23(18), 4850–4856 (2011).

37. Xu, W. et al. Recent advances in near-infrared luminescent nanothermometers, *Nanophotonics* 7(12), 1901–1919 (2018).

38. Chen H., Li B., Ren X., Li S., Ma Y., Cui S. & Gu Y. Multifunctional near-infrared-emitting nano-conjugates based on gold clusters for tumor imaging and therapy, *Biomaterials* 33(33):8461–8476 (2012). doi:10.1016/j.biomaterials.2012.08.034.

39. Zhang, F. et al. Bright quantum dots emitting at ~1,600 nm in the NIR-IIb window for deep tissue fluorescence imaging, *Proc. Nat. Acad. Sci.* 108(6), 2092–2097 (2011).

40. Liu, Y. et al. Cross-sensitivity-free luminescence thermometry via intramolecular excitation transfer. *Anal. Chem.* 93(24), 8360–8366 (2021).

41. Brites, C. D. S. Self-calibrated double luminescent thermometers through upconverting nanoparticles, *Front. Chem.* 7, 1–10 (2019).

42. Sahoo, M., Ghosh, K., Sahoo, S., Sahoo, P. K., Mathews, T. & Dhara, S. Determination of thermal conductivity of phase pure 10H-SiC thin films by non-destructive Raman thermometry, arXiv:2308.05437v1, 1–5 (2023).

Chapter 4

Metal-organic framework for luminescent thermometers

Alka Rani, Arpit Verma, and Bal Chandra Yadav

4.1 INTRODUCTION

Detecting illness and monitoring treatment effectiveness can be facilitated through the measurement of body temperature. A fever, indicated by an elevated temperature, serves as an important marker in this regard. Temperature measurement and monitoring constitute essential components of an effective food cold chain management system. This process involves both the measurement and recording of temperatures. Various instruments, including bimetal-style thermometers, thermocouples, thermistors, infrared thermometers, etc., can be employed to achieve precise temperature measurements in the condition of food safety and quality assurance [1]. Many sides of nanoscale electrical and photonic systems require accurate, noninvasive temperature measurement and mapping with a sub-micrometric scale. This entails being aware of heat dissipation, heat transport, heat transfer profiles, and thermal reactions [2].

For temperature measurements at sub-micrometer range, conventional temperature measuring instruments including pyrometers, thermocouples, bimetallic and liquid-filled thermometers are usually unsuitable [3]. Recent developments in thermoset technology include temperature-dependent configuration alterations of molecular spring superstructures. Ga-filled carbon/MnO nanotubes, which enable great spatial resolution and operation at sizes smaller than a few micrometers with luminescence based measurements, scanning thermal microscopy probes, and Raman and infrared spectroscopic probing [4–6]. Among noninvasive spectroscopic techniques for temperature evaluation, phosphor luminescence stands out as an accurate method because of its dependency on thermal signature, which includes peak energy, band shape, intensity, and the durability of excited states. This technique is known as thermographic phosphor thermometry [7]. This technology has been used to remotely monitor temperature in difficult environments, such as biological fluids, strong EM fields, and fast-moving objects, with short collecting intervals. It also provides excellent spatial resolution and detection sensitivity [8]. Conventional

DOI: 10.1201/9781032661537-4

techniques frequently fail to produce satisfying results when dealing with situations like temperature changes within cells, molecular temperature, microcircuits, and micro-fluids [9].

The development of optical thermometers for large regions or fluid samples has been aided by the intrinsic limitations of mechanical or electrical thermometers [10]. There is an increasing need for a new generation of nanoscale thermometers that are capable of measuring temperature distributions at the nanoscale with accuracy. Luminescence-based temperature sensors are becoming more and more common due to their superior spatial resolution, fast reaction times, and safety during remote handling [11]. Because luminescence probes are intrinsically flexible, luminosity thermometry can be easily adapted to assist theranostics and nanotheranostics. Luminescence probes can be used in luminescence bioimaging in addition to thermometry, and many of these probes may also be useful in magnetic resonance imaging and X-ray computed tomography [12].

In general, luminescence quantum yield OF organic compounds tends to decrease with increasing temperature. Some compounds, on the other hand, have a special behavior whereby their luminescence intensity either stays constant or even rises from lower temperatures to room temperature. One example of these compounds is twisted intermolecular charge transfer (TICT) compounds [11]. A colorimetric shift in luminescence is also present with this phenomenon, which results from an alteration in the thermal equilibrium between the emission of excited states in the surrounding environment and the excited states in TICT [13]. The practical implementation of TICT compounds in thermometers is difficult, despite their intriguing features. Most TICT compounds have only low quantum yields of luminescence, and at increasing temperatures, their luminescence decreases dramatically [4]. One possible way to improve the conductivity of these MOF systems is to think about including metal ions or metal complexes in the polymeric chain [14].

The most recent developments in luminous molecular thermometers based on the metal-organic framework are discussed in this chapter. The photoluminescence properties of MOFs have had a major impact on scientific literature during the last ten years [15]. This effect results from both their extensive use in solid-state lighting materials like LLPs and OLEDs as well as the quickness of the related procedures [16]. Because physical or chemical stimuli cause a noticeable shift in their luminescence emission, MOFs are useful for quickly identifying temperature or analyte changes. Since emissions in MOFs might come from different regions inside the MOF, the PL in MOFs recognizes the hybrid character of their structures. These might include charge transfer (CT) between ligands and metal ions, as well as ligand-centered (LC) and metal-centered (MC) emissions. Additionally, it's possible that these emissions include guest molecules, which is important for photoluminescent detection in MOFs [17].

4.2 TEMPERATURE SENSING BASED ON LUMINESCENCE

The best method for obtaining a temperature measurement from luminescence varies depending on the intended use. Instead of providing the temperature directly, luminescence thermometry provides an indicator (Q), which may include variables like a band shift, luminescence lifespan, or a ratio of emission intensities [18]. Since the emission qualities are intrinsic to the emitting center, the development of thermal probes at the nanoscale is not limited by any fundamental issues. Luminescence thermometry makes use of temperature-induced changes in emission using two methods: time-integrated scheme, which tracks changes in a given emission spectrum, and time-resolved scheme, which tracks changes in a particular transition [19]. Jio Feng et al. designed a luminous triaryl boron fluorescent thermometer in which a temperature gradient was applied to a vertical quartz tube containing DPTB-MOE solution by heating it from the top and chilling it from the bottom [20]. This led to the temperature distribution of the liquid fluid taking the shape of a color gradient, with blue at the top and green at the bottom. As the temperature rose, the intermediate zone showed a seamless shift from green to cyan to blue. By using a temperature-dependent CIE chromaticity diagram or direct eye observation, this color-changing pattern made it possible to estimate or quantify the ambient temperature. This visual aid for temperature distribution detection is a useful visualization approach [20]. Runowski et al. prepared YVO_4:Yb^{3+}-Tm^{3+} nanomaterial for a luminescent nanothermometer intended to operate at very high temperature [21]. In Figure 4.1a, the UC emission band of Tm^{3+} ions and the NIR emission band of Yb^{3+} ions are shown. They calculated the band IR with temperature for TCLs of Tm^{3+} and non-TCLs of Yb^{3+} and Tm^{3+}, finding that the LIR associated with TCLs increases from 0.001 to 0.125 with temperature, while the LIR associated with non-TCLs increases from 0.25 to 50, a 200-fold times increases [21].

4.3 LUMINESCENCE THERMOMETER FIGURES OF MERIT

4.3.1 Thermal sensitivity

The way the thermometric parameter responds to temperature changes is known as thermal sensitivity. When the sensor element is used, it is established by the rate at which the fluorescence intensity ratio varies with temperature [22]. Absolute thermal sensitivity (S_a) is given by the specific Eq. (4.1) [23] as below:

$$S_a = \frac{\partial \Delta}{\partial T} = \mathrm{FIR}\left(\frac{\partial E}{k_B T^2}\right) \tag{4.1}$$

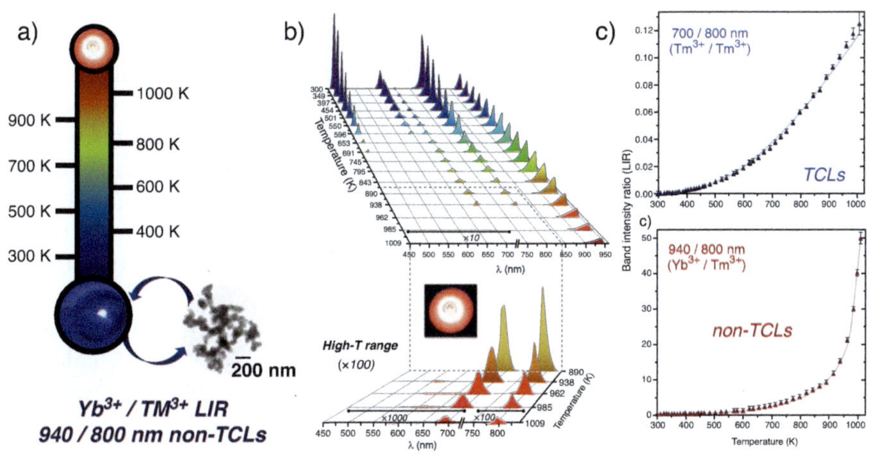

Figure 4.1   (a) Luminescent nanothermometer working at very high temperatures (b) the UC emission spectra for the synthesized YVO_4:Yb^{3+}- Tm^{3+} nanoparticles. (c) Luminescence intensity ratios for Tm^{3+} TCLs and non-TCLs of Yb^{3+} and Tm^{3+} as a change in temp. (Reprinted with permission from Ref. [21] American Chemical Society 2020.)

Regardless of the design or materials employed, relative thermal sensitivity (S_r), which represents the absolute change of FIR per degree of temperature as described in Eq. 4.2, is a useful tool for evaluating the performance of various thermometers.

$$S_r = \frac{1}{\Delta}\left|\frac{\partial \Delta}{\partial T}\right| = \frac{\partial E}{k_B T^2} \tag{4.2}$$

Thus parameter, denoted in units of (% K^{-1}) was introduced in 2007 within the realm of temperature optical sensors [24]. Since then, regardless of the type of thermometer, it has gained widespread acceptance as a figure of worth for comparison [25]. Although the absolute maximum sensitivity to compare host materials, S_r has the benefit of being autonomous of the kind of thermometer mechanical, electrical, or optical. Several thermometry techniques may be quantitatively compared because of their independence.

4.3.2 Temperature resolution and repeatability

As well as sensitivity, an essential system of measurement for evaluating the thermometer is temperature resolution (δT). It is calculated from Eq. (4.3)

$$\delta T = \frac{1}{S_r}\frac{\delta \Delta}{\Delta} \tag{4.3}$$

where $\delta\Delta/\Delta$ represents the relative error in the thermometric parameter measurements made using the experimental detection apparatus. The temperature resolution curve for the Rh101@UiO-67 thermometer was determined by Y. Zhou et al. inside the 20°C–60°C range, and it shows a resolution of better than 0.1°C on throughout the observed temperature span [26]. By analyzing the ratiometric luminescence response under heat cycling, the reversibility of the Rh101@UiO-67 thermometer was evaluated. The remarkable repeatability of the system was confirmed by the low variation in the Δ values at various temperatures throughout several test runs. Tifeng Xia et al. [27] revealed that the luminous thermometer's repeatability was evaluated using a series of heating-cooling cycles with temperatures ranging from 298 K to 473 K. When subjected to thermal cycling, calcined MOF thermometer exhibits a reversible luminescence response without any noticeable hysteresis caused by temperature variation. Throughout 12 consecutive heating-cooling cycles, the thermometric parameter Δ exhibits continuous reliability.

4.3.3 Response time

The amount of time it takes for a luminous thermometer to react to a step change in temperature by reaching 90% of its equilibrium fluorescence intensity is known as the reaction time [28]. A double exponential Eq. 4.4 may be used to fit the luminescence decline curves.

$$I(t) = A_1 \exp\left(-\frac{t}{\tau_1}\right) + A_2 \exp\left(-\frac{t}{\tau_2}\right) \tag{4.4}$$

The variables in this equation are I, the luminescence intensity, A_1 and A_2, time t, and the luminescence decay lifetimes of the two exponential components, τ_1 and τ_2, respectively [29]. The average lifetime τ, is determined by the following Eq. 4.5 [30].

$$\tau = \frac{\int I(t)\,t\,dt}{\int I(t)\,dt} \tag{4.5}$$

where $I(t)$ is the emission intensity at time t.

According to Francisco Sanchez et al., UiO-66-(OH)$_2$ MOF can identify temperature variations that cause luminescence intensity, which ranges from 30°C to 160°C [30]. They observed that, as shown in Figure 4.2 (a), the emission intensity of MOF gradually drops as the temperature rises. Figure 4.2(b) illustrates a linear response of MOF to temperature, where $I(t)$ and I_o values were recorded at 465 nm. Another crucial component of any LMOF being researched for use in a luminous thermometer is high

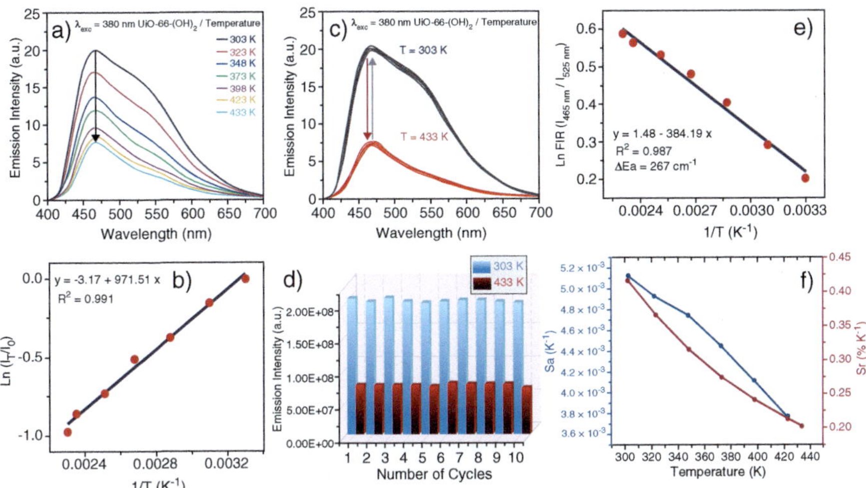

Figure 4.2 (a) Emission spectra of UiO-66-(OH)$_2$ in powder from collected at diff. increasing temperature (b) depiction of Arrhenius analysis (c) illustration of emission spectra of UiO-66-(OH)$_2$ (d) visualization of the emission intensity maxima observed at 465 nm (e) display of the Arrhenius analysis incorporating the FIR of the emission bands at 465 nm and 525 nm. (f) Graphical representation of the S_a and S_r vs. temperature. (Reprinted with permission from Ref. [31] American Chemical Society 2023.)

repeatability [31]. The emission intensity of UiO-66-(OH)$_2$ is reduced at 433 K and subsequently recovers at 303 K, as seen in Figure 4.2(c). Figure 4.2(d) displays the maximum emission intensity values at these temperatures throughout a 10-cycle period. The Arrhenius equation predicts activation energy (ΔE_a) of 267 cm^{-1} when FIR values are substituted for the usual intensity (Figure 4.2e). Plotting the values of S_a and S_r versus temperature, Figure 4.2(f) displays the greatest values of S_a and S_r.

4.4 LUMINESCENCE THERMOMETRY TECHNIQUE

By measuring temperature-dependent photoluminescence characteristics such as intensity, longevity, spectral location, and bandwidth, luminosity thermometry is used to determine temperature [32]. With short acquisition periods, this technique produces precise temperature readings with good spatial resolution. It shows great promise in a number of fields, including photonics, micro-and nanofluidic, nanomedicine, optoelectronics, and photonics [33]. Luminescence intensity changes are the reporting mechanism for many optical temperature sensors. For example, a rise in temperature may increase the rate of nonradiative decay, which may decrease luminosity [34]. The Majority of the time, luminescence temperature detection involves the use of luminous molecular probes, nanoparticles, or nanoaggregates. These nanosensors and probes can be put to the surface of the sample being studied, such as cells, or incorporated inside it. They are commonly applied as a thin layer to the surface, frequently inside a polymer host that holds the particles or probes [35].

4.4.1 Types of luminescence thermometry

4.4.1.1 Intensity-based luminescence thermometry

Intensity-based temperature sensors frequently encounter errors as a result of variations in probe concentration, inefficiencies in excitation or detection, and a lack of temperature specificity. Reducing these difficulties can be achieved by using ratiometric detecting systems, which depend on the ratio of different intensities to give better accuracy [36]. The two main categories of thermographic phosphor temperature sensing algorithms are decay times and intensity ratio algorithms [37]. While decay times techniques employ the temperature dependency of lifetime of an emitting level, intensity ratio processes use the intensity of one or more transitions directly to calculate the temperature. The strength of each transition is just proportionate to the total number of atoms in specific excited states at temperature T, as shown in Eq. 4.6.

$$I \propto gAhn\exp\left(-\frac{E}{k_B T}\right) \tag{4.6}$$

The degeneracy of the state is represented by g in this instance, along with the spontaneous emission rate (A), frequency (n), Plank and Boltzmann constants (h and k_B, respectively), and energy of the level (E) [9]. Tetrafluoride-based single-band ratiometric luminescence thermometer is a prominent example of an intensity-based luminescence thermometer. Two emissions bands from a tetrafluoride material one stimulated by Ground State Absorption (GSA) and the other by Stimulated State Absorption (SSA) are used in this thermometer to measure luminescence intensity ratio. Because the GSA and SSA excitation routes have differing degrees of temperature sensitivity, the intensity ratio varies with temperature. This thermometer has the benefit of functioning only in the infrared spectrum, which makes it appropriate for use in situations without access to visible light [38]. Based on the LIRESA/GSA temperature dependency, an LIR thermometry system is built. Figure 4.3(a) shows the schematic representation of the system. The single-channel luminescence detection technique and alternating excitations are integrated into this design. To switch between the two excitation lights, an optical switch with programmable control and fast reaction speed is used. The intensities of the excitation sources in Figure 4.3(a) are represented by I_{exc1} and I_{exc2}, which stand for the GSA and SSA processes of Tb^{3+} ions, respectively [39].

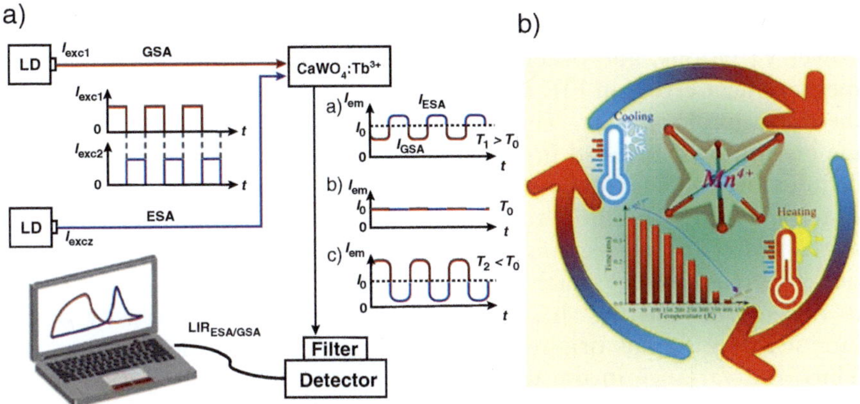

Figure 4.3 (a) Illustration of the LIR thermometry system. (Reprinted with permission from ref. [39] Optica 2021.) (b) Lifetime-based luminescent thermometry of $Sr_2InTa_{1-x}O_6$:Mn^{4+} material. (Reprinted with permission from Ref. [40] American Chemical Society 2022.)

4.4.1.2 Lifetime-based luminescence thermometry

Drifts in the optoelectronics system, such as adjustments to lights and detectors, might potentially negatively impact intensity-based measurements. On the other hand, lifetime data are more reliable since these problems do not impact them. Hongshang Peng et al. show the effect of temperature on NPS lifespan that can be clearly seen in Figure 4.3(b). The lifespan was shown to significantly decrease with temperature, resulting in a temperature sensitivity of $-2.2\%°C^{-1}$ between 25°C and 45°C. The decay-time figure shows a lesser sensitivity to temperature than the slope shown in the luminescence-intensity plot [3]. Wen Liu et al. constructed fluorescence lifetime-based luminescence material $Sr_2InTa_{1-x}O_6$:Mn^{4+} lead to a rapid decay of the lifetime from 0.403 to 0.008 ms with the increase in temperature [40].

4.4.1.3 Ratiometric luminescence thermometry

Rather than measuring the absolute intensity of a single emission, the ratiometric luminescence approach, also known as the fluorescence intensity ratio (FIR) technique, depends on calculating the relative intensity ratio between two separate emissions. This method reduces measurement errors brought on by variations in external variables like probe concentration and excitation power [32]. Numerous methods of measuring temperature have been studied, making use of the effects of temperature on a variety of characteristics, including FIR, lifespan, band shift, bandwidth, and more. Compared to other optical thermometry methods, FIR technology shows less reliance on the circumstances of measurement. Variations in air pressure, fluorescence detection losses, excitation source power variations, and other aspects are responsible for the reduction of measurement errors in FIR technology [41]. Two versions of FIR thermometric technology exist. The ratio of the same emission peak at various temperatures is included in the first. The ratio of several emission peaks or integrated intensity at a certain temperature is the second kind [42]. K. Maciejewska et al. report that when a metastable energy state acts both as the source for stokes emission and as the starting point for thermalization to higher energy states at higher temperatures, it becomes feasible to observe contrasting thermal trends in the emission intensity of two luminescence signals originating from a single type of dopant ions [43]. Thus, the variance in the temperature dependency of these two signals facilitates the advancement of a ratiometric luminescence thermometer with great thermal sensitivity [43]. Figure 4.4 illustrates how temperature affects the emission spectra of nanoparticles of $NaYF_4:Nd^{3+}$. Temperature causes a drop in the strength of the $4F_{3/2} \rightarrow 4I_{9/2}$ band and an increase in the $4F_{7/2}$ and $4S_{3/2} \rightarrow 4I_{9/2}$ bands. The concentration of Nd^{3+} has a significant impact on the integrated emission intensity of the $4F_{3/2} \rightarrow 4I_{9/2}$ band. The LIR is highly influenced by the concentration of Nd^{3+}. The LIR increases monotonically with temperature only when the concentration of Nd^{3+} is more than 25% [29].

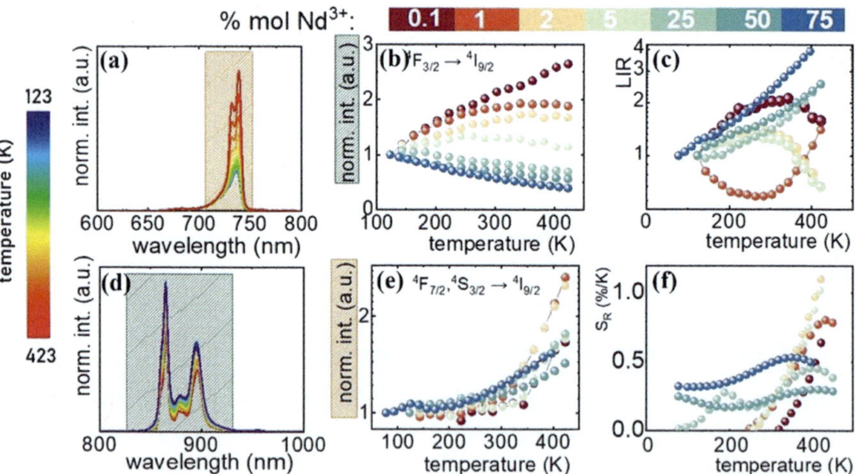

Figure 4.4 (a) The anti-stokes (b) stokes (c) the emission characteristics of $NaYF_4$:1%Nd³⁺ nanoparticles under 808 nm excitation reveal the effect of Nd³⁺ ion concentration on the thermal evolution of normalized integral emission intensities within the $4F_{3/2} \rightarrow 4I_{9/2}$ (d) and $4F_{7/2}$, $4S_{3/2} \rightarrow 4I_{9/2}$ (e) emission bands; the temperature-dependent LIR values of $NaYF_4$:Nd³⁺ nanoparticles (f) the corresponding S_r. (Reprinted with permission from Ref. [43] Nature 2023.)

4.4.1.4 Wavelength-based luminescence thermometry

Ratiometric luminous thermometers face several problems, the main ones being the requirement for precise control over components within the crystal lattice or for logical structural design at the molecular level. Since the emission wavelength of a wavelength-dependent luminous thermometer is independent of concentration, sample morphology, and measurement method, it may also provide accurate temperature detection [44]. Using a sinusoidally modulated light source, the indicator probe is excited in the frequency domain approach. Typically, the modulation frequencies range from 0.1 to 10 times the decay time. When the probe is stimulated sinusoidally, its emission matches the frequency of the excitation and exhibits an extra phase shift that depends on the lifespan. The following Eq. (4.7) may be used to calculate the decay period based on the phase shift (F).

$$t = \frac{\tan F}{2\pi f_{\mathrm{mod}}} \tag{4.7}$$

where f_{mod} is the modulation frequency of the excitation light [35].

4.4.2 The origin of luminescence in MOFs

The two primary kinds of luminescence are fluorescence and phosphorescence, which are frequently shown by the Jablonski Diagram in Figure 4.5(a) [45]. The diverse structural composition of MOFs, comprising a range of ligand molecules, inorganic ions or clusters, and guest molecular or ionic species, generates many pathways for the PL of MOFs. The processes depicted in Figure 4.5(b) include ligand-to-ligand charge transfer (LLCT), metal-to-metal charge transfer (MMCT), and emission arising from the interaction between the ligand and the metal center. Moreover, activities like guest-centered or sensitized emission that involve guest molecules inside MOF pores affect the overall photoluminescent behavior [46]. Organic linkers containing aromatic moieties or prolonged π structures are frequently utilized in the creation of porous MOFs due to their unbending molecular structure [47]. The π electrons in these linkers are crucial for luminescence, which may be divided into two categories: linker-based luminescence and LLCT [48]. Interestingly, Zn(II) and Cd(II) compounds frequently exhibit LMCT, but Cu(I) and Ag(I) compounds generally exhibit MLCT [49,50]. Liu et al. [51] used a number of electron-rich polycyclic aromatic hydrocarbons (PAHs) as guest molecules to construct a variety of fluorescent coordination polymers, as seen in Figure 4.5(c). Among these were triphenylene (green), coronene (yellow), and perylene (red). Figure 4.5(d) displays the 3D luminous structure.

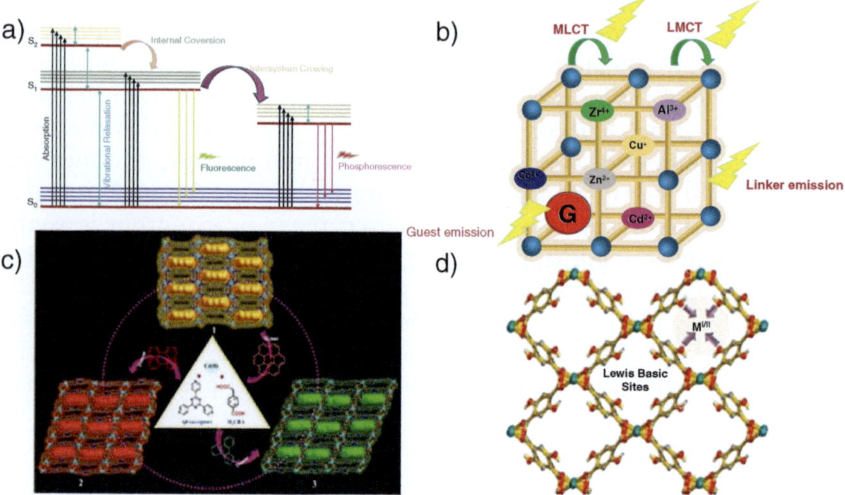

Figure 4.5 (a) Brief outline of Jablonski Diagram (b) schematic representation of diverse luminescence modes in MOFs, featuring inorganic metal clusters (multicolor spheres) interconnected by functional organic linkers (orange rods), and incorporating guest molecules (red sphere) (d) 3D luminescent context of [Mg (DHT) (DMF)$_2$]$_n$ showing Lewis basic pendant -OH groups accountable for sensing Cu(II) ion. (Reprinted with permission from Ref. [42] Springer 2019.)

4.5 EXAMPLES OF LUMINESCENT MOF THERMOMETERS

Luminescent MOFs have emerged as promising materials for luminescent thermometry because of their exceptional properties, such as high porosity, tenable luminescence properties, and thermal stability.

4.5.1 Intensity-based MOF thermometers

Zn-TDPAT, based on a new luminous MOF, was presented by Dingxuan Ma et al. as a sensor with dual functionality in detecting nitrobenzene and temperature changes [52]. They have recorded the Zn-TDPAT PL spectra at ambient temperature. Zn-TDPAT exhibits significant luminescence in the solid state, with a luminescence quantum yield of 0.213. Zn-TDPAT has such high luminous intensity because the highly conjugated systems in the ligand TDPAT6 are key sources of luminescence.

4.5.2 Ratiometric luminescent MOF thermometers

Luminescent metal-ligand complexes make up a significant class of molecular probes used in temperature detection. These complexes usually have lifetimes in the millisecond range, absorb visible light, show moderate brightness, and have large Stokes shifts. Because of their long lives, their luminescence, primarily phosphorescence often responds to oxygen levels [35]. For instance, Qian et al. [53] developed, the dual-emitting MOF/perylene composite, which allowed for the successful creation of a ratiometric MOF thermometer, and the luminous perylene dye inside the pores of a rare earth MOF. The exceptional sensitivity of this composite was made possible by the distinct energy transfer that occurs between perylene molecules and Eu^{3+} throughout the physiological temperature range. ZJU-88/perylene (0.1%) demonstrated a reducing emission intensity of perylene dye at 473 nm and a growing emission intensity of Eu^{3+} at 615 nm as the temperature rose from 293 K to 353 K. The luminescence intensities of the two emissions showed a significant linear connection. At 293 K, this thermometer's maximum sensitivity of 1.285 K^{-1} was greater than the previously published mixed Ln-MOF. In physiological circumstances, ZJU-88/perylene demonstrated remarkable stability and non-toxicity, suggesting potential use in biomedical situations requiring ratiometric luminescence temperature monitoring [54]. The temperature gradient within a copper pipe with one end flooded in liquid nitrogen (LN_2) was used in the first experiment on the materials (Figure 4.6a and b). The surface of the pipe was coated with $[EA]_2NaCr(HCOO)_6$ crystals using thermal paste. Thermometric calculations were achieved using the model derived from the published temperature-dependent luminescence properties. Using the PL found at 405 nm excitation wavelength, the integrated intensity ratio was calculated, and the findings were compared with the temperature relationship in the model. $[EA]_2NaCr(HCOO)_6$ features an alternating arrangement of octahedral CrO_6 and NaO_6 units in a crystal structure similar to perovskites. The 3-D metal format framework includes EA^+ cation-occupied voids (Figure 4.6(e)). Decay patterns were found for samples containing 21–100 mol% Cr^{3+}, as shown in Figure 4.6(f). Additionally, it demonstrates that the sample $[EA]_2NaCr_{0.21}A_{10.79}(HCOO)_6$ had the highest values of τ_1 and τ_2, which were found to be 1.35 and 2.26 ms, individually [55].

4.5.3 Mixed-lanthanide MOF thermometers

In luminous thermometers, Ln-MOFs are widely used as temperature probes. It is possible to modify Ln-MOFs to create ratiometric luminescence thermometers by adding certain guest molecules, bridging organic ligands, or carefully selecting metal ions. This makes it possible to adjust optical output wavelengths and response temperature ranges to meet particular needs [56]. There are several obstacles in the way of producing

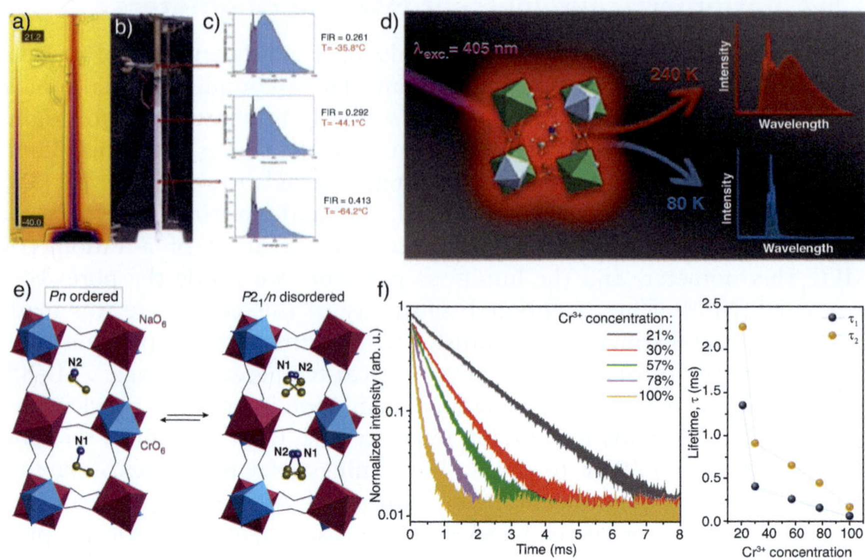

Figure 4.6 Exemplary thermometric system: (a) thermal image (b) image of the experimental system (c) recorded emission spectra for discrete points with considered FIR (d) MOF with perovskite structures with their luminescence (e) crystal structure of EANaCr ($[EA]_2NaCr(HCOO)_6$) in the high temp. $P2_{1/n}$ and low temp. P_n phases (f) decay profiles of $[EA]_2NaCrxAl_{1-x}(HCOO)_6$ measured at 77 K and changes of time parameters as a function of Cr^{3+} concentration. (Reprinted with permission from Ref. [50] American Chemical Society 2012.)

luminescence-based thermometers, especially when it comes to single-transition intensity measurements that are sensitive to changes in excitation power, material homogeneity, or sensor concentration [57,58]. Some of these issues have been addressed by the development of mixed-lanthanide MOFs, which function as ratiometric thermometers. The intensity ratio between the individual emissions of Tb^{3+} at 545 nm and Eu^{3+} at 613 nm is displayed by these MOFs, and significant linear correlations between 10 K and 300 K have been found. The ligands play a crucial role in the luminous characteristics of M'LnMOFs, affecting the development of crystal structures between the ligands and Ln [59–61].

The dual functionality of Ln-doped nanoparticles has been the subject of several investigations in the literature. These studies have looked into simultaneous optical heating and real-time temperature determination based on thermally linked excited states [62]. These studies include a wide range of host material selections, dopant ions, their concentrations, and particle shapes. But there is a problem that there is a greater chance of nonradiative depopulation of excited states, which can cause luminescence quenching and affect temperature reading accuracy. Therefore, in order to

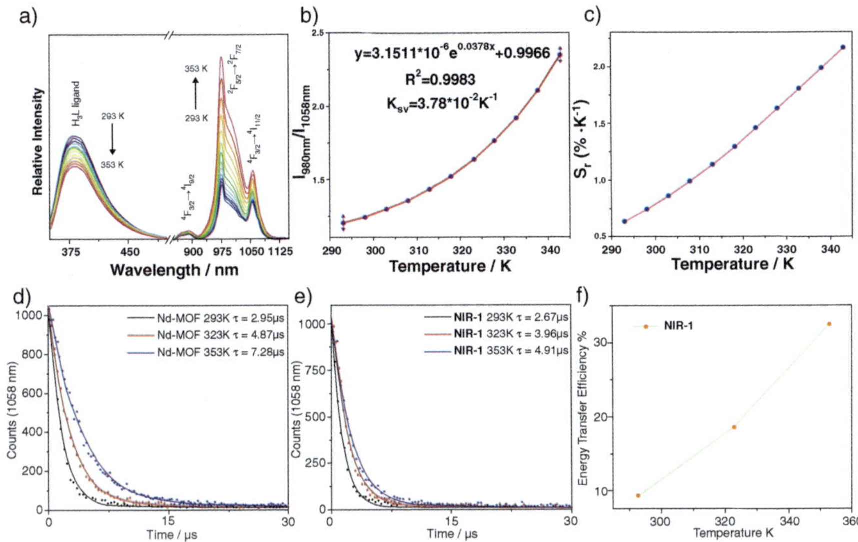

Figure 4.7 (a) Fluorescent spectra of NIR-1 at variable temp. with respect to wavelength (b) nonlinear association between FIR of Yb^{3+} and Nd^{3+} and temperature (c) relative sensitivity for NIR-1 at different temp. (d) Luminescent lifetime of Nd-MOF at diff. temperature (e) the solid luminescent lifetime of NIR-1 (f) energy transfer efficiency vs temp. dependence in NIR-1. (Reprinted with permission from Ref. [59] American Chemical Society 2018.)

overcome this problem, some research has concentrated on using core-shell and yolk-shell topologies to separate the heating functionality from luminescent thermometry [63]. Figure 4.7(a) shows the solid-state PL spectroscopy of the synthesized material (NIR Yb-Nd MOF) Nd-MOF@Yb-MOF@SiO_2@Fe_3O_4 (NIR-1) by Yu-Peng Jiang et al. [64] was carried out between 293 and 353 K in temperature. The measured relationship between Δ and temperature demonstrated a strong nonlinear link, as seen in Figure 4.7 (b and c). As the temperature climbed from 293 to 353 K, the lifespans of NIR-1 and Nd-MOF improved, respectively, from 2.76 to 4.91 μs and 2.95 to 7.28 μs (Figure 4.7(d and e)). With increasing temperature, Figure 4.5 f demonstrated a progressive improvement in energy transfer efficiency from Nd^{3+} to Yb^{3+}.

4.6 CONCLUSIONS

Finally, it should be noted that the use of metal-organic frameworks (MOFs) for luminous thermometry has become a vibrant and exciting topic with broad ramifications. This chapter illustrated the variety of bright signal temperature measurement techniques that MOFs can employ. These techniques include luminescent thermometry based on intensity, lifespan, wavelength, and ratiometric measurements. The source of luminescence in MOFs has been a major area of debate, elucidating the fundamental principles governing their efficacy as thermometric materials. Examining MOFs for bright thermometry presents exciting new avenues for research and development. Further work must focus on enhancing the synthesis methods to increase the luminous MOFs' repeatability and scalability. Subsequent endeavors might concentrate on investigating novel MOF designs and functionalization to enhance sensitivity and expand the scope of temperature regimes that are within reach. Researches on metal-organic frameworks for luminous thermometry is expected to continue to influence the field of temperature-sensing technologies.

CONFLICTS OF INTEREST

The authors declare that they have no known competing financial interests or personal relationships that could have appeared to influence the work reported in this paper.

ACKNOWLEDGMENTS

Ms. Alka Rani is thankful to the Council of Scientific & Industrial Research and University Grant Commission (CSIR-UGC), Government of India for financial support in the form of Junior Research Fellowship (NTA Ref. No.: 231610026817). Mr. Arpit Verma and Prof. B. C. Yadav acknowledge to Uttar Pradesh Council of Science and Technology, Lucknow for financial assistance in the form of project Ref: CST/D-2290.

REFERENCES

1. S.J. James, C. James, Chilling and freezing, *Food Safety Management*, (2023) 453–474.
2. S. Sadat, A. Tan, Y.J. Chua, P. Reddy, Nanoscale thermometry using point contact thermocouples, *Nano Letters* 10(7) (2010) 2613–2617.
3. H. Peng, M.I.J. Stich, J. Yu, L.N. Sun, L.H. Fischer, O.S. Wolfbeis, Luminescent europium (III) nanoparticles for sensing and imaging of temperature in the physiological range, *Advanced Materials* 22(6) (2010) 716–719.
4. J. Lee, A.O. Govorov, N.A. Kotov, Nanoparticle assemblies with molecular springs: A nanoscale thermometer, *Angewandte Chemie International Edition* 44(45) (2005) 7439–7442.
5. E. Saïdi, J. Labéguerie-Egéa, L. Billot, J. Lesueur, M. Mortier, L. Aigouy, Imaging joule heating in an 80 nm wide titanium nanowire by thermally modulated fluorescence, *International Journal of Thermophysics* 34(8–9) (2013) 1405–1412.

6. S.K. Singh, K. Kumar, S.B. Rai, Multifunctional Er^{3+}–Yb^{3+} codoped Gd_2O_3 nanocrystalline phosphor synthesized through optimized combustion route, *Applied Physics B* 94 (2009) 165–173.

7. A. Rabhiou, J. Feist, A. Kempf, S. Skinner, A. Heyes, Phosphorescent thermal history sensors, *Sensors Actuators A: Physical* 169(1) (2011) 18–26.

8. C.D.S. Brites, P.P. Lima, N.J. Silva, A. Millán, V.S. Amaral, F. Palacio, L.D. Carlos, Thermometry at the nanoscale, *Nanoscale* 4(16) (2012) 4799–4829.

9. C.D.S. Brites, P.P. Lima, N.J.O. Silva, A. Millan, V.S. Amaral, F. Palacio, L.D. Carlos, Lanthanide-based luminescent molecular thermometers, *New Journal of Chemistry* 35(6) (2011) 1177–1183.

10. J. Lee, N.A. Kotov, Thermometer design at the nanoscale, *Nano Today* 2(1) (2007) 48–51.

11. D. Braun, W. Rettig, Kinetic studies of twisted intramolecular charge transfer in highly viscous solvents as a function of pressure and temperature, *Chemical Physics* 180(2–3) (1994) 231–238.

12. H. Xing, W. Bu, S. Zhang, X. Zheng, M. Li, F. Chen, Q. He, L. Zhou, W. Peng, Y. Hua, Multifunctional nanoprobes for upconversion fluorescence, MR and CT trimodal imaging, *Biomaterials* 33(4) (2012) 1079–1089.

13. V.V. Volchkov, B.M. Uzhinov, Structural relaxation of excited molecules of heteroaromatic compounds, *High Energy Chemistry* 42 (2008) 153–169.

14. A. Verma, P. Chaudhary, R.K. Tripathi, A. Singh, B.C. Yadav, State of the art metallopolymer based functional nanomaterial for photodetector and solar cell application, *Journal of Inorganic and Organometallic Polymers and Materials* 32(8) (2022) 2807–2826.

15. A. Verma, P. Chaudhary, R. K. Tripathi, B. C. Yadav, Flexible, environmentally-acceptable and long-durable-energy-efficient novel WS_2–polyacrylamide MOFs for high-performance photodetectors, *Materials Advances* 3(9) (2022) 3994–4005.

16. E. San Sebastian, A. Rodríguez-Diéguez, J. M. Seco, J. Cepeda, Coordination polymers with intriguing photoluminescence behavior: the promising avenue for greatest long-lasting phosphors, *European Journal of Inorganic Chemistry* 2018(20–21) (2018) 2155–2174.

17. P. Leo, D. Briones, J. A. García, J. Cepeda, G. Orcajo, G. Calleja, A. Rodriguez-Dieguez, F. Martínez, Strontium-based MOFs showing dual emission: luminescence thermometers and toluene sensors, *Inorganic Chemistry* 59(24) (2020) 18432–18443.

18. M. D. Dramićanin, Trends in luminescence thermometry, *Journal of Applied Physics* 128(4) (2020) 040902.

19. C. D. S. Brites, S. Balabhadra, L. D. Carlos, Lanthanide-based thermometers: at the cutting-edge of luminescence thermometry, *Advanced Optical Materials* 7(5) (2019) 1801239.

20. J. Feng, K. Tian, D. Hu, S. Wang, S. Li, Y. Zeng, Y. Li, G. Yang, A triarylboron-based fluorescent thermometer: sensitive over a wide temperature range, *Angewandte International Edition Chemie* 50(35) (2011) 8072–8076.

21. M. Runowski, P. Woźny, N. Stopikowska, I. R. Martín, V. Lavín, S. Lis, Luminescent nanothermometer operating at very high temperature—sensing up to 1000 K with upconverting nanoparticles (Yb^{3+}/Tm^{3+}), *ACS Applied Materials Interfaces* 12(39) (2020) 43933–43941.

22. S. A. Wade, S. F. Collins, G. W. Baxter, Fluorescence intensity ratio technique for optical fiber point temperature sensing, *Journal of Applied Physics* 94(8) (2003) 4743–4756.

23. S. F. Collins, G. W. Baxter, S. A. Wade, T. Sun, K. Grattan, Z. Y. Zhang, A.W. Palmer, Comparison of fluorescence-based temperature sensor schemes: theoretical analysis and experimental validation, *Journal of Applied Physics* 84(9) (1998) 4649–4654.

24. V. K. Rai, Temperature sensors and optical sensors, *Applied Physics B* 88 (2007) 297–303.

25. D. Manzani, J. F.D.S. Petruci, K. Nigoghossian, A. A. Cardoso, S. J. L. Ribeiro, A portable luminescent thermometer based on green up-conversion emission of Er^{3+}/Yb^{3+} co-doped tellurite glass, *Scientific Reports* 7(1) (2017) 41596.

26. Y. Zhou, D. Zhang, J. Zeng, N. Gan, J. Cuan, A luminescent lanthanide-free MOF nanohybrid for highly sensitive ratiometric temperature sensing in physiological range, *Talanta* 181 (2018) 410–415.

27. T. Xia, Y. Cui, Y. Yang, G. Qian, A luminescent ratiometric thermometer based on thermally coupled levels of a Dy-MOF, *Journal of Materials Chemistry C* 5(21) (2017) 5044–5047.

28. A. Rani, A. Verma, A. Singh, B. C. Yadav, Monitoring of UV-A radiation by TiO_2/CdS nanohybrid along with the high on-off ratio, *Sensors Actuators A: Physical* 367 (2024) 115060.

29. F. Chi, B. Jiang, Z. Zhao, Y. Chen, X. Wei, C. Duan, M. Yin, W. Xu, Multimodal temperature sensing using Zn_2GeO_4: Mn^{2+} phosphor as highly sensitive luminescent thermometer, *Sensors Actuators B: Chemical* 296 (2019) 126640.

30. J. Yu, L. Sun, H. Peng, M. I. Stich, Luminescent terbium and europium probes for lifetime based sensing of temperature between 0 and 70°C, *Journal of Materials Chemistry A* 20(33) (2010) 6975–6981.

31. F. Sánchez, M. Gutiérrez, A. Douhal, Taking advantage of a luminescent ESIPT-based Zr-MOF for fluorochromic detection of multiple external stimuli: acid and base vapors, mechanical compression, and temperature, *ACS Applied Materials Interfaces* 15(48) (2023), 56587–56599.

32. Y. Cheng, Y. Gao, H. Lin, F. Huang, Y. Wang, Strategy design for ratiometric luminescence thermometry: circumventing the limitation of thermally coupled levels, *Journal of Materials Chemistry C* 6(28) (2018) 7462–7478.

33. C. D. S. Brites, R. Marin, M. Suta, A. N. Carneiro Neto, E. Ximendes, D. Jaque, L. D. Carlos, Spotlight on luminescence thermometry: basics, challenges, and cutting-edge applications, *Advanced Materials* 35(36) (2023) 2302749.

34. P. Löw, B. Kim, N. Takama, C. Bergaud, High-spatial-resolution surface-temperature mapping using fluorescent thermometry, *Small* 4(7) (2008) 908–914.

35. X.-D. Wang, O.S. Wolfbeis, R. J. Meier, Luminescent probes and sensors for temperature, *Chemical Society Reviews* 42(19) (2013) 7834–7869.

36. E. J. McLaurin, L. R. Bradshaw, D. R. Gamelin, Dual-emitting nanoscale temperature sensors, *Chemistry of Materials* 25(8) (2013) 1283–1292.

37. A. L. Heyes, On the design of phosphors for high-temperature thermometry, *Journal of Luminescence* 129(12) (2009) 2004–2009.

38. K. Trejgis, K. Ledwa, A. Bednarkiewicz, L. Marciniak, A single-band ratiometric luminescent thermometer based on tetrafluorides operating entirely in the infrared region, *Nanoscale Advances* 4(2) (2022) 437–446.

39. Z. Yuan, P. Lixin, T. Peng, Z. Zhiguo, Luminescence intensity ratio thermometry based on combined ground and excited states absorptions of Tb^{3+} doped $CaWO_4$, *Optics Express* 29(14) (2021) 22805–22812.

40. W. Liu, D. Zhao, R.-J. Zhang, Q.-X. Yao, S.-Y. Zhu, Fluorescence lifetime-based luminescent thermometry material with lifetime varying over a factor of 50, *Inorganic Chemistry* 61(41) (2022) 16468–16476.

41. Q. Wang, M. Liao, Q. Lin, M. Xiong, Z. Mu, F. Wu, A review on fluorescence intensity ratio thermometer based on rare-earth and transition metal ions doped inorganic luminescent materials, *Journal of Alloys Compounds* 850 (2021) 156744.

42. Q. Chi, Z. Gao, C. Zhang, T. Zhang, Y. Cui, X. Wang, Q. Lei, Microstructures and energy storage property of sandwiched BZT-BCT@ Fe3O4/polyimide composites, *Journal of Materials Science: Materials in Electronics* 30 (2019) 1–8.

43. K. Maciejewska, L. Marciniak, The role of Nd3+ concentration in the modulation of the thermometric performance of Stokes/anti-Stokes luminescence thermometer in $NaYF_4$: Nd^{3+}, *Scientific Reports* 13(1) (2023) 472.

44. D.-N. Song, D.-J. Zhang, Y.-L. Wang, J.-J. Wang, X.-S. Xing, Z.-Y. Lv, F. Liu, J.-X. Han, R.-C. Zhang, S.-J. Liao, Luminescent thermochromic silver iodides as wavelength-dependent thermometers, *Inorganic Chemistry* 59(18) (2020) 13067–13077.

45. M. Pamei, A. Puzari, Luminescent transition metal-organic frameworks: an emerging sensor for detecting biologically essential metal ions, *Nano-Structures Nano-Objects* 19 (2019) 100364.

46. J. Dong, D. Zhao, Y. Lu, W.-Y. Sun, Photoluminescent metal-organic frameworks and their application for sensing biomolecules, *Journal of Materials Chemistry A* 7(40) (2019) 22744–22767.

47. A. K. Shukla, V. Verma, P. Goriyan, A. Rani, A. Verma, A. Singh, B. C. Yadav, R. K. Baimuratova, A. V. Andreeva, G. I. Dzhardimalieva, $Zr_6O_4(OH)_4$ based metal-organic frameworks for the enhanced chemiresistive sensing of ethanol, *Journal of Inorganic Organometallic Polymers Materials* (2024) 1–16. doi:10.1007/s10904-023-02986-1.

48. Z. Hu, B. J. Deibert, J. Li, Luminescent metal–organic frameworks for chemical sensing and explosive detection, *Chemical Society Reviews* 43(16) (2014) 5815–5840.

49. J.-C. Dai, X.-T. Wu, Z.-Y. Fu, C.-P. Cui, S.-M. Hu, W.-X. Du, L.-M. Wu, H.-H. Zhang, R.-Q. Sun, Synthesis, structure, and fluorescence of the novel cadmium (II)–trimesate coordination polymers with different coordination architectures, *Inorganic Chemistry* 41(6) (2002) 1391–1396.

50. G. A. Senchyk, V. O. Bukhan'ko, A. B. Lysenko, H. Krautscheid, E. B. Rusanov, A. N. Chernega, M. Karbowiak, K. V. Domasevitch, AgI/VV heterobimetallic frameworks generated from novel-type $\{Ag_2\ (VO_2F_2)_2\ (triazole)_4\}$ secondary building blocks: a new aspect in the design of SVOF hybrids, *Inorganic Chemistry* 51(15) (2012) 8025–8033.

51. X.-T. Liu, B. Zhao, Y.-H. Zhang, S.-S. Chen, J. Zhu, Z. Chang, X.-H. Bu, Structure and emission modulation of a series of Cd(ii) luminescent coordination polymers through guest dependent donor–acceptor interaction, *Crystal Growth Design* 19(2) (2019) 1391–1398.

52. D. Ma, B. Li, X. Zhou, Q. Zhou, K. Liu, G. Zeng, G. Li, Z. Shi, S. Feng, A dual functional MOF as a luminescent sensor for quantitatively detecting the concentration of nitrobenzene and temperature, *Chemical Communications* 49(79) (2013) 8964–8966.

53. Y. Cui, R. Song, J. Yu, M. Liu, Z. Wang, C. Wu, Y. Yang, Z. Wang, B. Chen, G. Qian, Dual-emitting MOF⊃ dye composite for ratiometric temperature sensing, *Advanced Materials* 27(8) (2015) 1420–1425.

54. Y. Li, Temperature and humidity sensors based on luminescent metal-organic frameworks, *Polyhedron* 179 (2020) 114413.

55. A. Kabański, M. Ptak, D. Stefańska, Metal–organic framework optical thermometer based on Cr^{3+} ion luminescence, *ACS Applied Materials Interfaces* 15(5) (2023) 7074–7082.

56. T. Feng, Y. Ye, X. Liu, H. Cui, Z. Li, Y. Zhang, B. Liang, H. Li, B. Chen, A robust mixed-lanthanide polyMOF membrane for ratiometric temperature sensing, *Angewandte Chemie International Edition* 132(48) (2020) 21936–21941.

57. Y. Salinas, R. Martínez-Máñez, M. D. Marcos, F. Sancenón, A. M. Costero, M. Parra, S. Gil, Optical chemosensors and reagents to detect explosives, *Chemical Society Reviews* 41(3) (2012) 1261–1296.

58. S.-N. Zhao, G. Wang, D. Poelman, P. V. D. Voort, Luminescent lanthanide MOFs: a unique platform for chemical sensing, *Materials* 11(4) (2018) 572.

59. X.-D. Zhu, K. Zhang, Y. Wang, W.-W. Long, R.-J. Sa, T.-F. Liu, J. Lü, Fluorescent metal–organic framework (MOF) as a highly sensitive and quickly responsive chemical sensor for the detection of antibiotics in simulated wastewater, *Inorganic Chemistry* 57(3) (2018) 1060–1065.

60. J. Feng, K. Tian, D. Hu, S. Wang, S. Li, Y. Zeng, Y. Li, G. Yang, A triarylboron-based fluorescent thermometer: sensitive over a wide temperature range, *Angewandte Chemie International Edition* 50(35) (2011) 8072–8076.

61. J. Parikh, B. Mohan, K. Bhatt, N. Patel, S. Patel, A. Vyas, K. Modi, Effect of temperature on metal-organic frameworks chemical sensors detection properties, *Microchemical Journal* 184 (2023) 108156.

62. L. Marciniak, K. Kniec, K. Elzbieciak, A. Bednarkiewicz, Non-plasmonic NIR-activated photothermal agents for photothermal therapy, *Near Infrared-Emitting Nanoparticles for Biomedical Applications* 1 (2020) 305–347.

63. K. Elzbieciak-Piecka, L. Marciniak, Optical heating and luminescence thermometry combined in a Cr^{3+}-doped $YAl_3(BO_3)_4$, *Scientific Reports* 12(1) (2022) 16364.

64. Y.-P. Jiang, X.-H. Fang, Q. Wang, J.-Z. Huo, Y.-Y. Liu, X.-R. Wang, B. Ding, Near-infrared magnetic core-shell nanoparticles based on lanthanide metal-organic frameworks as a ratiometric felodipine sensing platform, *Communications Chemistry* 6(1) (2023) 96.

Chapter 5

Quantum dots for luminescence thermometers

Gurpreet Kaur, Daljit Kaur, Vinita Sharma, and Deepak Kumar

5.1 INTRODUCTION

Temperature is a potential physical parameter like length, mass, and time to be examined. The measurement of temperature, i.e., thermometry, is therefore crucial since the increase or decrease of its value from its optimum value affects the environment and our daily life activities. It is anticipated that temperature sensors will account for up to 75%–80% of the global sensor market. Thermometers can be classified as liquid-filled glass devices, thermocouples, and optical sensors. The Zeroth law in thermodynamics discusses the temperature as a state parameter, and this lays the foundation for novel technology. The development of this field from contact mode liquid column thermometers to nanoprobe optical thermometers has been quite long. The first alcohol thermometer is thought to have been invented in 1641 by Galileo. In 1665, Robert William Boyle underlined the significance of thermometer scale comparability. E. Halley, a British scientist, utilized mercury as a thermometric liquid for the first time in 1693. In 1708, Daniel Gabriel Fahrenheit presented the design of mercury filled in glass as a potential thermometer [1,2]. In 1742, Anders Celsius introduced the well-known Celsius scale. Absolute and relative thermometers are the two types of basic thermometers. Absolute primary thermometers measure thermodynamic temperature using Boltzmann constant. Relative primary thermometers, on the other hand, measure temperature indirectly by employing temperature fixed points to derive the parameters in the equation of state. Acoustic gas thermometry (AGT), constant-volume gas along with refractive index thermometry, etc. are some of the ways of temperature measurement. Thermometers can be classified in various ways as contact or noncontact thermometers, direct or inferential thermometer, manually operated or automatic [3]. Further, the arrival of fluorescent lamp started the new field of luminescence thermometry. Bradley investigated the thermographic phosphors combined with the binders along with ceramic materials to assess temperature of the surface in 1953, which is considered to be the application of luminescence thermometry [4]. Luminescence/optical thermometry is a fascinating field that explores the relationship between

DOI: 10.1201/9781032661537-5

temperature and light emission. By harnessing the unique properties of certain materials, researchers can measure temperature with remarkable precision and accuracy, offering a promising alternative to conventional thermometers. In this chapter, we will delve into the principles and applications of luminescence thermometry; shedding light on its potential to revolutionize temperature sensing technology. Optical thermometry relies on the phenomenon of luminescence, which is basically emission of the light from a material under test after it has undergone absorption of energy. This energy can come from various sources such as heat or light excitation. By examining a material's luminescent properties such as quantum dots, it is possible to determine the temperature of that material because temperature directly affects this energy absorption and subsequent light emission. With advancements in materials science and nanotechnology, this noncontact method of luminescence thermometry has the potential to be applied in various fields such as nanotechnology, medicine, environmental and industrial processes due to its versatility and accuracy. The range of sensing of luminescence thermometers lies between 299 and 480 K depending upon the material used and sensing luminescence techniques like ratiometric intensity, bandshift, band shape, etc.

5.2 MATERIALS USED FOR LUMINESCENCE THERMOMETRY

There are different materials whose luminescence profile can directly give a measure of their temperature as mentioned here.

  i. Rare-earth lanthanide doped nanomaterials
  ii. Semiconducting quantum dots
  iii. Metal-organic frameworks
  iv. Dyes and functional organic polymers

5.3 MECHANISM OF TEMPERATURE SENSING IN LUMINESCENCE THERMOMETRY

The discipline, where the interaction of electromagnetic (EM) waves is dealt with in relation with matter, is known as spectroscopy. Every material has certain specified levels of energy derived from Schrodinger's equation of motion applied to different motions of the molecules and atoms. The energy differences between these levels and allowed transitions of electrons from one level to another are also governed by laws of quantum mechanics. Whenever a photon of an EM wave has energy in line with energy difference for a required transition between two levels, the photon transfers its energy to the electron of the material system and further exciting it to higher level of energy. Thereafter, the excited system returns to its ground energy state, losing its energy either by radiative or non-radiative processes. The term "Luminescence" was first introduced by a German physicist, Eilhard Wiedemann, in 1888, derived from the phenomenon of observing light from a substance apart from thermal radiation or incandescence [5]. Luminescence arises due to the emission of visible light radiation from a substance under test when the electrons undergo transitions between different vibrational-electronic energy levels. Luminescence can be understood as various types like photoluminescence, chemiluminescence, bioluminescence, ionoluminescence, cathodoluminescence, mechanoluminescence, piezoluminescence, thermoluminescence, electroluminescence, sonoluminescence and tribo- or fractoluminescence in view of the luminescent material being excited by light, chemical reaction, biological mechanism, ion beam, electron beam, mechanical rubbing or grinding, dynamic pressure, heat, electric field or current, sound and breaking of chemical bond or fracture of crystals respectively. The radiative transitions can result in various processes [1] in luminescence like fluorescence, phosphoresecence, delayed luminescence, internal conversion, vibrational relaxation, quenching, intersystem crossing which are presented in Figure 5.1 as Jablonski diagram. Photoluminescence begins with photon absorption (A), a quick radiative process lasting 10^{-15} seconds transferring energy to ground state (S_0) electrons to proceed to states $(S_1, S_2, ...)$ having higher energy and a

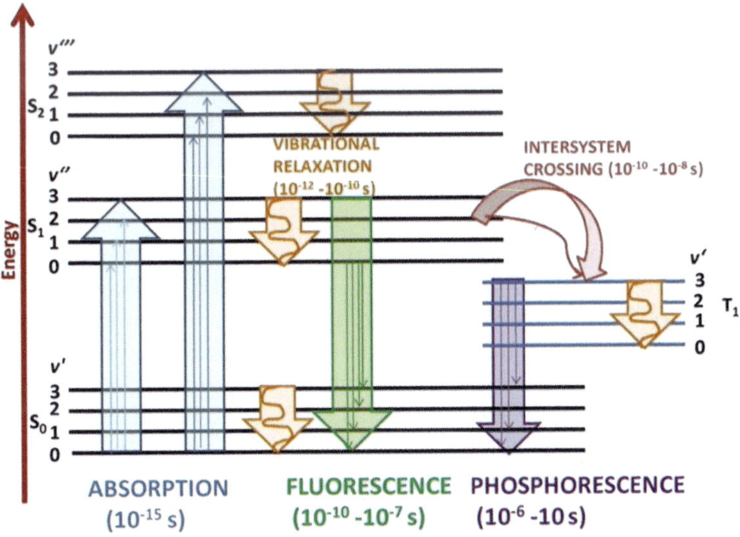

Figure 5.1 Jablonski diagram showing various processes like absorption, fluorescence, phosphorescence, intersystem crossing, vibration relaxation involved in the interaction of electromagnetic radiation energy with the matter [1].

vibrationally excited state (v). They interact with phonons and this is followed by undergoing the vibrational relaxation (VR) to the lowest vibrational state of excited electronic state, losing energy in non-radiative way. They also lose energy in non-radiative manner through internal conversion (IC). Both operations have a time duration of 10^{-12} seconds. The electrons then de-excite in a radiative way to their ground state and emit lower energy photons than the photons that provided excitation, while the other electrons lose energy through some non-radiative process (as luminescence quenching). Fluorescence (FL) is a slow process of around 10^{-9}–10^{-7} seconds and is considered as an allowed electronic transition [6]. When electrons in an excited single state are undergoing intersystem crossing (ISC) to a state with a different spin multiplicity (triplet state, T), phosphorescence (PHOS) and delayed luminescence (DL) come into existence. This is a transition of around 10^{-8}–10^{-3} seconds which is disallowed. However, the electron-phonon coupling makes it weakly permitted. In this instance, electrons can lose energy either through phosphorescence or in a non-radiative way through quenching. Electrons can also cross back (ISC) to first excited singlet state before de-exciting in radiative manner to ground electronic state. This leads to a phenomenon known as delayed luminescence.

Luminescence in solids is classified as intrinsic or extrinsic. Extrinsic luminescence is caused by electronic transitions of impurities inserted (typically purposefully) into the host medium, whereas intrinsic luminescence is caused by electronic transitions corresponding to intrinsic host medium.

An inherent band-to-band luminescence is produced when an electron corresponding to conduction band recombines with a hole corresponding to valence band, which occurs in semiconducting quantum dots and nanoparticles. Intrinsic luminescence among the organic solids is induced as a consequence of the electronic transitions occurring between the lowest unoccupied molecular orbital (LUMO) and the highest occupied molecular orbital (HOMO) as well as intermolecular complexes in the excited states [7]. The luminescence measurements can be understood as steady-state luminescence and time-resolved. Further, alterations in the magnitudes of temperature can affect numerous aspects related to luminescence including the position and intensities of bands, the shapes of emission spectra, etc. [8]. The various ways of temperature read-out in luminescence thermometry are based on the variation of observable parameters such as intensity, bandshift, bandwidth, polarization, ratiometric, luminescence kinetics, and single band ratiometric, as depicted in Figure 5.2, and are followed by discussion in sub-sections.

5.3.1 Temperature sensing by intensity of emission band

Some compounds that are non-luminescent at ambient temperature may exhibit intense luminescence at cryogenic temperatures; while others like Dy^{3+} doped $Y_3Al_5O_{10}$, are of luminescent nature even at extremely high

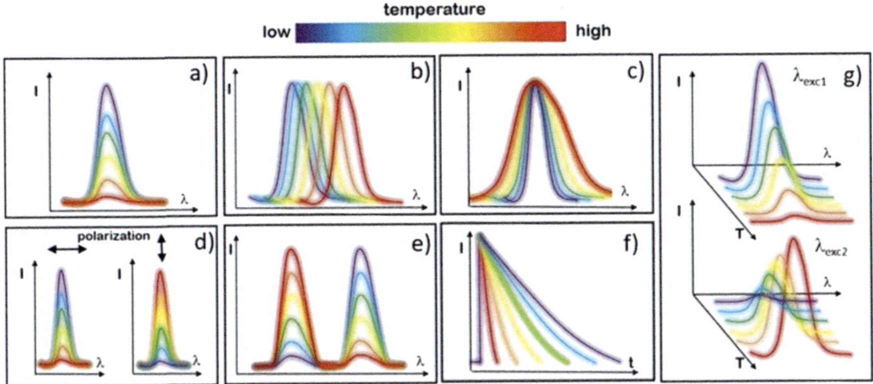

Figure 5.2 Temperature read-out routes in luminescence thermometry based on (a–c) intensity, bandshift, and the bandwidth, respectively, while (d–g) on polarization, ratiometric, luminescence kinetics and the single band ratiometric, respectively [8]. (Reprinted from Ref. [8] © 2022 The Author(s). Published by Elsevier B.V., under Creative Commons CC-BY license, https://creativecommons.org/licenses/by/4.0/.)

magnitudes of temperatures (>1000 K). The changes in lanthanide ion transition energies are less than $0.1\,cm^{-1}/K$, which results from the variations happening in the crystal field acting on them. Kusama et al. observed the effect from the line-shift of emission as $Eu^{3+}\,^5D_0 \rightarrow ^7F_0$ (in Y_2O_2S) that happens when the temperature increases from $-16°C$ to $72°C$ [9]. The discovery was examined by the appearance of Nd^{3+} (in LaF_3) 863 nm emission peak changes to longer values of wavelengths when the temperature increases at a rate of 0.007 nm/K [10]. The maximum wavelength of the charge transfer band in trivalent lanthanide ion materials has significant temperature dependence, reaching 0.6 nm/K for Eu^{3+} doped Y_2O_3, and can be used to measure temperatures. The change in the intensity of a single emission band has been quite high (2%/K) for quantum dots (QDs) and organic dyes/polymers.

5.3.2 Temperature sensing by band shifting of emission band

Semiconducting QDs have exceptional temperature dependence on the maxima of emission band rather than on the intensities of the emission bands. This feature makes them a better candidate for nanothermometry as the measurement of band maximum is done with greater accuracy than the intensity. To mention, the emission band maxima are redshifted by 0.193 nm/°C rise in temperature [11] in the context of CdTe QDs and double hydroxide layers, along with other reports on CdSe QDs [12] in this direction.

5.3.3 Temperature sensing by ratiometric intensity of emission bands

The most commonly employed temperature reading procedure in contemporary luminescence thermometry practice is one based on determining the ratio of intensity magnitudes of different bands corresponding to a sample [13] and applied for numerous cases [14].

5.3.4 Temperature sensing by temporal dependence of emission band (lifetime of an excited state)

This is one of the potential ways toward sensing of temperature. The size of the QDs has a great impact on the lifetime values as reported in research work based on CdTe QDs [15] along with the impact of temperature.

5.4 THERMOMETRY AT THE NANOSCALE

The task of the temperature measurement in functional structures with characteristic dimensions in nanometer range using traditional procedures is extremely challenging in view of spatial resolution issue. The nanoscale temperature probes, nanoparticles or nanocolloids endorse the size-dependent optical activity which is used to monitor temperature. The dynamic feature of nanoscale assemblies is crucial in view of reversible operation of a thermometer and not as a one-shot measurement [16]. For casting dynamic nanoscale assemblies, polymers are proven as an excellent material allowing superstructures from particles, the logical choice behind these structures. Due to the drastic alteration in physicochemical and thermodynamic properties at the nanoscale, the evolution of nanothermometers requires materials with novel physical properties [17,18]. Further, the advanced nanothermometers can be employed in different scenarios, which is potentially apt for biological applications due to their biocompatibility. The nanoscale thermometry is substantially employed in the fields of thermoelectricity, nanofluidics [19], etc. With the advent of technology, the field of scientific research requires the development of temperature sensors which can perform accurately for sub-micrometer spatial resolution. Therefore, the development of novel temperature-sensing elements is important for scientific community. In this context, the nanoluminescence temperature probes can play a significant role. Luminescence thermometry might facilitate the sensitivity magnitude of >1% K^{-1} within acquisition times <1 ms along with resolution <10 µm [20,21]. In addition, the luminescent characteristics for the nanoparticles are correlated with temperature and can be considered as a potential nanothermometer [20]. The main drawbacks are associated with designing materials of thermometers such as their surface profile, irregularity in shape, etc. [22], which can be explored with advancements in new platforms like carbon nanotubes [23,24].

5.5 QUANTUM DOTS: IMPLICATIONS IN THERMOMETRY

The burning desire to precisely measure the temperature of nanoscale systems has resulted in the potential procedures of thermometry at quantum scales. Quantum dots (QDs) are extremely small crystals with the size of the order of a few nanometers i.e. 2–10 nm or 10–50 atoms exhibit unusual properties, unlike an ordinary bulk semiconductor, which are generally macroscopic objects. The bandgap in relation with a spherical quantum dot shows a proportional behavior as $(1/R^2)$ with R denotes radius, thus the energy gap of excitons in QDs is strongly size dependent. Figure 5.3 depicts that the emission wavelengths are strongly dependent on the size of quantum dots, which influence the optical and electrical properties and pave the way for smart applications [17,25]. Owing to their size, QDs are expected to reduce phonon contributions and can be utilized as scatterers for phonons. The reduction in phonon heat transport across the device supports good electrical conduction, suitable for promising candidate as futuristic thermoelectric devices with high efficiency [17]. Recently, Coulomb-coupled configurations have been exploited for triple quantum dot thermometry applications [26] along with enhancements in sensitivity.

Here, the temperature evaluation in relation with reservoir (target) can be carried out in terms of voltage or the electric current, which is obtained across the terminals in electrical isolation from target reservoir (Figure 5.4). In addition, thermoelectric energy harvesting devices are probed in view of quantum dots configurations [27].

Figure 5.3 The size of quantum dots in relevance to emission wavelength, i.e. quantum size effects [25].

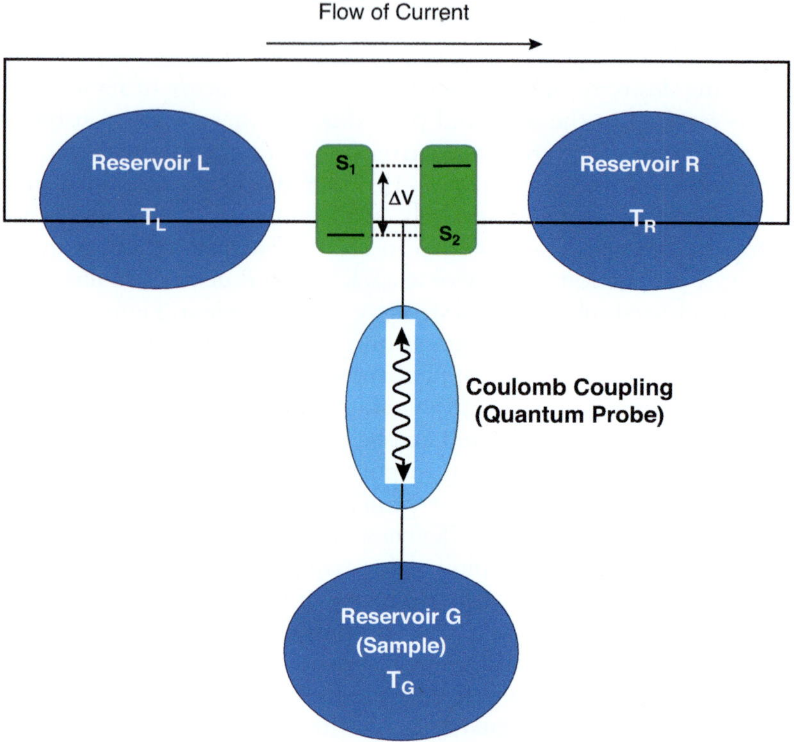

Figure 5.4 Schematic layout for Coulomb-coupled quantum dots as a thermometer [26].

Further, the choice of lanthanide candidates in view of applications in ratiometric near-infrared (NIR) thermometery can be done by the intensities corresponding to emissions/transitions between the excited and ground energy states. Figure 5.5 depicts the energy diagrams with the NIR emission transitions of Nd, Er, and Ho ions with excitation around 536.8 (corresponding to all three ions) and 649.7 nm (corresponding to Er and Ho) [28].

5.6 SINGLE QUANTUM DOTS AS TEMPERATURE MARKERS

Single quantum dots are emerging as a subject of great interest. The alteration in strain or shape results in additional continuous variation of the Eigen energies from dot to dot. Due to their exciting features, a great effort is currently underway to fabricate nanoscale QDs [29]. Recently, CdSe quantum dots [12,30] have been explored to sense variation in temperature from 24.4 to 43.6 (measured in °C) under noncontact system configuration via

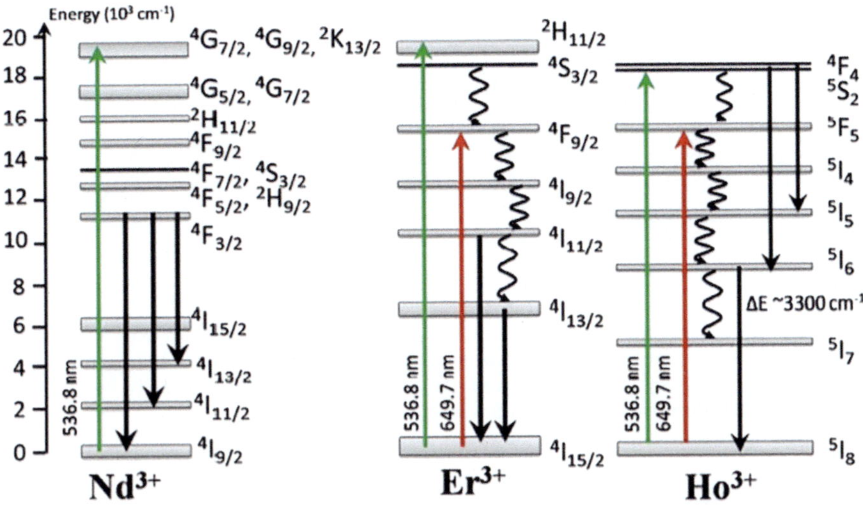

Figure 5.5 Illustrating energy levels diagrams of Nd³⁺ (left), Er³⁺ (middle) and Ho³⁺ (right) and their respective emission transitions [28]. (Reprinted from Ref. [28] © 2019 Elsevier B.V. All rights reserved.)

the study of spectral shift. Around 10 single dots are monitored with excitation achieved by 532 nm wavelength. Further, Grundmann et al. studied the origin of luminescence in single quantum dot made of InAs employing the molecular beam epitaxy technique. The study is carried out till a temperature of 50 K [31]. In addition, Zn-In-S quantum dots are prepared to build temperature-sensing element along with dopants of copper and manganese ions [32]. These efforts can pave the way for applications in sectors of optical thermometry [33].

5.7 QUANTUM DOTS IN VIEW OF TEMPERATURE-DEPENDENT PHOTOLUMINESCENCE

Information regarding the photoluminescence (PL) profile of a quantum dot is of immense help to understand its excited electronic state. A plethora of temperature-mediated PL studies on numerous quantum dot systems [34–36] including water-soluble CdSe/ZnS core/shell, CdSe clusters, $CH_3NH_3PbBr_3$ perovskite, etc. are being reported so far targeting versatile functionalities as temperature sensors and light-emitting diodes. In this direction, Ji et al. studied Ag_2Se quantum dots [37] of sizes 3.2, 3.7, and 4.2 nm within a temperature range of 80–360 K as depicted by their photoluminescence spectral characteristics (Figure 5.6a–c). Their raw products are extracted employing methanol and purified forms are dispersed for testing by Omni-λ300 monochromotor employing tetrachloroethylene. It can

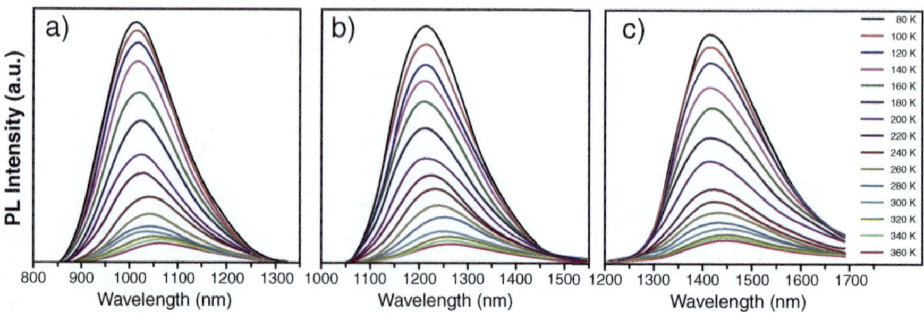

Figure 5.6 (a–c) Photoluminescence spectral details of Ag$_2$Se quantum dots with variation in temperature corresponding to size of particles as 3.2, 3.7, and 4.2 nm respectively [37]. (Reprinted (adapted) with permission from Ref. [37] © 2015, American Chemical Society.)

be observed from the spectra (Figure 5.6) that there exists a reduction in photoluminescence intensity, a shift in energy of emission peak along with broadening with an increase in temperature. In addition, the peak position remains constant at low temperatures, while it moves toward longer wavelengths (redshifted) when the temperature exceeds 200 K. These results can be comprehended in terms of the interaction between the exciton and acoustic phonon, which can pave the way for realization of novel sensing strategies and solar cell applications followed by accelerating the advancements in quantum dots made from semiconductors.

5.8 LUMINESCENCE NANOTHERMOMETER BASED ON POLYMER-ENCAPSULATED QUANTUM DOTS

Encapsulation of quantum dots in polymeric forms has a paramount significance to tailor their surface topology. This has brought potential examples [38,39] in relation with potential luminescence characteristics as hydroxide-based composites, amalgamation with polyethylene glycol, polyvinyl pyrrolidone, etc. for imaging, temperature, and microfluidic applications [40,41]. Concerning this, Liu et al. realized a luminescence nanothermometer [16] with polymer encapsulation of the CdSeS/ZnS quantum dots employing precipitation with ethanol. Their photoluminescence information is monitored from 400 to 700 nm with the help of a spectrophotometer and control of temperature is achieved via incubator and thermocouple. Figure 5.7a depicts the photographs of prepared quantum dots in water as observed under natural light (left) and 365 nm wavelength (right) revealing their transparent behavior and green luminescence respectively. Their photoluminescence intensity spectra are graphed with

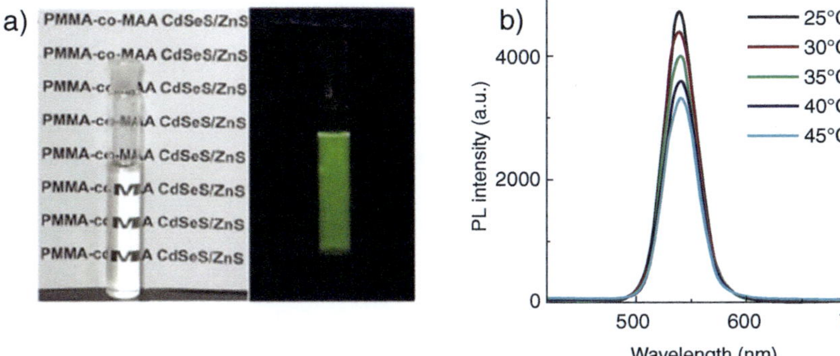

Figure 5.7 (a) Polymer-encapsulated quantum dot in aqueous solution under natural light (left) and ultraviolet irradiation (right) and (b) variation of its photoluminescence with temperature from 25°C to 45°C [16]. (Reprinted from Ref. [16] © 2015, The Author(s), licensed under a Creative Commons Attribution 4.0 International License, http://creativecommons.org/licenses/by/4.0/.)

wavelength in Figure 5.7b, which show a decreasing behavior with temperature ranging from 25°C to 45°C along with almost negligible emission shift. These results reveal that quantum dots under consideration have no photodegradtion and can be comprehended in terms of non-radiative decay mechanism in relevance to surface state. Such polymer-encapsulated quantum dots-based thermometers are highly beneficial as temperature sensors for biological applications [42].

5.9 PERSPECTIVES IN QUANTUM DOT MATERIALS FOR LUMINESCENCE THERMOMETER

Quantum dots are one of the highly exploited light-emitting platforms to actualize efficient nanothermometers among several other luminescent systems including organic dyes, nanoparticles with doping of lanthanide as well as non-luminescent labels. Still, there are technical issues to be addressed as instability, blinking effect, toxic behavior for biological environment, agglutination, etc. for their wide range of applications. These factors might be eradicated to an appreciable extent via coating their surface with the help of inert materials and adopting emission measurements for a large density of particles to minimize blinking. Also, there is an ample scope to bring advancements in sectors of single molecular thermometry in view of spatial and temperature resolution and insensitivity issues toward pH, etc. This can be of great advantage to build intracellular temperature probes and we can study further cell-related phenomena via employing polymers with quantum dots due to their biocompatible and nontoxic behavior. In addition, their quantum yield [43,44] can be optimized via utilizing proper capping ligands and layering the core with a material having larger value of energy band gap. Moving ahead, layered double hydroxide-based quantum dot systems might serve effectively to minimize aggregation along with manipulating the chemical stability to be beneficial for display applications. Still, this field is in its infancy and with judicious choice of these potential strategies and developments in chemical processing techniques [45–55], we can gain prominent momentum to transform the prototypes of thermometers based on quantum dots into real products.

5.10 CONCLUSIONS

Here, we have highlighted the recent progress in quantum dots for new generation luminescence thermometers. These can be attributed to their tunable band gap assisted by a potential control over the size of particles in view of quantum confinement effect. Further, the emission characteristics of quantum dots in terms of intensity, spectral peak location, decay lifetime, band shape, bandwidth, etc. prove to be helpful to analyze the evolution of their thermal profile. Such temperature read-out procedures can foster their applications in the interdisciplinary fields of optics, medical science, and electronics.

REFERENCES

1. Dramićanin, M., 2018. *Luminescence Thermometry: Methods, Materials, and Applications*. Woodhead Publishing.
2. McGee, T.D., 1988. *Principles and Methods of Temperature Measurement*. John Wiley & Sons.
3. Camuffo, D. and Bertolin, C., 2012. The earliest temperature observations in the world: the Medici Network (1654–1670). *Climatic Change*, 111, pp. 335–363.
4. Bradley III, L.C., 1953. A temperature-sensitive phosphor used to measure surface temperatures in aerodynamics. *Review of Scientific Instruments*, 24(3), pp. 219–220.
5. Williams, F., 1978. Overview and trends of luminescent research. In: Di Bartolo, B., Godberg, V., Pacheco, D. (eds) *Luminescence of Inorganic Solids* (pp. 1–13). Boston, MA: Springer US. https://doi.org/10.1007/978-1-4684-3375-3_1.
6. Lakowicz, J.R. (eds), 2006. Fluorescence Sensing. In Principles of Fluorescence Spectroscopy (pp. 623–673). Springer, Boston, MA. https://doi.org/10.1007/978-0-387-46312-4_19.
7. Lenhardt, L. and Dramićanin, M.D., 2016. PARAFAC: a tool for the analysis of phosphor mixture luminescence. *Journal of Luminescence*, 170, pp. 136–140.

8. Marciniak, L., Kniec, K., Elżbieciak-Piecka, K., Trejgis, K., Stefanska, J. and Dramićanin, M., 2022. Luminescence thermometry with transition metal ions. A review. *Coordination Chemistry Reviews*, 469, p. 214671.

9. Kusama, H., Sovers, O.J. and Yoshioka, T., 1976. Line shift method for phosphor temperature measurements. *Japanese Journal of Applied Physics*, 15(12), p. 2349.

10. Rocha, U., Jacinto da Silva, C., Ferreira Silva, W., Guedes, I., Benayas, A., Martinez Maestro, L., Acosta Elias, M., Bovero, E., van Veggel, F.C., Garcia Sole, J.A. and Jaque, D., 2013. Subtissue thermal sensing based on neodymium-doped LaF_3 nanoparticles. *ACS Nano*, 7(2), pp. 1188–1199.

11. Liang, R., Tian, R., Shi, W., Liu, Z., Yan, D., Wei, M., Evans, D.G. and Duan, X., 2013. A temperature sensor based on CdTe quantum dots–layered double hydroxide ultrathin films via layer-by-layer assembly. *Chemical Communications*, 49(10), pp. 969–971.

12. Li, S., Zhang, K., Yang, J.M., Lin, L. and Yang, H., 2007. Single quantum dots as local temperature markers. *Nano Letters*, 7(10), pp. 3102–3105.

13. Dramićanin, M.D., 2016. Sensing temperature via downshifting emissions of lanthanide-doped metal oxides and salts. A review. *Methods and Applications in Fluorescence*, 4(4), p. 042001.

14. Gavrilović, T.V., Jovanović, D.J., Lojpur, V. and Dramićanin, M.D., 2014. Multifunctional Eu^{3+}-and Er^{3+}/Yb^{3+}-doped $GdVO_4$ nanoparticles synthesized by reverse micelle method. *Scientific Reports*, 4(1), p. 4209.

15. Haro-González, P., Martínez-Maestro, L., Martín, I.R., García-Solé, J. and Jaque, D., 2012. High-sensitivity fluorescence lifetime thermal sensing based on CdTe quantum dots. *Small*, 8(17), pp. 2652–2658.

16. Liu, H., Fan, Y., Wang, J., Song, Z., Shi, H., Han, R., Sha, Y. and Jiang, Y., 2015. Intracellular temperature sensing: an ultra-bright luminescent nanothermometer with non-sensitivity to pH and ionic strength. *Scientific Reports*, 5(1), p. 14879.

17. Yue, Y. and Wang, X., 2012. Nanoscale thermal probing. *Nano Reviews*, 3(1), p. 11586.

18. Tang, L., Zhang, Y., Liao, C., Guo, Y., Lu, Y., Xia, Y. and Liu, Y., 2022. Temperature-dependent photoluminescence of CdS/ZnS core/shell quantum dots for temperature sensors. *Sensors*, 22(22), p. 8993.

19. Menges, F., Mensch, P., Schmid, H., Riel, H., Stemmer, A. and Gotsmann, B., 2016. Temperature mapping of operating nanoscale devices by scanning probe thermometry. *Nature Communications*, 7(1), p. 10874.

20. Brites, C.D., Marin, R., Suta, M., Carneiro Neto, A.N., Ximendes, E., Jaque, D. and Carlos, L.D., 2023. Spotlight on luminescence thermometry: basics, challenges, and cutting-edge applications. *Advanced Materials*, 35(36), p. 2302749.

21. Dramićanin, M.D., 2020. Trends in luminescence thermometry. *Journal of Applied Physics*, 128(4), p. 040902.

22. Kuzubasoglu, B.A. and Bahadir, S.K., 2020. Flexible temperature sensors: a review. *Sensors and Actuators A: Physical*, 315, p. 112282.

23. Gao, Y. and Bando, Y., 2002. Carbon nanothermometer containing gallium. *Nature*, 415(6872), pp.599–599.

24. Brites, C.D., Lima, P.P., Silva, N.J., Millán, A., Amaral, V.S., Palacio, F. and Carlos, L.D., 2012. Thermometry at the nanoscale. *Nanoscale*, 4(16), pp. 4799–4829.

25. Ren, D., Wang, B., Hu, C. and You, Z., 2017. Quantum dot probes for cellular analysis. *Analytical Methods*, 9(18), pp. 2621–2632.

26. Dhongade, S.G., Haque, A.A., Roy, S.S. and Singha, A., 2022. Non-local triple quantum dot thermometer based on Coulomb-coupled systems. *Scientific Reports*, 12(1), p. 15842.

27. Sothmann, B., Sánchez, R. and Jordan, A.N., 2014. Thermoelectric energy harvesting with quantum dots. *Nanotechnology*, 26(3), p. 032001.

28. Avram, D., Colbea, C., Florea, M. and Tiseanu, C., 2019. Highly-sensitive near infrared luminescent nanothermometers based on binary mixture. *Journal of Alloys and Compounds*, 785, pp. 250–259.

29. Hoffmann, E.A. and Linke, H., 2009. Nanoscale thermometry with a quantum dot. *Journal of Low Temperature Physics*, 154, pp. 161–171.

30. Bukowski, T.J. and Simmons, J.H., 2002. Quantum dot research: current state and future prospects. *Critical Reviews in Solid State and Material Sciences*, 27(3–4), pp. 119–142.

31. Grundmann, M., Christen, J., Ledentsov, N.N., Böhrer, J., Bimberg, D., Ruvimov, S.S., Werner, P., Richter, U., Gösele, U., Heydenreich, J. and Ustinov, V.M., 1995. Ultranarrow luminescence lines from single quantum dots. *Physical Review Letters*, 74(20), p. 4043.

32. Cao, S., Zheng, J., Zhao, J., Yang, Z., Shang, M., Li, C., Yang, W. and Fang, X., 2016. Robust and stable ratiometric temperature sensor based on Zn–In–S quantum dots with intrinsic dual-dopant ion emissions. *Advanced Functional Materials*, 26(40), pp. 7224–7233.

33. Quintanilla, M. and Liz-Marzan, L.M., 2018. Guiding rules for selecting a nanothermometer. *Nano Today*, 19, pp. 126–145.

34. Liu, T.C., Huang, Z.L., Wang, H.Q., Wang, J.H., Li, X.Q., Zhao, Y.D. and Luo, Q.M., 2006. Temperature-dependent photoluminescence of water-soluble quantum dots for a bioprobe. *Analytica Chimica Acta*, 559(1), pp. 120–123.

35. Biju, V., Makita, Y., Sonoda, A., Yokoyama, H., Baba, Y. and Ishikawa, M., 2005. Temperature-sensitive photoluminescence of CdSe quantum dot clusters. *The Journal of Physical Chemistry B*, 109(29), pp. 13899–13905.

36. Woo, H.C., Choi, J.W., Shin, J., Chin, S.H., Ann, M.H. and Lee, C.L., 2018. Temperature-dependent photoluminescence of $CH_3NH_3PbBr_3$ perovskite quantum dots and bulk counterparts. *The Journal of Physical Chemistry Letters*, 9(14), pp. 4066–4074.

37. Ji, C., Zhang, Y., Zhang, T., Liu, W., Zhang, X., Shen, H., Wang, Y., Gao, W., Wang, Y., Zhao, J. and Yu, W.W., 2015. Temperature-dependent photoluminescence of Ag_2Se quantum dots. *The Journal of Physical Chemistry C*, 119(24), pp. 13841–13846.

38. Cho, S., Kwag, J., Jeong, S., Baek, Y. and Kim, S., 2013. Highly fluorescent and stable quantum dot-polymer-layered double hydroxide composites. *Chemistry of Materials*, 25(7), pp. 1071–1077.

39. Sabah, A., Tasleem, S., Murtaza, M., Nazir, M. and Rashid, F., 2020. Effect of polymer capping on photonic multi-core-shell quantum dots CdSe/CdS/ZnS: impact of sunlight and antibacterial activity. *The Journal of Physical Chemistry C*, 124(16), pp. 9009–9020.

40. Fan, Y., Liu, H., Han, R., Huang, L., Shi, H., Sha, Y. and Jiang, Y., 2015. Extremely high brightness from polymer-encapsulated quantum dots for two-photon cellular and deep-tissue imaging. *Scientific Reports*, 5(1), p. 9908.

41. Tohgha, U.N., Alvino, E.L., Jarnagin, C.C., Iacono, S.T. and Godman, N.P., 2019. Electrowetting behavior and digital microfluidic applications of fluorescent, polymer-encapsulated quantum dot nanofluids. *ACS Applied Materials & Interfaces*, 11(31), pp. 28487–28498.

42. Fanizza, E., Zhao, H., De Zio, S., Depalo, N., Rosei, F., Vomiero, A., Curri, M.L. and Striccoli, M., 2020. Encapsulation of dual emitting giant quantum dots in silica nanoparticles for optical ratiometric temperature nanosensors. *Applied Sciences*, 10(8), p. 2767.

43. Jaque, D. and Vetrone, F., 2012. Luminescence nanothermometry. *Nanoscale*, 4(15), pp. 4301–4326.

44. Efros, A.L. and Brus, L.E., 2021. Nanocrystal quantum dots: from discovery to modern development. *ACS Nano*, 15(4), pp. 6192–6210.

45. Fahad, S., Li, S., Zhai, Y., Zhao, C., Pikramenou, Z. and Wang, M., 2024. Luminescence-based infrared thermal sensors: comprehensive insights. *Small*, 20(3), p. 2304237.

46. Kumar, D., Jain, R., Shahjahan, Banerjee, S., Prabhu, S.S., Kumar, R., Azad, A.K. and Roy Chowdhury, D., 2020. Bandwidth enhancement of planar terahertz metasurfaces via overlapping of dipolar modes. *Plasmonics*, 15, pp. 1925–1934.

47. Li, L., Wang, W., Luk, T.S., Yang, X. and Gao, J., 2017. Enhanced quantum dot spontaneous emission with multilayer metamaterial nanostructures. *ACS Photonics*, 4(3), pp. 501–508.

48. Kumar, D., Gupta, M., Srivastava, Y.K., Devi, K.M., Kumar, R. and Chowdhury, D.R., 2022. Photoinduced dynamic tailoring of near-field coupled terahertz metasurfaces and its effect on Coulomb parameters. *Journal of Optics*, 24(4), p. 045101.

49. Kumar, D., Devi, K.M., Kumar, R. and Chowdhury, D.R., 2021. Dynamically tunable slow light characteristics in graphene based terahertz metasurfaces. *Optics Communications*, 491, p. 126949.

50. Singh, G., Fisch, M. and Kumar, S., 2016. Emissivity and electrooptical properties of semiconducting quantum dots/rods and liquid crystal composites: a review. *Reports on Progress in Physics*, 79(5), p. 056502.

51. Mohan Rao, S.J., Kumar, D., Kumar, G. and Chowdhury, D.R., 2017. Modulating the near field coupling through resonator displacement in planar terahertz metamaterials. *Journal of Infrared, Millimeter, and Terahertz Waves*, 38, pp. 124–134.

52. Kala, A., Inam, F.A., Biehs, S.A., Vaity, P. and Achanta, V.G., 2020. Hyperbolic metamaterial with quantum dots for enhanced emission and collection efficiencies. *Advanced Optical Materials*, 8(15), p. 2000368.

53. Rao, S.J.M., Kumar, D., Kumar, G. and Chowdhury, D.R., 2016. Probing the near-field inductive coupling in broadside coupled terahertz metamaterials. *IEEE Journal of Selected Topics in Quantum Electronics*, 23(4), pp. 1–7.

54. Banerjee, S., Amith, C.S., Kumar, D., Damarla, G., Chaudhary, A.K., Goel, S., Pal, B.P. and Chowdhury, D.R., 2019. Ultra-thin subwavelength film sensing through the excitation of dark modes in THz metasurfaces. *Optics Communications*, 453, p. 124366.

55. Karim, H., Delfin, D., Chavez, L.A., Delfin, L., Martinez, R., Avila, J., Rodriguez, C., Rumpf, R.C., Love, N. and Lin, Y., 2017. Metamaterial based passive wireless temperature sensor. *Advanced Engineering Materials*, 19(5), p. 1600741.

Chapter 6

Rare earth-doped phosphors for luminescence-based optical thermometers

Ashvini Pusdekar, Nilesh Ugemuge, and Sanjay J. Dhoble

6.1 INTRODUCTION

Demand of temperature sensor in various scientific fields and industries increasing rapidly due to accuracy and consistency in measurements. High-accuracy temperature sensors are required for regulated manufacturing, scientific research, medical diagnostics, and production safety [1,2]. Temperature is considered as the most important thermodynamic parameter and is indispensable in many fields, such as industrial manufacturing, scientific research, and other scientific fields. Temperature enables the diagnosis of medical conditions of patients. Increased temperature is the indication of many types of inflammation and diseases; due to this reason, it is important to measure temperature with great accuracy [3]. In industry, temperature regulates the industrial manufacturing process and equipment diagnostics to avoid accidents [4]. Conventionally, temperature is measured by using thermometers, and thermocouples need direct thermal transmission and the subsequent heat balance between the sensor and the measured object, which has limitations in contactless situation, dynamic systems, and small-scale objects [5,6]. Due to these important reasons, many efforts have been devoted by several researchers to develop advanced contactless thermometry technology. In this technology, the measurement of temperature of an object can be directly detected by non-contact temperature sensors using optical parameters, for instance, single fluorescence intensity, FIR, peak wavelength, emission bandwidth, fluorescence lifetime, etc. [7]. Optical thermometry built on the FIR mode has gained attention due to several advantages including high precision, high signal yield and accuracy, adaptability to tough environments, remote detection, and fast response [8].

FIR temperature thermometric technology is realized through the diverse luminescent intensities of two important thermally coupled levels (TCLs) at ambient temperature, which is unaffected by spectrum losses and fluctuations of pumping intensity. The TCLs are two adjacent energy levels with a

DOI: 10.1201/9781032661537-6

very trivial energy gap (ΔE); ΔE is about 200–2000 cm^{-1} [9]. On the basis of ΔE, temperature measurement with high spatial resolution is enormously significant.

Recently, lanthanide ions incorporating optical temperature sensors with non-contact detection mode have attracted extensive attention in various fields ranging from medical science to industries [10] on account of their characteristics like contactless measurement, un-injurious, and being able to work in high-temperature conditions which avoid the defects that arise in earlier technologies such as long responsive time and imprecision [11].

Several efforts of researchers have been concentrated on the research of temperature-dependent upconversion (UC) phosphors. The outcome of these attempts creates sensors for temperature measurement that are extremely attractive for their possible applications in several fields such as electrical power stations, biological environments, coal mines, and oil refineries [12]. As the center of luminescence, it is particularly attractive that trivalent rare earth (RE) ions have rich energy levels, which are located in the wide wavelength range from ultraviolet to infrared [13]. Oxide materials generally exhibit greater chemical stability and eco-friendly features. Therefore, several oxide-based UC phosphors have been widely investigated in the last two decades.

There are two major advantages of studying RE-doped phosphor materials for optical thermometry, namely temperature sensing range and sensitivity. To fulfill these requirements, lanthanide-doped phosphors are recommended as desired fluorescent thermometers [14,15]. Fluorescent nano-thermometers show numerous temperature-dependent fluorescence parameters which are usually employed to explore temperature, such as decay lifetime, FIR, emission peak position, emission bandwidth, and emission peak intensity [16]. Among these parameters based on FIR of two independent emission peaks for ratiometric detection, which can avoid some disturbances in measurements such as the fluctuation of excitation power and the non-radiative relaxation prompted by the local environment of the lattice. Furthermore, lanthanide ions are considered as an excellent material due to their large Stokes shift, long fluorescence lifetime, and sharp emission profiles [17].

Most of the temperature-sensing measurements are attained by a single UC or DC process or simultaneously in combination of both processes. Among these processes, UC fluorescent materials have attracted the attention of researchers because of measurements of temperature by gathering the FIR analogous to the upper energy levels of RE ions; for instance, the excited states $^2H_{11/2}$ and $^4S_{3/2}$ of Er^{3+} ions can be used [18]. Yb^{3+} ions are typically selected as a sensitizer in the UC process to improve the fluorescence intensity of trivalent lanthanide (Ln^{3+}) ions because of their simple energy levels, which are responsible to improve the absorption of energy [19].

In this chapter, we have presented several aspects of RE-based thermometers having a broader range of measurements and a higher relative sensitivity to other thermometry. In addition to this, the fundamental properties of lanthanide ions used in thermometry, principles of measurements, and adequacies and inadequacies of various types of optical thermometry are also described.

6.2 FUNDAMENTAL ASPECTS OF OPTICAL TEMPERATURE SENSING

Generally, optical temperature sensing depends on the temperature dependence of luminescence properties of un-doped and RE-doped phosphor materials, such as FIR, fluorescence intensity, and fluorescence lifetime. Very few hosts, such as Vanadate, Tungstate, and Molybdate, produce luminescence under typical UV-excitation. Phosphor which gives luminescence comprises hosts, activators, and/or sensitizers. Ideal host matrix can provide suitable crystal field environment for lanthanide ions to yield fluorescence. The activator Ln^{3+} ions as a luminescence center possesses peculiar luminescence characteristics due to their electronic structure. For the enhancement of probability of emission, the sensitizer ions are used which impact on local crystal field and help for effective energy transfer. Efficiency of phosphors depends on host material and activators, particularly when sensitizers are present. Due to these reasons, thermometry primarily depends upon the appropriate luminescence centers like Ln^{3+} ions due to abundant 4f energy levels. The apparent variation in temperature has an innumerable impact on the intensity of RE ions. Primarily, temperature changes leading to the variation in the coupling states in between the luminescence center arise from Ln^{3+} ions and afterward lattice field created due to host matrix, which results in variation in terms of physical process associated with phonon. After this apparent changes in lattice, with the increase of temperature, the probability of non-radiative relaxation of trivalent lanthanide ions in the excited state under lattice vibration increases, which results in decrease in fluorescence intensity of Ln^{3+} ions. The relationship between the number of ions in the excited state and the observed temperature can be employed in thermometry. Additionally, RE ions characteristically feature low molar extinction coefficients and long luminescence lifetimes, mostly ascribing to the Laporte prohibited nature of the f-f transitions [20].

Advancing from unique energy level scheme of lanthanide ions, the Ln^{3+} ions are typically chosen as the luminescence centers in several optical applications. Fluorescence temperature sensors are mostly divided into three common types: (i) based on fluorescence intensity, (ii) fluorescence decay lifetime, and (iii) FIR. The principles of working and applicability of these sensors are explained below.

6.2.1 Fluorescence intensity method

This type of temperature measurement is known as direct temperature measurement method. In this method, the absolute luminescence intensity and temperature, along with their relationship, are taken into consideration. Important parameters affecting the fluorescence intensity can be mathematically expressed as follows:

$$I = 2.3k\varphi I_0 \varepsilon l c \tag{7.1}$$

where I – emitted fluorescence intensity; I_0 – excitation light intensity; k – constant; φ – efficiency of energy transfer (ET); ε – the molar absorbance; l – absorption path length; c – fluorescent ions concentration. As long I as I_0 and c are determined, only I is depending on φ. The ET efficiency largely includes between host and Ln^{3+} ions, and the non-radiative transition probability of Ln^{3+} ions, which are affected by temperature.

Normally, with rising temperature, both the parameter increases gradually. This method of measurement is direct and simple; however, absolute fluorescence intensity is mainly influenced by the fluctuation of the excitation power of laser and environmental factors. Along with these parameters, the effect of thermal quenching on the value of I should also be taken into consideration.

6.2.2 Fluorescence decay lifetime method

Under excitation through laser, the molecules present in the material absorb energy from the laser, and transitions occur from the ground to the excited level and then return to ground level with typical radiative transition along with some non-radiative transitions. When excitation ceases, then I of the molecule declines exponentially [21]:

$$I_t = I_0 \exp\left(-t/\tau\right) \tag{7.2}$$

where I_0 – the intensity at $t=0$, τ – lifetime and defined by $I(\tau)=I_0/e$.

As per theory of semiconductor physics, the intensity of lattice vibration rises with increasing temperature of the system, which leads to rise of the number of atoms involved in absorption process and eventually lessens the quenching process of light. Therefore, temperature determines the speed of light quenching process, i.e. size of τ and the ambient temperature at that time can be found by measuring the value of τ.

6.2.3 Fluorescence intensity ratio method

This method is extensively studied and technologically important in various thermometry applications. Mostly, available optical thermometers are designed through fluorescence intensity ratio (FIR) technique based on Ln^{3+}

doped phosphors [22]. In the process of photoluminescence, the atoms present in the TCLs are arranged in a DC or UC process, and the atomic arrangements in the TCLs can be altered by altering the ambient temperature around the phosphor materials. In case of FIR technology, it uses temperature-dependent emission intensities of two TCLs of the Ln^{3+} ions, such as Er^{3+}, Nd^{3+}, Tm^{3+}, Ho^{3+}, etc. [23–26]. Ln^{3+} ions provide a large number of TCLs which are suitable for the detection of temperature. Generally used TCLs of Ln^{3+} ions are (ΔE between 200 and 2000 per cm). The accurate measurement of temperature on the basis of emission intensity, could be powerfully affected by an external factor [27]. Thermometry based on FIR technique is helpful to reduce the dependence of measurement conditions. Also, it has several advantages such as high sensitivity, there is no requirement for contact, rapid response, and high spatial and temperature resolutions [28].

6.3 LUMINESCENT THERMOMETERS BASED ON TRANSITION METAL (TM) IONS

Luminescent materials incorporated with transition metal ions are used as one of the probes for temperature measurements in thermometry. TM ion-doped phosphor gives emission in visible and NIR region. These emission and absorption bands of materials are highly vulnerable to temperature variations. Sensitive temperature can be measured by peculiar characteristics of TM ion emission intensities, excited state lifetimes, emission bandwidths, band shifts, and the ratio of intensities between several emission bands in phosphors [29].

The materials in which TM ions are doped, luminescence arises due to the d-d electronic transitions. The d-orbitals, unlike the 4f orbitals in trivalent lanthanides, are not shielded by the outer orbitals. The spectroscopic properties of TM ions are based on the crystal field strength of the host matrix [30,31]. Therefore, emission peak intensity and the rate of temperature quenching of the luminescence can be modified by changing the type of host matrix [32–34]. These properties of TM ions and host matrix help to the development of sensitive luminescence thermometers [35–37].

Although there are several advantages of TM ion-activated thermometry, but still it is not efficiently recommended for typical thermometry applications; therefore, lanthanide ions-based thermometry is preferred and explored by several researchers.

6.4 LUMINESCENT THERMOMETERS BASED ON RE IONS

Traditional contact thermometers have exposed many limitations, such as the inability to measure fast-moving objects, corrosive environments, and nano-sized or sub-micro-meter objects. RE ions-based thermometry is a relatively novel approach for the measurement of remote temperature by using variations in intensities of emission peaks due to temperature fluctuations [38,39]. Also, in this type of thermometry, the temperature is measured by analyzing the luminescence parameter of the phosphor in physical contact with the object for which we are measuring temperature [40,41]. Luminescent materials with RE^{3+} ions have an abundant luminescent resource on account of the availability of abundant energy states [42]. For example, under UV excitation, Eu^{3+} ion is considered as red-emitting activator originating from $^5D_0$–^7F_J ($J=0$, 1, 2, 3, 4) electronic transitions [43].

RE ions have analogous chemical properties. They also have unique physical properties of light, magnetism, electricity, and heat because of their distinctive structures. The multiple number of 4f electrons in each RE electronic shell structure makes them unique from one another, and the peculiar luminescence properties of these materials are primarily caused by electron transitions in the 4f orbital of RE ions. In RE^{3+}-doped phosphors, optical transitions mainly occur in the 4f orbitals that are well-insulated from their surrounding environment by fully filled outer 5s and 5p orbitals. As a result, the majority of RE^{3+} ions f-f transitions are less impacted by their surroundings.

6.5 ADVANTAGES OF LUMINESCENT THERMOMETERS BASED ON RE IONS OVER TRANSITION METAL (TM) IONS

Research in the field of phosphor sensors has shown that they have significant advantages compared to conventional sensors in terms of their properties like greater sensitivity, freedom from electromagnetic interference, long path monitoring, and independence of compatibility with electronic devices [44]. The following are the major advantages of RE^{3+}-based optical thermometry:

1. Contactless operation, high spatial resolution, and rapid response, which are useful wherever contact measurement is not feasible [45,46].
2. High radiative probabilities for the emitting levels to obtain high emission intensities [47].
3. Various optical thermometry techniques have been studied in the past few decades via the luminescence of probes such as metal–ligand complexes phosphor thermometry and quantum dots.
4. Among RE^{3+}, praseodymium, neodymium, europium, holmium, and erbium have been used for optical temperature sensing because they generate emissions that are dependent on temperature owing to the presence of TCLs.

6.6 DIFFERENT ROUTES OF MATERIAL SYNTHESIS

6.6.1 Solid-state reaction

Solid-state reaction (SSR) is a simple route of synthesis of materials in a pure form. In this method, more than one stages of calcination, in inert or air atmosphere is used. This method consumes more energy for the preparation of due to requirement of long time and high-temperature calcination periods in order to obtain phase pure materials. On account of its adequacies over very few inadequacies, it is considered as most preferred method of synthesis.

In the first stage, precursors are mixing thoroughly by using ball-milling or grounded with mortar and pestle for a certain amount of time, usually, in the presence of ethanol or acetone followed by heat treatment [48]. Because of easy steps, SSR is mostly used for large-scale industrial applications. However, consumption of high-energy and heat treatment results in formation of non-uniform particles, which are the major inadequacies of SSR.

Several researchers have synthesized the phosphors for optical thermometry by using SSR route such as $GdNbO_4:Tm^{3+}/Yb^{3+}$ [49], $BaLaZnSbO_6:Mn^{4+}$ [50], $Er^{3+}/Ho^{3+}-Yb^{3+}$-doped $LaNbO4$ [51], Er^{3+}/Yb^{3+}-codoped $Al_2Mo_3O_{12}$ [52], Yb^{3+}, Er^{3+}, Tm^{3+} co-doped $K_3Gd(PO_4)_2$ [53], $Er^{3+}/Tm^{3+}/Yb^{3+}$ tri-doped bismuth lanthanum tungstate phosphors [54], Eu^{3+} doped $Ba_2La_6Y_2(SiO_4)_6O_2$ [55], $GdNbO_4:Er^{3+}/Yb^{3+}$ [56]. The phosphor shows excellent thermometric performances including strong UCL, good thermal stability, appropriate sensitivity, and a wide sensitivity response range.

6.6.2 Hydrothermal method

Hydrothermal synthesis is the process of materials preparation under hydrothermal conditions that occurs at high pressure and low temperature. It is mostly used in the synthesis process to obtain phase pure materials used in many technological fields such as ceramics, biomedical, and electronics. This process requires lower temperature than the SSR process, but it needs typical conditions, like high pressure and appropriate temperature with single synthesis stage, to manufacture products with desired properties. This is one of the promising methods with well-controlled reaction parameters (temperature, time, pH, and surfactant). This method is the most commonly applied research method as it may yield metal oxides with numerous morphologies depending on the organic templates. Besides, each parameter of the synthesized conditions has a considerable influence on the properties of the resulting materials [57,58].

This method has several advantages such as mild reaction conditions, cheaper experimental arrangement, low reaction temperatures, high purity compounds, narrow particle size distribution, controllable morphology, smaller and better nucleation control, low energy consumption, uniform particle distribution, and selective size growth [59].

Several researchers have synthesized the phosphors for optical thermometry by using hydrothermal method such as $CaWO_4:Eu^{3+}$ [60], $ZnTiO_3$ [61], $Rb_2ZrCl_6:Te^{4+}$ [62], $LaOF:Yb^{3+},Tm^{3+}$ [63], $CaWO_4:Eu^{3+}$ [64], $CeO_2:Eu^{3+}/Sm^{3+}$ [65], $NaYF_4:0.2Yb^{3+}/0.02Ho^{3+}$ [66], $Zn_2(OH)BO_3:Pb^{2+}$ [67], $Ca_3(PO_4)_2:Tm^{3+}/Yb^{3+}$ [68], $Rb_2ScCl_5 \cdot H_2O:Te^{4+}$ [69].

6.6.3 Sol-gel method

Sol-gel is mostly used as wet chemistry method of synthesis to obtain well-controlled stoichiometry, phase pure, nano, and micro size particles at relatively low costs and low temperatures, in comparison with SSR method [70,71]. The sol-gel method provides promising access to high-performance nano-structured tungsten materials with both high strength and ductility and can be easily scaled for industrial production. Sol-gel is a facile fabrication technique for producing metal complexes with certain physicochemical properties. In the sol-gel process, colloidal or gel can be prepared by mixing liquid reactants. The process is based on the transfer of the produced sol into gel. Initial stage of this process is connected with the formation of a highly dispersed colloidal sol of a certain chemical composition. In this method, mostly acetates or nitrates of lithium and manganese compounds are used as spinel's precursors, while citric acid is extensively used as a chelating agent [72,73]. Solution of ammonia is used during the precursor's homogenization to maintain a neutral pH. With only the requirement of chemical reactants, sol-gel technique offers very simple and low-cost synthesis to fabricate tungstate in micro and nano crystalline structure by fine-tuning sol-gel parameters. The advantage of the developed sol-gel technology is the possibility of manufacturing tungstate powders in a nanocrystalline state not only at elevated temperatures but at room temperatures as well.

Several researchers have synthesized the phosphors for optical thermometry by using sol-gel method such as $Na_2YbMg_2(VO_4)_3$ Eu^{3+} [74], Er^{3+}/Yb^{3+}-codoped $Gd_2Mo_3O_{12}$ [75], Yb^{3+}- and Er^{3+}-doped $Ca_3Sc_2Si_3O_{12}$ [76], $Y_2O_3:Nd^{3+}$ [77], $ZnAl_2O_4:Eu^{3+}$ [78], Er^{3+}/Li^+ co-doped Y_2O_3 [79], Er^{3+}/Yb^{3+} co-doped yttrium niobate [80], $Gd_2O_3:x\%Nd^{3+}/Yb^{3+}(0.2 \leq x \leq 0.8)$ [81].

6.6.4 Co-precipitation method

Co-precipitation method is promising among the diverse methods of synthesis because of their technical simplicity, flexibility, versatility, and scalable production of phase pure materials with uniform particle distribution [82]. It is considered as good synthesis approach for the preparation of many inorganic compounds. Co-precipitation method exhibits several advantages, such as being inexpensive, having short reaction time, being simple, effective, having low energy consumption, and providing good yields [83].

Several researchers have synthesized the phosphors for optical thermometry by using Co-precipitation method such Yb^{3+}, Er^{3+} co-doped $KBi(MoO_4)_2$ [84], $YPO_4:Tm^{3+},Yb^{3+}$ [85], $Ca_3Bi(PO_4)_3:Er^{3+}$ [86], $NaYF_4:Tm^{3+}$, Er^{3+}; Yb^{3+}, Ho^{3+}, Nd^{3+} [87], $La_2MgTiO_6:Nd^{3+}$ [88], $Ba_2MgWO_6:Er^{3+}$ [89], $SrLaAlO_4:Er^{3+}$ [90], and α-$(La,Yb,Er)_2W_2O_9$ [91].

6.6.5 Combustion method

Combustion synthesis (CS) is energy and time-efficient method of material preparation in comparison with the solid-state reaction method, hydrothermal synthesis, co-precipitation, sono-chemical method, microwave-assisted sol–gel synthesis, etc. There are two types of CS such as solution combustion synthesis (SCS) and Self-propagating High-temperature Synthesis (SHS). In one of the types of CS, a solution mixture of metal precursor and an organic fuel such as urea is first dehydrated, then ignited and finally brought to combustion. Important feature of CS is the exothermicity nature of the combustion reaction, which provides the energy requirement for the synthesis of material in this method.

The high combustion temperature arises because of the exothermicity of the reaction is helpful to improve the crystallinity of the products. SCS method has been proven to be simple, efficient, cost- and time-effective which overcomes the material preparation difficulties by making use of exothermicity of the self-sustaining chemical reaction itself to drive the reaction. In SCS, almost all precursors are used in the liquid phase to have controlled accuracy and uniform mixing. In the combustion process, during preparation of precursor, heat energy is liberated by the redox exothermic reaction at a relative low igniting temperature required between metal nitrates and fuels [92]. Also, in the SCS method, a large number of gaseous products can be generated which will produce a large number of pores in the oxide and inhibit their agglomeration, leading to a large specific surface area of the resulting material [93]. Self-propagating High-temperature Synthesis (SHS) is also considered as one type of CS. Execution of this reaction is based on the exothermic redox reaction between metal nitrates and an appropriate fuel.

Several researchers have synthesized the phosphors for optical thermometry by using combustion method such YOF: Er^{3+} [94], Dy^{3+}/Eu^{3+} co-doped $SrMoO_4$ [95], $Tm^{3+}/Yb^{3+}:Y_2O_3$ and $Tm^{3+}/Yb^{3+}/Gd^{3+}:Y_2O_3$ [96]. Therefore, our research opens up new avenues for the development of color-tunable luminescent materials for various optoelectronic and temperature-sensing applications.

6.7 RARE-EARTH ION-DOPED PHOSPHORS FOR OPTICAL THERMOMETRY

Trivalent lanthanide ion-doped phosphors prepared by various methods are used in optical temperature sensors. Mostly Ln^{3+} ions such as Er^{3+}, Tm^{3+}, Ho^{3+}, Nd^{3+}, Dy^{3+}, and Eu^{3+}, are used as luminescence centers (or activators) in thermometry, as shown in Figure 6.1. Phosphors are excited with a UV, Visible, and NIR light source, and the emission in the ultraviolet, visible, and NIR regions. For optical thermometry, the TCL should satisfy some of the conditions that depend strongly on the host matrix into which the Ln^{3+} ions are doped. The factors affecting the TCL of Ln^{3+} ions are as follows:

1. To prevent significant overlap between the two fluorescence wavelengths, the TCL separation should be greater than $200\,cm^{-1}$; conversely, it should be less than $2000\,cm^{-1}$ to prevent an excessively tiny population at the upper energy level for the temperature measurement range.
2. The radiative transitions from the upper level should outnumber its non-radiative transitions in order to obtain an adequate fluorescence intensity from the upper-level transition.

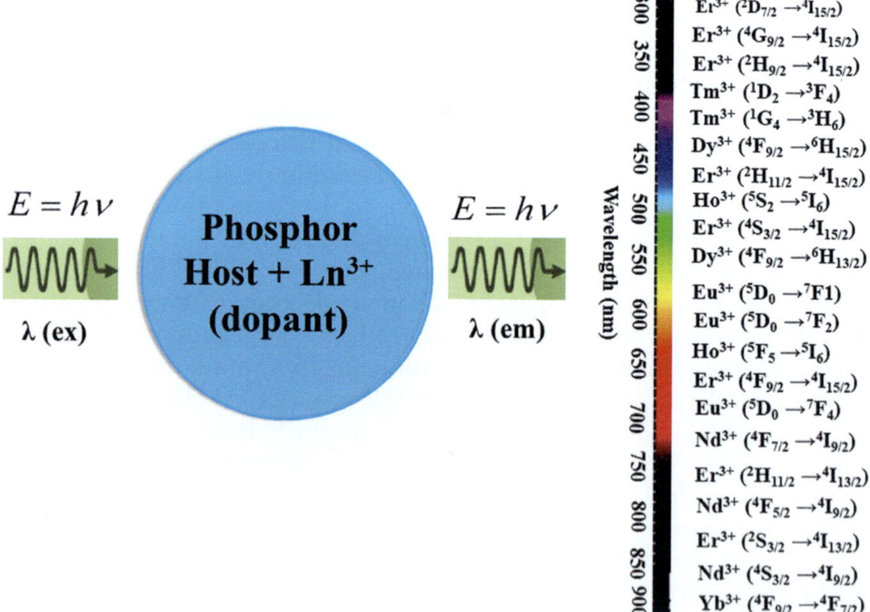

Figure 6.1 Transitions of Ln^{3+} ion-doped phosphors for optical thermometry.

Various TCL proven by experiment are listed as follows: Er^{3+}:$^2H_{11/2}$ and $^4S_{3/2}$, and $^4D_{7/2} \rightarrow {}^4G_{9/2}$; Tm^{3+}:$^3F_{2,3}$ and $^3H_4$, and $^1G_4$(a) and $^1G_4$(b); Ho^{3+}:$^5S_2$ and $^5F_4$, $^5F_{2,3}$ and $^3K_8$, and $^5G_6$ and $^5F_1$; Nd^{3+}:$^4F_{5/2}$ and $^4F_{3/2}$, $^4F_{7/2}$ and $^4F_{3/2}$, and $^4F_{7/2}$ and $^4F_{5/2}$; Dy^{3+}:$^4I_{5/2}$ and $^4F_{9/2}$; and Eu^{3+}:$^5D_1$ and $^5D_0$. TCLs are populated by the UC and DC processes of the above-mentioned Ln^{3+} ions. Optical thermometry has been achieved by analyzing the temperature-dependent luminescence properties originating from the transitions from the TCL to the other levels.

6.7.1 Er³⁺-doped phosphors

When excited in the near-infrared (NIR) region, trivalent erbium ion (Er^{3+}) exhibits strong green UC emission from two TCLs, namely $^2H_{11/2}$ and $^4S_{3/2}$. This makes Er^{3+} stand out among the several activator ions used in FIR-based optical temperature sensors. One of the most significant activator ions for FIR-based optical temperature sensors is widely acknowledged to be Er^{3+} [97,98].

The PLE and PL spectra of the $(Y_{0.975}Er_{0.025})_2O_3$ are depicted in Figure 6.2. The excitation is monitored at 563 nm. The excitation spectra have a very strong narrow line at 380 nm, which is assigned to the hypersensitive transition $^4I_{15/2} \rightarrow {}^4G_{11/2}$ of erbium ion. The other lines are observed at 258, 368, and 410 nm related to transitions from $^4I_{15/2}$ state to the $^4D_{5/2}$,

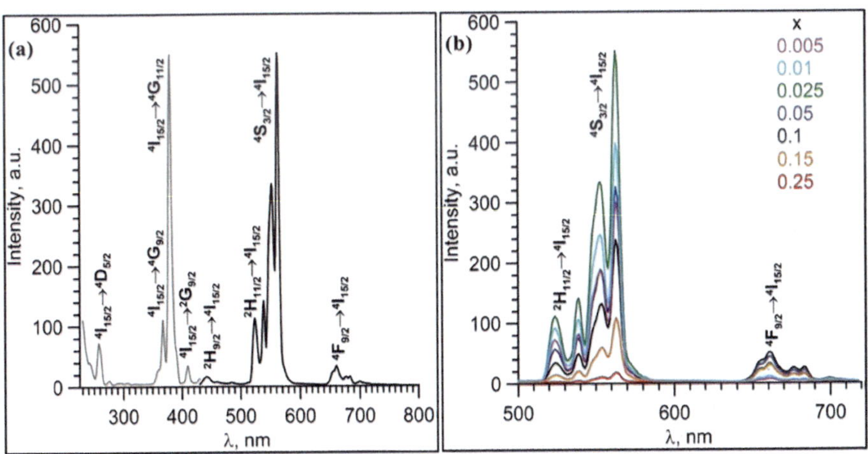

Figure 6.2 (a) Excitation (λ_{em}=563 nm) and luminescence (λ_{ex}=380 nm) spectra of $(Y_{0.975}Er_{0.025})_2O_3$ oxide (b) luminescence spectra (λ_{ex}=380 nm) of $(Y_{1-x}Er_x)_2O_3$ oxides with x=0.005–0.25. (Adopted with permission from Ref. [99] © Elsevier, 2023.)

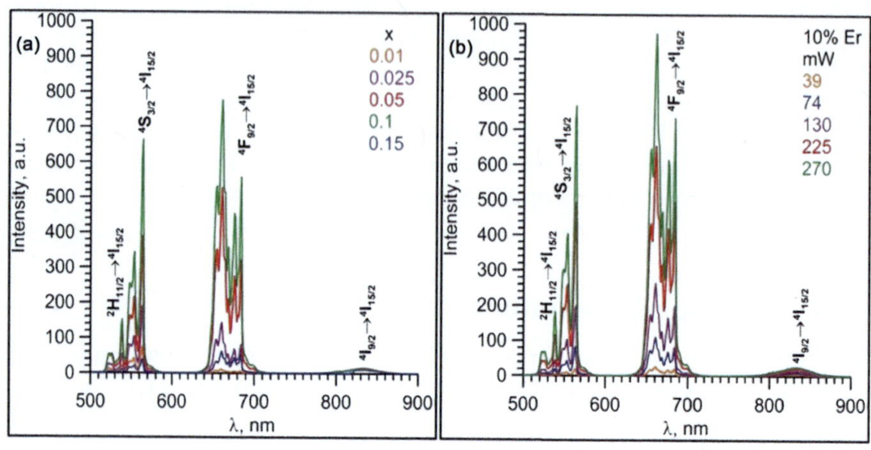

Figure 6.3 (a) UC luminescence spectra of $(Y_{1-x}Er_x)_2O_3$ oxides with $x=0.01-0.15$. (b) UC luminescence spectra ($\lambda_{ex}=980\,nm$) of $(Y_{0.9}Er_{0.1})_2O_3$ oxide measured at different excitation power. (Adopted with permission from Ref. [99] © Elsevier, 2023.)

$^4G_{9/2}+^2G_{7/2}$, and $^2G_{9/2}$ states of Er^{3+} ion, respectively. When the excitation is taken under UV radiation at 380 nm, the following transitions of erbium ion are observed in the emission spectrum: $^2H_{9/2}\rightarrow^4I_{15/2}$ (442 nm – blue line), $^2H_{11/2}\rightarrow^4I_{15/2}$ (524 and 539 nm – green lines), $^4S_{3/2}\rightarrow^4I_{15/2}$ (553 and 563 nm – green lines), $^4F_{9/2}\rightarrow^4I_{15/2}$ (655, 661, 676, and 683 nm – red lines). Also, in the PL spectra, the green lines are more intense compared with the red line irrespective of the concentration of erbium ions in oxides (Figure 6.2).

Figure 6.3a represents the UC luminescence spectra excited at 980 nm under IR laser excitation. The upconversion lines of erbium ions are observed in visible as well as NIR region. The lines in the visible region are linked with the transition of erbium ions are $^2H_{11/2}\rightarrow^4I_{15/2}$ (522, 526, and 538 nm), $^4S_{3/2}\rightarrow^4I_{15/2}$ (548, 554, and 563 nm), and $^4F_{9/2}\rightarrow^4I_{15/2}$ (655, 661, 669, 676, and 683 nm). The line in the NIR region is observed at 832 nm associated with $^4I_{9/2}\rightarrow^4I_{15/2}$ transition (Figure 6.3b).

In recent years, some Er^{3+}-activated phosphors have been developed such as $Y_2Mo_4O_{15}:Er^{3+}/Yb^{3+}$ (YMO) [100], $NaYb(MoO_4)_2:Er^{3+}$ [101], nanosized $(Y_{1-x}Er_x)_2O_3$ oxides [99], Er^{3+} and Yb^{3+} co-doped $SrTiO_3$ [102], Yb^{3+}, Er^{3+} co-doped $KBi(MoO_4)_2$ nanomaterials [103] has great application potential in optical thermometry.

6.7.2 Yb³⁺–Tm³⁺ co-doped phosphors

Among all the RE ions, Tm^{3+} ion is such ion that exhibits emission from near infrared (NIR) to ultraviolet (UV) excited under 980 nm laser diode. It was revealed that the energy levels of Tm^{3+} ion were insufficient for effectively

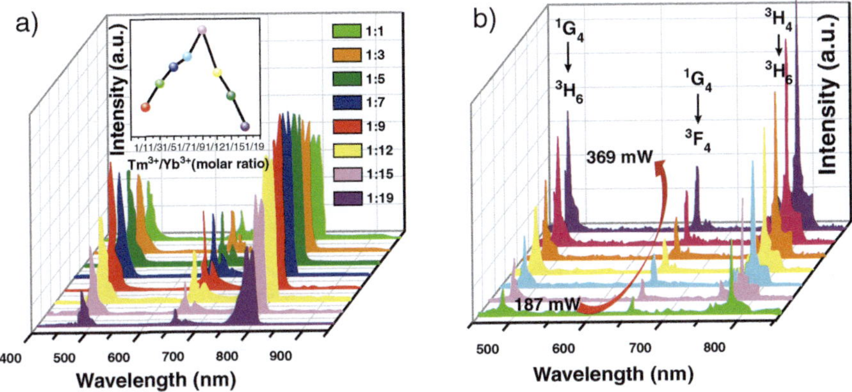

Figure 6.4 (a) UC spectra of CaZnOS:Tm³⁺/Yb³⁺ with different molar ratio and (b) the UC intensity as a function of laser power (Tm³⁺/Yb³⁺ = 1:9). (Adopted with permission from ref. [99] © Elsevier, 2023. Adopted with permission from Ref. [107] © Elsevier, 2021.)

absorbing wavelengths of 980 nm. An effective sensitizer is needed. So, the Yb^{3+} ion is considered as a suitable candidate in this regard. Yb^{3+} ion have a large absorption cross section [104]. It enhances the intensity of Tm^{3+} ion emission through energy transfer processes that are both cooperative and sequential. Recent years have seen a large amount of research on Tm^{3+}/Yb^{3+}-based upconversion luminescence in a variety of host materials [105,106].

Figure 6.4a represents the upconversion spectra of CaZnOS: Tm/Yb excited at 980 nm with various molar ratio. The inserted figure shows that the strongest UC intensity was reached when the molar ratio of Tm/Yb was 1:9. In Figure 6.4b, the integral emission intensity shows a predictable increase with the increasing of the power density. Emission spectrum consists of narrow bands which are observed in the range 500–800 nm, corresponding to the f-f transitions. The bands observed at 483, 652, and 792 nm were assigned to transitions $^1G_4{\rightarrow}^3H_6$, $^1G_4{\rightarrow}^3F_4$, and $^3H_4{\rightarrow}^3H_6$, respectively [108].

In recent years, some Er^{3+}-activated phosphors have been developed such as $Tm^{3+}/Yb^{3+}:Na_3GdV_2O_8$ (NGVO) phosphors [109], $Tm/Yb:SrF_2$ [110], Tm^{3+}/Yb^{3+} co-doped CaZnOS phosphors [107], Well-crystallized $LiLa(MoO_4)_2$ co-doped with 20% Yb^{3+} and 0.5% Tm^{3+} [111] has great application potential in optical thermometry.

6.7.3 Nd³⁺-doped phosphors

Of all the available RE ions, due to the strong emission at 1060 nm, Nd^{3+} ion is one of the most effective ions for solid-state lasers [112]. In order to develop high-power NIR solid-state lasers, trivalent neodymium (Nd^{3+})

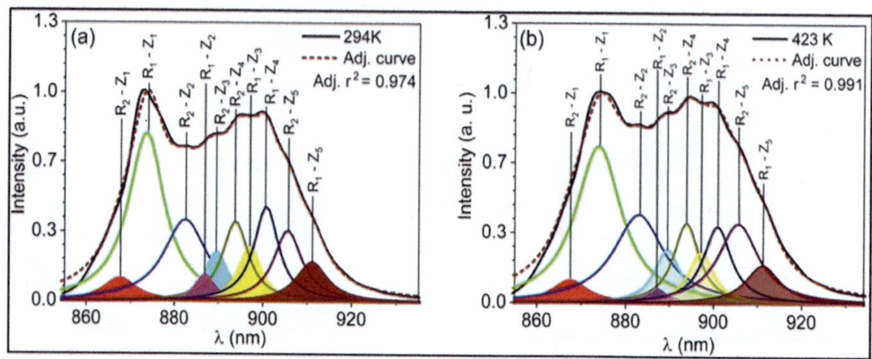

Figure 6.5 Deconvolution of the band $^4F_{3/2} \rightarrow {}^4I_{9/2}$ (a) at 294 K and (b) 423 K. (Adopted with permission from Ref. [114] © Elsevier, 2020.)

ion-doped variety of crystals and glasses were extensively studied under 808 and 885 nm laser diode excitation [113].

LiBaPO$_4$:Nd^{3+}phosphor emission band within BW I recorded at 294 and 423 K is shown in Figure 6.5. Observed emission is associated with $^4F_{3/2} \rightarrow {}^4I_{9/2}$ transition of Nd^{3+} ions. Because of host crystalline field, the energy level $^4F_{3/2}$ can split in two Stark's sublevels, named as R_1 and R_2, while the ground state, $^4I_{9/2}$ is divided into five different sublevels such as Z_1, Z_2, Z_3, Z_4, and Z_5. In recorded spectra, splitting is prominently observed after deconvolution of the energy band. Ten narrow emission lines, which are connected to the changes between those Stark's sublevels, are superposed to form the emission band.

In recent years, some Nd^{3+}-activated phosphors have been developed such as LuVO$_4$:Nd^{3+} [115], Nd^{3+}-doped Yb$_2$Mo$_3$O$_{12}$ phosphor [116], Neodymium-doped LiBaPO$_4$ phosphors [114], CaWO$_4$:Nd^{3+} [117]. This work offers a prospective strategy to provide precise and effective optical thermometry, including at ultralow temperatures.

6.7.4 Dy^{3+}-doped phosphors

Dy^{3+}-doped phosphors attracted particular interest because of their importance in visible lasers and light-emitting materials. Two prominent emission bands, $^4F_{9/2} \rightarrow {}^6H_{15/2}$ (blue) and $^4F_{9/2} \rightarrow {}^6H_{13/2}$ (yellow), correspond to the transitions in the Dy^{3+} luminescence spectrum. In Dy^{3+} activated phosphors, white emission can be produced by a suitable combination of blue and yellow emission. The yellow emission, represented by the hypersensitive transition $^4F_{9/2} \rightarrow {}^6H_{13/2}$, is highly dependent on host material. Therefore, the Dy^{3+}-doped phosphors can produce tunable emission by choosing the right host materials [118,119].

Figure 6.6a represents the PLE of LiCa$_2$Mg$_2$V$_3$O$_{12}$:0.08Dy^{3+} phosphor recorded at fixed emission wavelength of 576 nm and Figure 6.6b

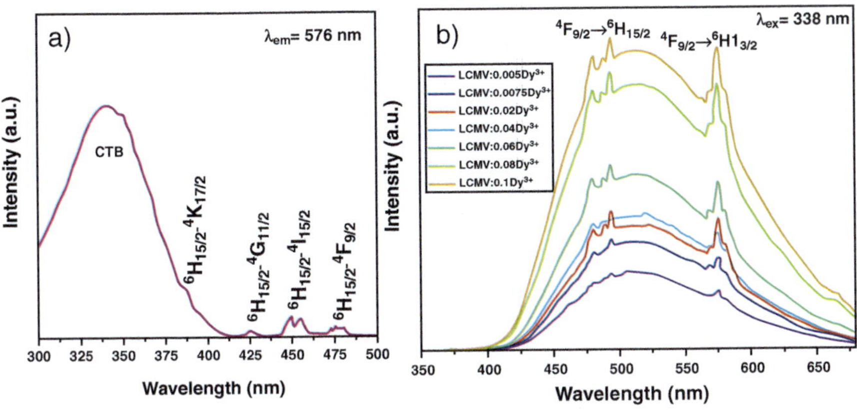

Figure 6.6 (a) PLE of $LiCa_2Mg_2V_3O_{12}$:0.08Dy^{3+} phosphor (b) PL spectra of $LiCa_2Mg_2V_3O_{12}$:xDy^{3+} (x=0.005–0.1) phosphor. (Adopted with permission from Ref. [120] © Elsevier, 2023.)

represents the PL spectra of $LiCa_2Mg_2V_3O_{12}$:xDy^{3+} (x=0.005–0.1) phosphor at an excitation of 338 nm respectively. The excitation spectra have a wide band in the range of 300–400 nm and have sufficient absorption in the UV region. The other excitation peaks observed at 385, 425, 449, and 476 nm are assigned to the $^6H_{15/2}{\rightarrow}^4K_{17/2}$, $^6H_{15/2}{\rightarrow}^4G_{11/2}$, $^6H_{15/2}{\rightarrow}^4I_{15/2}$ and $^6H_{15/2}{\rightarrow}^4F_{9/2}$ transitions of Dy^{3+} [121]. In addition to the broadband spectra of $LiCa_2Mg_2V_3O_{12}$, the PL Spectra of $LiCa_2Mg_2V_3O_{12}$: xDy^{3+} (x=0.005–0.1) consists of two dominant peaks at 494 and 576 nm. The peaks at 494 (blue) and 576 nm (yellow) are attributed to the transitions $^4F_{9/2}{\rightarrow}^6H_{15/2}$ and $^4F_{9/2}{\rightarrow}^6H_{13/2}$, respectively [122].

In recent years, some Dy^{3+}-activated phosphors have been developed such as Dy^{3+}-doped $LiCa_2Mg_2V_3O_{12}$ phosphors [120], Dy^{3+}-doped $CaWO_4$ [123], $CaLa_4(SiO_4)_3O$: Dy^{3+} [124], $GdPO_4$:1%Dy^{3+} phosphor [125], Dy^{3+}-doped $CaWO_4$ [126]. The optical properties of the phosphor suggest that it finds applications in the fields of solid-state lighting, temperature sensing, optoelectronic devices, and W-LEDs.

6.7.5 Eu^{3+}-doped phosphors

Eu^{3+} ion is a commonly used dopant in many host matrices such as vanadate, oxyfluoride, tungstate, and silicate [127–130]. The excitation of Eu^{3+} ion is in the NUV region. The narrow line emission is in the orange-red region because of $^5D_0{\rightarrow}^7F_J$ (J=1–4) transitions. The strongest red emission between 610 and 630 nm ($^5D_0{\rightarrow}^7F_2$) meets the requirements of WLEDs. It is well known that the luminescent performance is heavily dependent on the host material [131,132].

Figure 6.7a represents the photoluminescence excitation (PLE) and emission spectra of NYbM:$x$$Eu^{3+}$ (x=1.0, 0.9, 0.8, 0.4, and 0.2) at room

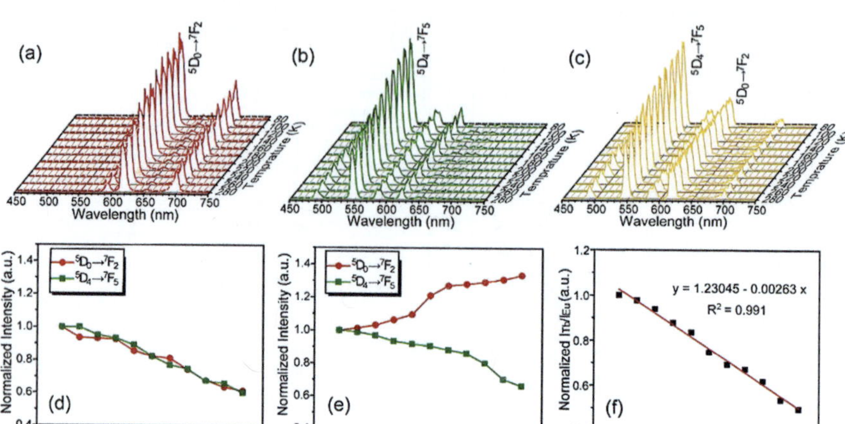

Figure 6.7 (a) PLE spectra of NYbM:1.0Eu^{3+} samples monitored at the 610 and 706 nm emission wavelength, emission spectra of (b) NYbM:xEu^{3+} (x = 1.0, 0.9, 0.8, 0.4, and 0.2). (Adopted with permission from Ref. [74] © Elsevier, 2024.)

temperature. The excitation spectra consist of three peaks in the range of 200–600 nm, one is strong broadband at 334 nm, other two peaks are weak but sharp at 394 and 465 nm and are assigned to the $^7F_0 \rightarrow {}^5L_6$ and $^7F_0 \rightarrow {}^5D_2$ electronic transitions of Eu^{3+} ions, respectively [133,134].

Figure 6.7b displays the PL spectra of NYbM: xEu^{3+} (x = 0–1.0) samples under 355 nm excitation at room temperature. The emission spectra contain broad band host luminescence (covering the 400–720 nm range, centered at ~550 nm and owing to $^3T_{1,2} \rightarrow {}^1A_1$ transitions of [VO$_4$]$^{3-}$) and some sharp characteristic emission of Eu^{3+} centered at 594, 612, 654, and 708 nm, corresponding to $^5D_0 \rightarrow {}^7F_J$ (J = 1–4) transitions, respectively [135,136]. The emission peak is located at 612 nm with a narrow band, which is far more dominant than others. The emission intensity of Eu^{3+} increases drastically while the emissions of [VO$_4$]$^{3-}$ groups decrease with the Eu^{3+} concentrations.

In recent years, some Eu^{3+}-activated phosphors have been developed such Sr$_3$V$_2$O$_8$:2%Eu^{3+} material [137], Eu^{3+}-doped Na$_2$YbMg$_2$(VO$_4$)$_3$ red-emitting phosphors [138], YVO$_4$: Eu^{3+} phosphor [139], Eu^{3+}-doped Ca$_3$LiZnV$_3$O$_{12}$ phosphors [140], CaWO$_4$:Eu^{3+} phosphors [141]. This phosphor has potential applications in the lighting field and optical thermometry.

6.8 CONCLUSIONS

In summary, lanthanide-doped phosphor-based optical thermometers offer a wide range of advantages over other types of thermometer materials, which is a motivation for large-scale research. The potential of these materials is reflected in the rapidly increasing research and development of not only new thermometer materials but also their practical applications in sensing and thermal imaging. Despite the extensive literature on this topic, it can be assumed that the coming years will bring many interesting and scientifically inspiring discoveries in this area.

REFERENCES

1. M. I. Tiwana, S. J. Redmond, N. H. Lovell, A review of tactile sensing technologies with applications in biomedical engineering. *Sens. Actuat. A Phys.* 179 (2012) 17–31, https://doi.org/10.1016/j.sna.2012.02.051.
2. I. Mordor, Temperature Sensors Market - Growth, Trends, and Forecast (2020–2025), 2019. [Online]. Available: https://www.mordorintelligence. com/industry-reports/temperature-sensors-market-industry.
3. G. Wasner, J. Schattschneider, K. Heckmann, C. Maier, R. Baron, Vascular abnormalities in reflex sympathetic dystrophy (CRPS I): Mechanisms and diagnostic value. *Brain* 124 (3) (2001) 587–599, https://doi.org/10.1093/brain/124.3.587.
4. M. Aldén, A. Omrane, M. Richter, G. Särner, Thermographic phosphors for thermometry: A survey of combustion applications. *Prog. Energy Combust. Sci.* 37 (4) (2011) 422–461, https://doi.org/10.1016/j.pecs.2010.07.001.
5. R. Shi, L. Lin, P. Dorenbos, H. Liang, Development of a potential optical thermometric material through photoluminescence of Pr^{3+} in La_2MgTiO_6. *J. Mater. Chem. C* 5 (41) (2017) 10737–10745, https://doi.org/10.1039/C7TC02661G.
6. X. D. Wang, O. S. Wolfbeis, R. J. Meier, Luminescent probes and sensors for temperature. *Chem. Soc. Rev.* 42 (2013) 7834–7869, https://doi.org/10.1039/C3CS60102A.
7. M. K. Mahata, K. Kumar, V. K. Rai, Er^{3+}–Yb^{3+} doped vanadate nanocrystals: A highly sensitive thermographic phosphor and its optical nanoheater behaviour. *Sens. Actuators B* 209 (2015) 775–780, https://doi.org/10.1016/j.snb.2014.12.039.
8. A. H. Khalid, K. Kontis, Thermographic phosphors for high temperature measurements: Principles, current state of the art and recent applications. *Sensors* 8 (2008) 5673–5744, https://doi.org/10.3390/s8095673.
9. X. F. Wang, Q. Liu, Y. Y. Bu, C. S. Liu, T. Liu, X. H. Yan, Optical temperature sensing of rare-earth ion doped phosphors. *RSC Adv.* 5 (2015) 86219–86236, https://doi.org/10.1039/C5RA16986K.
10. J. Cao, X. Li, Z. Wang, Y. Wei, L. Chen, H. Guo, Optical thermometry based on up-conversion luminescence behavior of self-crystallized K_3YF_6:Er^{3+} glass ceramics. *Sens. Actuators B* 224 (2016) 507–513, https://doi.org/10.1016/j.snb.2015.10.087.
11. M. Back, E. Trave, J. Ueda, S. Tanabe, Ratiometric optical thermometer based on dual near-infrared emission in Cr^{3+}-doped bismuth-based gallate host. *Chem. Mater.* 28 (2016) 8347–8356, https://doi.org/10.1021/acs.chemmater.6b03625.
12. B. Dong, B. S. Cao, Y. Y. He, Z. Liu, Z. P. Li, Z. Q. Feng, Temperature sensing and in vivo imaging by molybdenum sensitized visible upconversion luminescence of rare-earth oxides. *Adv. Mater.* 24 (2012) 1987–1993, https://doi.org/10.1002/adma.201200431.

13. H. Dong, L. D. Sun, C. H. Yan, Energy transfer in lanthanide upconversion studies for extended optical applications. *Chem. Soc. Rev.* 44 (2015) 1608–1634, https://doi.org/10.1039/C4CS00188E.

14. Y. Cui, H. Xu, Y. Yue, Z. Guo, J. Yu, Z. Chen, J. Gao, Y. Yang, G. Qian, B. Chen, A luminescent mixed-lanthanide metal-organic framework thermometer. *J. Am. Chem. Soc.* 134 (2012) 3979–3982, https://doi.org/10.1021/ja2108036.

15. R. F. D'Vries, S. Álvarez-García, N. Snejko, L. E. Bausá, E. Gutiérrez-Puebla, A. de Andrés, M. Á. Monge, Multimetal rare earth MOFs for lighting and thermometry: Tailoring color and optimal temperature range through enhanced disulfobenzoic triplet phosphorescence. *J. Mater. Chem. C* 1 (2013) 6316–6325, https://doi.org/10.1039/C3TC30858H.

16. M. Peng, Z. Pei, G. Hong, Q. Su, The reduction of Eu^{3+} to Eu^{2+} in $BaMgSiO_4$:Eu prepared in air and the luminescence of $BaMgSiO_4$:Eu^{2+} phosphor. *J. Mater. Chem.* 13 (2003) 1202–1205, https://doi.org/10.1039/B211624C.

17. W. Jiang, Z. Bian, C. Hong, C. Huang, A mild liquid reduction route toward uniform blue-emitting $EuCl_2$ nanoprisms and nanorods. *Inorg. Chem.* 50 (2011) 6862–6864, https://doi.org/10.1021/ic200993w.

18. M. A. R. C. Alencar, G. S. Maciel, C. B. de Araújo, A. Patra, Er^{3+}-doped $BaTiO_3$ nanocrystals for thermometry: Influence of nano environment on the sensitivity of a fluorescence-based temperature sensor. *Appl. Phys. Lett.* 84 (2004) 4753–4755, https://doi.org/10.1063/1.1760882.

19. Y. M. Yang, Z. Y. Li, J. Y. Zhang, Y. Lu, S. Q. Guo, Q. Zhao, X. Wang, Z. J. Yong, H. Li, J. P. Ma, Y. Kuroiwa, C. Moriyoshi, L. L. Hu, L. Y. Zhang, L. R. Zheng, H. T. Sun, X-ray-activated long persistent phosphors featuring strong UVC afterglow emissions. *Light Sci. Appl.* 7 (2018) 88–99, https://doi.org/10.1038/s41377-018-0089-7.

20. F. Wang, R. Deng, J. Wang, Q. Wang, Y. Han, H. Zhu, X. Chen, X. Liu, Tuning upconversion through energy migration in core-shell nanoparticles. *Nat. Mater.* 10 (2011) 968–973, https://doi.org/10.1038/nmat3149.

21. Y. Kitaoka, K. Nakamura, T. Akiyama, T. Ito, M. Weinert, A.J. Freeman, Excited Cr impurity states in Al_2O_3 from constraint density functional theory. *Phys. Rev. B* 87 (2013) 205113, https://doi.org/10.1103/PhysRevB.87.205113.

22. M. Xu, X. Zou, Q. Su, W. Yuan, C. Cao, Q. Wang, X. Zhu, W. Feng, F. Li, Ratiometric nanothermometer in vivo based on triplet sensitized upconversion. *Nat. Commun.* 9 (2018) 2698, https://doi.org/10.1038/s41467-018-05160-1.

23. M. Runowski, P. Woźny, S. Lis, V. Lavín, I. R. Martín, Optical vacuum sensor based on lanthanide upconversion – Luminescence thermometry as a tool for ultralow pressure sensing. *Adv. Mater. Technol.* 5 (2020) 1901091, https://doi.org/10.1002/admt.201901091.

24. A. Yakovliev, T. Y. Ohulchanskyy, R. Ziniuk, T. Dias, X. Wang, H. Xu, G. Chen, J. Qu, A. S. L. Gomes, Noninvasive temperature measurement in dental materials using Nd^{3+}, Yb^{3+} doped nanoparticles emitting in the near infrared region. *Part. Syst. Charact.* 37 (2020) 1900445, https://doi.org/10.1002/ppsc.201900445.

25. X. Tu, J. Xu, M. Li, T. Xie, R. Lei, H. Wang, S. Xu, Color-tunable upconversion luminescence and temperature sensing behavior of Tm^{3+}/Yb^{3+} codoped $Y_2Ti_2O_7$ phosphors. *Mater. Res. Bull.* 112 (2019) 77–83, https://doi.org/10.1016/j.materresbull.2018.12.008.

26. N. Wang, Z. Fu, Y. Wei, T. Sheng, Investigation for the upconversion luminescence and temperature sensing mechanism based on $BiPO_4$:Yb^{3+}, RE^{3+} (RE^{3+}= Ho^{3+}, Er^{3+} and Tm^{3+}). *J. Alloys Compd.* 772 (2019) 371–380, https://doi.org/10.1016/j.jallcom.2018.09.070.

27. J. S. Zhong, D. Q. Chen, Y. Z. Peng, Y. D. Lu, X. Chen, X. Y. Li, Z. G. Ji, A review on nanostructured glass ceramics for promising application in optical thermometry. *J. Alloys Compd.* 763 (2018) 34–48, https://doi.org/10.1016/j.jallcom.2018.05.348.

28. M. Ding, M. Xu, D. Chen, A new non-contact self-calibrated optical thermometer based on $Ce^{3+} \rightarrow Tb^{3+} \rightarrow Eu^{3+}$ energy transfer process. *J. Alloys Compd.* 713 (2017) 236–247, https://doi.org/10.1016/j.jallcom.2017.04.202.

29. L. Marciniak, K. Kniec, K. Elżbieciak-Piecka, K. Trejgis, J. Stefanska, M. Dramićanin, Luminescence thermometry with transition metal ions: A review. *Coordination Chem. Rev.* 469 (2022) 214671, https://doi.org/10.1016/j.ccr.2022.214671.

30. M. G. Brik, Ab-initio studies of the electronic and optical properties of Al_2O_3:Ti^{3+} laser crystals, *Phys. B Condens. Matter* 532 (2018) 178–183, https://doi.org/10.1016/j.physb.2017.02.032.

31. D. Plessers, M. Bols, H. M. Rhoda, A. J. Heyer, E. I. Solomon, B. Sels, R. Schoonheydt, Single site spectroscopy of transition metal ions and reactive oxygen complexes in zeolites, *Reference Module in Chemistry, Molecular Sciences and Chemical Engineering*, Elsevier, 2021.

32. L. Marciniak, K. Trejgis, Luminescence lifetime thermometry with Mn^{3+}–Mn^{4+} co-doped nanocrystals. *J. Mater. Chem.* C 6 (26) (2018) 7092–7100, https://doi.org/10.1039/C8TC01981A.

33. M. Shi, L. Yao, S. Yu, Y. Dong, Q. Shao, Enhancing the temperature sensitivity of Cr^{3+} emissions by modification of the host's composition for fluorescence thermometry applications. *Dalt. Trans.* 51 (2) (2022) 587–593, https://doi.org/10.1039/d1dt03480d.

34. K. Elzbieciak, L. Marciniak, The impact of Cr^{3+} doping on temperature sensitivity modulation in Cr^{3+} doped and Cr^{3+}, Nd^{3+} co-doped $Y_3Al_5O_{12}$, $Y_3Al_2Ga_3O_{12}$, and $Y_3Ga_5O_{12}$ nanothermometers. *Front. Chem.* 6 (2018) 424, https://doi.org/10.3389/fchem.2018.00424.

35. W. Piotrowski, M. Kuchowicz, M. Dramićanin, L. Marciniak, Lanthanide dopant stabilized Ti^{3+} state and supersensitive Ti^{3+}-based multiparametric luminescent thermometer in $SrTiO_3$:Ln^{3+} (Ln^{3+}=Lu^{3+}, La^{3+}, Tb^{3+}) nanocrystals. *Chem. Eng. J.* 428 (2022) 131165, https://doi.org/10.1016/j.cej.2021.131165.

36. F. Li, J. Cai, F. F. Chi, Y. Chen, C. Duan, M. Yin, Investigation of luminescence from LuAG:Mn^{4+} for physiological temperature sensing. *Opt. Mater. (Amst)* 66 (2017) 447–452, https://doi.org/10.1016/j.optmat.2017.02.054.

37. K. Trejgis, M. D. Dramićanin, L. Marciniak, Highly sensitive multiparametric luminescent thermometer for biologically-relevant temperatures based on Mn^{4+}, Ln^{3+} co-doped $SrTiO_3$ nanocrystals. *J. Alloy. Compd.* 875 (2021) 159973, https://doi.org/10.1016/j.jallcom.2021.159973.

38. C. D. S. Brites, P. P. Lima, N. J. O. Silva, A. Millán, V. S. Amaral, F. Palacio, L. D. Carlos, Thermometry at the nanoscale. *Nanoscale* 4 (16) (2012) 4799, https://doi.org/10.1039/c2nr30663h.

39. M. Suta, A. Meijerink, A theoretical framework for ratiometric single ion luminescent thermometers – Thermodynamic and kinetic guidelines for optimized performance. *Adv. Theory Simul.* 3 (12) (2020) 2000176, https://doi.org/10.1002/adts.v3.1210.1002/adts.202000176.

40. J. Zhou, B. del Rosal, D. Jaque, S. Uchiyama, D. Jin, Advances and challenges for fluorescence nanothermometry. *Nat. Methods* 17 (10) (2020) 967–980, https://doi.org/10.1038/s41592-020-0957-y.

41. D. Jaque, F. Vetrone, Luminescence nanothermometry. *Nanoscale* 4 (15) (2012) 4301–4326, https://doi.org/10.1039/C2NR30764B.

42. K. Lenczewska, Y. Gerasymchuk, N. Vu, N. Q. Liem, G. Boulon, D. Hreniak, The size effect on the energy transfer in Bi^{3+}–Eu^{3+} co-doped $GdVO_4$ nanocrystals. *J. Mater. Chem.* C 5 (12) (2017) 3014–3023, https://doi.org/10.1039/C6TC04660F.

43. A. Escudero, C. Carrillo-Carrion, M. V. Zyuzin, S. Ashraf, R. Hartmann, N. O. Nunez, M. Ocana, W. J. Parak, Synthesis and functionalization of monodisperse near-ultraviolet and visible excitable multifunctional Eu^{3+}, Bi^{3+}:$REVO_4$ nanophosphors for bioimaging and biosensing applications. *Nanoscale* 8 (2016) 12221–12236, https://doi.org/10.1039/C6NR03369E.

44. X. Wang, Q. Liu, Y. Bu, C.-S. Liu, T. Liu, X. Yan, Optical temperature sensing of rare-earth ion doped phosphors. *RSC Adv.* 5 (105) (2015) 86219–86236, https://doi.org/10.1039/C5RA16986K.

45. C. D. Brites, K. Fiaczyk, J. F. Ramalho, M. Sójka, L. D. Carlos, E. Zych, Widening the temperature range of luminescent thermometers through the intra- and interconfigurational transitions of Pr^{3+}. *Adv. Opt. Mater.* 6 (10) (2018) 1701318, https://doi.org/10.1002/adom.201701318.

46. X. F. Wang, Y. Wang, J. Marques-Hueso, X. H. Yan, Improving optical temperature sensing performance of Er^{3+} doped Y_2O_3 microtubes via co-doping and controlling excitation power. *Sci. Rep.* 7 (1) (2017) 758, https://doi.org/10.1038/s41598-017-00838-w.

47. D. Manzani, J. Flávio da Silveira Petruci, K. Nigoghossian, A. Alves Cardoso, S. J. L. Ribeiro, A portable luminescent thermometer based on green upconversion emission of Er^{3+}/Yb^{3+} co-doped tellurite glass. *Sci. Rep.* 7 (1) (2017) 41596, https://doi.org/10.1038/srep41596.

48. B. Li, X. Wei, Z. Chang, X. Chen, X. Z. Yuan, H. Wang, Facile fabrication of $LiMn_2O$ microspheres from multi-shell MnO_2 for high-performance lithium-ion batteries. *Mater. Lett.* 135 (2014) 75–78, https://doi.org/10.1016/j.matlet.2014.07.117.

49. M. M. Upadhyay, N. K. Mishra, K. Kumar, Upconversion luminescence based temperature sensing properties and anti-counterfeiting applications of $GdNbO_4$:Tm^{3+}/Yb^{3+} phosphor. *Spectrochim. Acta A Mol. Biomol. Spectrosc.* 304 (2024) 123333, https://doi.org/10.1016/j.saa.2023.123333.

50. J. Wang, X. Zhao, M. Song, J. Zhao, J. Xue, Efficient lifetime-based optical thermometry using $BaLaZnSbO_6$:Mn^{4+} red-emitting phosphors. *J. Lumines.* 263 (2023) 120019, https://doi.org/10.1016/j.jlumin.2023.120019.

51. X. Yang, Y. Zhu, T. Li, S. Long, B. Wang, High-accuracy dual-mode optical thermometry based on up-conversion luminescence in Er^{3+}/Ho^{3+}-Yb^{3+} doped $LaNbO_4$ phosphors. *Ceramics Int.* 49 (2023) 21932–21940, https://doi.org/10.1016/j.ceramint.2023.04.017.

52. H. Lv, L. Liu, D. Wang, Z. Mai, F. Yan, G. Xing, P. Du, Enhanced upconversion emission in Er^{3+}/Yb^{3+}-codoped $Al_2Mo_3O_{12}$ microparticles via doping strategy: Towards multimode visual optical thermometer. *J. Lumin.* 252 (2022) 119333, https://doi.org/10.1016/j.jlumin.2022.119333.

53. X. Yin, Q. Xiao, L. Lv, X. Wu, X. Dong, Y. Fan, N. Zhou, X. Luo, Winning color-tunable upconversion luminescence and high-sensitive optical thermometry in $K_3Gd(PO_4)_2$:Yb^{3+},Er^{3+},Tm^{3+}, *Spectrochim. Acta A Mol. Biomol. Spectrosc.* 291 (2023) 122324, https://doi.org/10.1016/j.saa.2023.122324.

54. K. Pavani, A. J. Neves, R. J. B. Pinto, C. S. R. Freire, M. J. Soares, M. P. F. Graça, K. Upendra Kumar, S. K. Jakka, $BiLaWO_6$: $Er^{3+}/Tm^{3+}/Yb^{3+}$ phosphor: Study of multiple fluorescence intensity ratiometric thermometry at cryogenic temperatures. *Ceram. Int.* 48 (2022) 31344–31353, https://doi.org/10.1016/j.ceramint.2022.06.316.

55. J. Cheng, J. Cheng, S. Zhang, C. Ma, X. Bian, Z. Zhai, Photoluminescence, site occupation and optical thermometry of $Ba_2La_6Y_2(SiO_4)_6O_2$:Eu^{3+} phosphors with high thermal stability. *J. Lumin.* 252 (2022) 119265, https://doi.org/10.1016/j.jlumin.2022.119265.

56. M.-H. Liu, T.-T. Li, D.-L. Zhang, Synthesis and spectroscopic properties of Er^{3+}/Yb^{3+}-codoped $GdNbO_4$ phosphor for thermometry and marine safety protection. *Mater. Res. Bull.* 154 (2022) 111944, https://doi.org/10.1016/j.materresbull.2022.111944.

57. J. Wu, F. Duan, Y. Zheng, Y. Xie, Synthesis of Bi_2WO_6 nanoplate-built hierarchical nest-like structures with visible-light-induced photocatalytic activity. *J. Phys. Chem. C* 111 (2007) 12866–12871, https://doi.org/10.1021/jp073877u.

58. L. Zhang, W. Wang, Z. Chen, L. Zhou, H. Xu, W. Zhu, Fabrication of flower-like Bi_2WO_6 superstructures as high performance visible-light driven photocatalysts. *J. Mater. Chem.* 17 (2007) 2526–2532, https://doi.org/10.1039/B616460A.

59. T. Li, Y. Shen, S. Zhao, X. Zhong, W. Zhang, C. Han, D. Wei, D. Meng, Y. Ao, Sub-ppm level NO_2 sensing properties of polyethyleneimine-mediated WO_3 nanoparticles synthesized by a one-pot hydrothermal method. *J. Alloys Compd.* 783 (2019) 103–112, https://doi.org/10.1016/j.jallcom.2018.12.287.

60. H. Xiao, Q. Meng, Eu^{3+} doped $CaWO_4$ nanophosphor for high sensitivity optical thermometry. *Spectrochim. Acta A Mol. Biomol. Spectrosc.* 305 (2024) 123542, https://doi.org/10.1016/j.saa.2023.123542.

61. F. Hu, D. Zhang, Y. Wu, C. Sun, C. Xu, Q. Wang, Y. Xie, Q. Shi, S. Li, K. Wang, Optical behaviors of Mn^{4+}-modified cubic type $ZnTiO_3$:Eu^{3+} nanocrystals: Application in optical thermometers based on fluorescence intensity ratio and lifetime. *Spectrochim. Acta A Mol. Biomol. Spectrosc.* 304 (2024) 123339, https://doi.org/10.1016/j.saa.2023.123339.

62. J. Liu, Q. Hu, H. Xu, H. Yu, B. Du, Q. Han, W. Wu, Vacancy-ordered Te^{4+}-doped Rb_2ZrCl_6 double perovskite microcrystals for solid-state lighting and non-contact optical thermometry. *Appl. Phys. Lett.* 123 (2023) 111903, https://doi.org/10.1063/5.0165439.

63. S. Tamboli, G. B. Nair, A. K. Sharma, S. J. Dhoble, H. C. Swart, Structural, optical and temperature-sensing properties of $LaOF$:Yb^{3+}, Tm^{3+} nanophosphor prepared by microwave-assisted hydrothermal synthesis. *Mater. Res. Bull.* 169 (2024) 112503, https://doi.org/10.1016/j.materresbull.2023.112503.

64. H. Xiao, Q. Meng, C. Wang, Optical thermometry based on fluorescence intensity ratio of doped ions and matrix in CaWO$_4$:Eu^{3+} phosphors. *J. Lumin.* 263 (2023) 119975, https://doi.org/10.1016/j.jlumin.2023.119975.

65. K. K. Gupta, H. Darsono, A. Chafidz, C.-H. Lu, Synthesis and spectroscopic characterization of Ce$_{0.90}$Eu$_{0.07}$Sm$_{0.03}$O$_2$ phosphors for luminescence-based thermometry applications. *Optical Mater.* 142 (2023) 114107, https://doi.org/10.1016/j.optmat.2023.114107.

66. S. S. Nanda, P. Nayak, S. Pattnaik, V. K. Rai, S. Dash, Simultaneous influence of Mg^{2+} and Sc^{3+} co-doping on upconversion luminescence and optical thermometry in β-NaYF$_4$:Yb^{3+}/Ho^{3+} microphosphor. *J. Alloys Compd.* 934 (2023) 167732, https://doi.org/10.1016/j.jallcom.2022.167732.

67. M. Yılmaz, E. Alp, Low-cost Zn$_2$(OH)BO$_3$:Pb^{2+} phosphor for large-scale thermometric applications. *J. Alloys Compd.* 934 (2023) 167865, https://doi.org/10.1016/j.jallcom.2022.167865.

68. K. M. Krishna, M. M. Upadhyay, V. Kesarwan, K. Kumar, Blue frequency upconversion emission and optical thermometry of Ca$_3$(PO$_4$)$_2$:Tm^{3+}/Yb^{3+}. *Optik Int. J. Light Electron Optics* 273 (2023) 170477, https://doi.org/10.1016/j.ijleo.2022.170477.

69. L. K. Wu, H. Y. Sun, L. H. Li, R. F. Li, H. Y. Ye, J. R. Li, Te^{4+}-doping rubidium scandium halide perovskite single crystals enabling optical thermometry. *J. Phys. Chem.* C 126 (51) (2022) 21689–21698, https://doi.org/10.1021/acs.jpcc.2c07553.

70. L. J. Fu, H. Liu, C. Li, Y. P. Wu, E. Rahm, R. Holze, H. Q. Wu, Electrode materials for lithium secondary batteries prepared by sol–gel methods. *Prog. Mater. Sci.* 50 (2005) 881–928, https://doi.org/10.1016/j.pmatsci.2005.04.002.

71. T.-F. Yi, X.-G. Hu, C.-S. Dai, K. Gao, Effects of different particle sizes on electrochemical performance of spinel LiMn$_2$O$_4$ cathode materials. *J. Mater. Sci.* 42, 11 (2007) 3825–3830, https://doi.org/10.1007/s10853-006-0460-6.

72. B. Ebin, G. Lindbergh, S. Gürmen, Preparation and electrochemical properties of nanocrystalline LiB$_x$Mn$_{2-x}$O$_4$ cathode particles for Li-ion batteries by ultrasonic spray pyrolysis method. *J. Alloys Compd.* 620 (2015) 399–406, https://doi.org/10.1016/j.jallcom.2014.09.098.

73. Z. Quan, S. Ohguchi, M. Kawase, H. Tanimura, N. Sonoyama, Preparation of nanocrystalline LiMn$_2$O$_4$ thin film by electrodeposition method and its electrochemical performance for lithium battery. *J. Power Sources* 244 (2013) 375–381, https://doi.org/10.1016/j.jpowsour.2012.12.087.

74. W. Wang, L. Li, J. Xie, Q. Cao, Y. Li, X. Shen, Y. Pan, Non-concentration quenching phosphor Na$_2$YbMg$_2$(VO$_4$)$_3$:Eu^{3+} for WLEDs and optical thermometry applications. *J. Lumin.* 266 (2024) 120289, https://doi.org/10.1016/j.jlumin.2023.120289.

75. H. Gou, Q. Wu, L. Luo, W. Li, P. Du, Reengineering the thermometric behaviors of Er^{3+}/Yb^{3+}-codoped Gd$_2$Mo$_3$O$_{12}$ microparticles via dual-mode luminescence manipulation. *Ceram. Int.* 49 (2023) 38297–38304, https://doi.org/10.1016/j.ceramint.2023.09.162.

76. J. Hong, F. Liu, M. D. Dramićanin, L. Zhou, M. Wu, The upconversion luminescence of Ca$_3$Sc$_2$Si$_3$O$_{12}$:Yb^{3+}, Er^{3+} and its application in thermometry. *Nanomaterials* 13 (2023) 1910, https://doi.org/10.3390/nano13131910.

77. A. S. Laia, M. A. Gomes, A. Brandão-Silva, Y. Xing, G. S. Maciel, Z. S. Macedo, M. E.G. Valerio, J. J. Rodrigues, M. A. R. C. Alencar, Nd^{3+} doped Y_2O_3 micro- and nanoparticles: A comparative study on temperature sensing and optical heating performance within the 1st biological window. *Optical Mater.* 142 (2023) 114126, https://doi.org/10.1016/j.optmat.2023.114126.

78. S.-H. Yang, S.-M. Liao, Y.-Y. Tsai, C.-H. Wang, C.-C. Ho, Optical thermometry based on $ZnAl_2O_4$:Eu^{3+} with carbon dots incorporation. *J. Alloys Compd.* 933 (2023) 167761, https://doi.org/10.1016/j.jallcom.2022.167761.

79. Y. Yang, Y. Deng, L. Zhang, Y. He, S. Zhao, T. Xiao, S. Dang, Y. Bai, Fluorescence enhancement and inverse Boltzmann distribution in Li^+/Er^{3+} co-doped Y_2O_3 nanocrystals. *Ceram. Int.* 48 (2022) 27295–27301, https://doi.org/10.1016/j.ceramint.2022.07.010.

80. F. H. Borges, J. C. Martins, F. J. Caixeta, L. D. Carlos, R. A. S. Ferreira, R. R. Gonçalves, Luminescent thermometry based on Er^{3+}/Yb^{3+} co-doped yttrium niobate with high NIR emission and NIR-to-visible upconversion quantum yields. *J. Lumin.* 248 (2022) 118986, https://doi.org/10.1016/j.jlumin.2022.118986.

81. Y. Zhang, P. Wang, H. Wang, X. Zheng, Y. Guo, N. Zhang, H. Liu, Visible and near-infrared luminescence properties of Nd^{3+}/Yb^{3+} co-doped Gd_2O_3 phosphors for highly sensitive optical thermometry. *Dalton Trans.* 51 (2022) 10612–10622, https://doi.org/10.1039/D2DT01271E.

82. W. D. S. Pereira, J. C. Sczancoski, E. Longo, Tailoring the photoluminescence of $BaMoO_4$ and $BaWO_4$ hierarchical architectures via precipitation induced by a fast precursor injection. *Mater. Lett.* 293 (2021) 129681, https://doi.org/10.1016/j.matlet.2021.129681.

83. A. H. Marincaş, F. Goga, S. A. Dorneanu, P. Ilea, Review on synthesis methods to obtain $LiMn_2O_4$-based cathode materials for Li-ion batteries. *J. Solid State Electrochem.* 24 (2020) 473–497, https://doi.org/10.1007/s10008-019-04467-3.

84. S. K. Samal, Pushpendra, J. Yadav, B. S. Naidu, Upconversion properties of Er, Yb co-doped $KBi(MoO_4)_2$ nanomaterials for optical thermometry. *Ceram. Int.* 49 (2023) 20051–20060, https://doi.org/10.1016/j.ceramint.2023.03.127.

85. H. Thakur, A. K. Gathania, S. Kachhap, S. K. Singh, R. K. Singh, Coprecipitation synthesis, structural, optical properties, and thermometry application of Tm^{3+}/Yb^{3+} co-doped YPO_4 phosphor. *J. Lumin.* 254 (2023) 119513, https://doi.org/10.1016/j.jlumin.2022.119513.

86. M. K. Sahu, M. Jayasimhadri, D. Haranath, Temperature-dependent photoluminescence and optical thermometry performance in $Ca_3Bi(PO_4)_3$:Er^{3+} phosphors. *Solid State Sci.* 131 (2022) 106956, https://doi.org/10.1016/j.solidstatesciences.2022.106956.

87. X. Yuan, E. Cui, K. Liu, Y. Jiang, X. Yang, J. Tang, L. Yang, X. Liao, Y. Zhao, W. Sun, Y. Liu, J. Liu, Lanthanide-doped $NaYF_4$ near-infrared-II nanothermometers for deep tissue temperature sensing. *Ceram. Int.* 48 (2022) 35141–35149, https://doi.org/10.1016/j.ceramint.2022.08.110.

88. T. H. Q. Vu, B. Bondzior, D. Stefanska, P. J. Deren, Influence of temperature on near-infrared luminescence, energy transfer mechanism and the temperature sensing ability of La_2MgTiO_6:Nd^{3+} double perovskites. *Sens. Actuators A* 317 (2021) 112453, https://doi.org/10.1016/j.sna.2020.112453.

89. Q. Vu, B. Bondzior, D. Stefańska, P. J. Dereń, An Er^{3+} doped Ba_2MgWO_6 double perovskite: A phosphor for low-temperature thermometry. *Dalton Trans.* 51 (20) (2022) 8056–8065, https://doi.org/10.1039/D2DT00554A.

90. C. Yue, Y. Pu, D. Zhu, Q. Yan, A novel green-emitting $SrLaAlO_4$:Er^{3+} phosphor synthesized by co-precipitation method for w-LEDs and optical thermometry. *J. Mater. Sci. Mater. Electron.* 32 (2021) 4228–4238, https://doi.org/10.1007/s10854-020-05167-9.

91. X. Wang, X. Feng, C. Gong, M. Sun, C. Wang, Q. Wang, J.-G. Li, Coprecipitation synthesis of α-$(La,Yb,Er)_2W_2O_9$ tungstate as a novel upconversion phosphor for optical thermometry. *J. Asian Ceramic Soc.* 9 (4) (2021) 1419–1428, https://doi.org/ 10.1080/21870764.2021.1989750.

92. X. Lou, D. Chen, Synthesis of $CaWO_4$: Eu^{3+} phosphor powders via a combustion process and its optical properties. *Mater. Lett.* 62 (10–11) (2008) 1681–1684, https://doi.org/10.1016/j.matlet.2007.09.066.

93. F. T. Li, J. Ran, M. Jaroniec, S. Z. Qiao, Solution combustion synthesis of metal oxide nanomaterials for energy storage and conversion. *Nanoscale* 7 (42) (2015) 17590–17610, https://doi.org/10.1039/C5NR05299H.

94. N. Rakov, F. Matias, G. S. Maciel, M. Xiao, The near-infrared emission of Er^{3+}-doped yttrium oxy-fluoride ceramic powders: Temperature reading using uncoupled and thermally coupled energy levels. *J. Lumin.* 263 (2023) 119991, https://doi.org/10.1016/j.jlumin.2023.119991.

95. V. Chauhan, P. Dixit, P. K. Pandey, S. Chaturvedi, P. C. Pandey, Emission color tuning and dual-mode luminescence thermometry design in Dy^{3+}/Eu^{3+} co-doped $SrMoO_4$ phosphors. *Methods Appl. Fluoresc.* 12 (1) (2023) 015002, https://doi.org/10.1088/2050–6120/acf97b.

96. M. M. Upadhyay, K. Kumar, Gd^{3+} ion induced UV upconversion emission and temperature sensing in Tm^{3+}/Yb^{3+}:Y_2O_3 phosphor. *J. Rare Earths* 41 (2023) 1295–1301, https://doi.org/10.1016/j.jre.2022.07.005.

97. J. Zhang, J. Chen, Y. Zhang, Temperature-sensing luminescent materials $La_{9.67}Si_6O_{26.5}$: Yb^{3+}–Er^{3+}/Ho^{3+} based on pump-power-dependent upconversion luminescence. *Inorg. Chem. Front.* 7 (2020) 4892–4901, https://doi.org/10.1039/D0QI01058H.

98. Y. Zi, Y. Cun, X. Bai, Z. Xu, A. A. Haider, J. Qiu, Z. Song, A. Huang, J. Zhu, Z. Yang, Negative lattice expansion-induced upconversion luminescence thermal enhancement in novel Na_2MoO_4:Yb^{3+}, Er^{3+} transparent glass ceramics for temperature sensing applications. *J. Mater. Chem. C* 11 (2023) 1541–1549, https://doi.org/10.1039/D2TC05009A.

99. I. V. Baklanova, V. N. Krasilnikov, A. P. Tyutyunnik, Y. V. Baklanova, Simple express precursor synthesis of erbium-doped yttrium oxide luminescent material for optical thermometry. *Optical Mater.* 143 (2023) 114232, https://doi.org/10.1016/j.optmat.2023.114232.

100. K. Saidi, C. Hernández-Álvarez, M. Runowski, M. Dammak, I. R. Martín, Ultralow pressure sensing and luminescence thermometry based on the emissions of Er^{3+}/Yb^{3+} codoped $Y_2Mo_4O_{15}$ phosphors. *Dalton Trans.* 52 (2023) 14904–14916, https://doi.org/10.1039/D3DT02613B.

101. M. Wang, S. Dinga, C. Zhanga, H. Rena, Y. Zou, X. Tanga, Q. Zhang, Pure-green upconversion emission and high-sensitivity optical thermometry of Er^{3+}-doped stoichiometric $NaYb(MoO_4)_2$. *Ceram. Int.* 49 (2023) 37661–37669, https://doi.org/10.1016/j.ceramint.2023.09.092.

102. I. Kumar, A. Kumar, S. Kumar, G. B. Nair, H. C. Swart, A. K. Gathania, Simultaneous realization of FIR-based multimode optical thermometry and photonic molecular logic gates in Er^{3+} and Yb^{3+} co-doped $SrTiO_3$ phosphor. *Phys. Scr.* 98 (2023) 105532, https://doi.org/10.1088/1402–4896/acfa2b.

103. S. K. Samal, Pushpendra, J. Yadav, B. S. Naidu, Upconversion properties of Er, Yb co-doped $KBi(MoO_4)_2$ nanomaterials for optical thermometry. *Ceram. Int.* 49 (2023) 20051–20060, https://doi.org/10.1016/j.ceramint.2023.03.127.

104. R. S. Yadav, S. J. Dhoble, S. B. Rai, Enhanced photoluminescence in Tm^{3+}, Yb^{3+}, Mg^{2+} tri-doped $ZnWO_4$ phosphor: Three photon upconversion, laser induced optical heating and temperature sensing. *Sens. Actuat. B Chem.* 273 (2018) 1425–1434, https://doi.org/10.1016/j.snb.2018.07.049.

105. W. Ge, M. Xu, J. Shi, J. Zhu, Y. Li, Highly temperature-sensitive and blue upconversion luminescence properties of $Bi_2Ti_2O_7$:Tm^{3+}/Yb^{3+} nanofibers by electrospinning. *Chem. Eng. J.* 391 (2020) 123546, https://doi.org/10.1016/j.cej.2019.123546.

106. H. Song, C. Wang, Q. Han, X. Tang, W. Yan, Y. Chen, J. Jiang, T. Liu, Highly sensitive Tm^{3+}/Yb^{3+} codoped $SrWO_4$ for optical thermometry. *Sens. Actuat. A* 271 (2018) 278–282, https://doi.org/10.1016/j.sna.2018.01.037.

107. W. Gao, W. Ge, J. Shi, X. Chen, Y. Li, A novel upconversion optical thermometers derived from non-thermal coupling levels of CaZnOS:Tm/Yb phosphors. *J. Solid State Chem.* 297 (2021) 122063, https://doi.org/10.1016/j.jssc.2021.122063.

108. R. Lisiecki, J. Komar, M. Berkowski, W. Ryba-Romanowski, Effect of temperature on up-conversion phenomena in $Gd_3(Al,Ga)_5O_{12}$ crystals co-doped with Yb^{3+} and Tm^{3+}. *J. Lumin.* 216 (2019) 116721, https://doi.org/10.1016/j.jlumin.2019.116721.

109. K. Saidi, M. Dammak, K. Soler-Carracedo, I. R. Martín, A novel optical thermometry strategy based on emission of Tm^{3+}/Yb^{3+} codoped $Na_3GdV_2O_8$ phosphors. *Dalton Trans.* 51 (2022) 5108–5117, https://doi.org/10.1039/D1DT03747A.

110. N. Rakov, S. A. Vieira, A. S. L. Gomes, Tm^{3+}/Yb^{3+} co-doped SrF_2 up-conversion phosphors for non-invasive optical thermometry: Ratiometric approach using thermal and non-thermal coupled fluorescent emission bands. *Appl. Phys. A* 127 (2021) 1–10, https://doi.org/10.1007/s00339-021-05085-5.

111. W. Hu, F. Hu, X. Li, H. Fang, L. Zhao, Y. Chen, C.-K. Duan, M. Yin, Optical thermometry of a Tm^{3+}/Yb^{3+} Co-doped $LiLa(MoO_4)_2$ up-conversion phosphor with a high sensitivity. *RSC Adv.* 6 (2016) 84610–84615, https://doi.org/10.1039/C6RA18432D.

112. C. Madhukar Reddy, N. Vijaya, B. Deva Prasad Raju, NIR fluorescence studies of neodymium ions doped sodium fluoroborate glasses for 1.06 µm laser applications. *Spectrochim. Acta A* 115 (2013) 297–304, https://doi.org/10.1016/j.saa.2013.06.029.

113. A. Agnesi, P. Dallocchio, F. Pirzio, G. Reali, Compact sub-100-fs Nd: Silicate laser. *Opt. Commun.* 282 (10) (2009) 2070–2073, https://doi.org/10.1016/j.optcom.2009.02.016.

114. A. S. Laia, D. A. Hora, M. V. D. S. Rezende, Y. Xing, J. J. Rodrigues, G. S. Maciel, M. A. R. C. Alencar, Comparing the performance of Nd^{3+}-doped $LiBaPO_4$ phosphors as optical temperature sensors within the first biological window exploiting luminescence intensity ratio and bandwidth methods. *J. Lumin.* 227 (2020) 117524, https://doi.org/10.1016/j.jlumin.2020.117524.

115. A. A. Kalinichev, E. V. Afanaseva, E. Y. Kolesnikov, I. E. Kolesnikov, Boltzmann-type cryogenic ratiometric thermometry based on Nd^{3+}-doped $LuVO_4$ phosphors. *J. Mater. Chem. C* 11 (2023) 12234–12242, https://doi.org/10.1039/D3TC02043F.

116. L. Chen, H. Chen, G. Bai, X. Yang, H. Xie, S. Xu, Near-infrared excitation and emitting thermometer based on Nd^{3+} doped ytterbium molybdate with thermally enhanced emissions. *J. Lumin.* 228 (2020) 117655, https://doi.org/10.1016/j.jlumin.2020.117655.

117. Q. Chang, X. Zhou, X. Zhou, L. Chen, G. Xiang, S. Jiang, L. Li, Y. Li, X. Tang, Strategy for optical thermometry based on temperature-dependent charge transfer to the Eu^{3+} 4f-4f excitation intensity ratio in $Sr_3Lu(VO_4)_3$:Eu^{3+} and $CaWO_4$:Nd^{3+}. *Optics Lett.* 45 (13) (2020) 3637–3640, https://doi.org/10.1364/OL.396456.

118. D. D. Ramteke, R. S. Gedam, H. C. Swart, Physical and optical properties of lithium borosilicate glasses doped with Dy^{3+} ions. *Phys. B Condens. Matter.* 535 (2018) 194–197, https://doi.org/10.1016/j.physb.2017.07.035.

119. S. Damodaraiah, V. Reddy Prasad, S. Babu, Y. C. Ratnakaram, Structural and luminescence properties of Dy^{3+} doped bismuth phosphate glasses for greenish yellow light applications. *Opt. Mater.* 67 (2017) 14–24, https://doi.org/10.1016/j.optmat.2017.03.023.

120. R. Raji, P. S. Anjana, N. Gopakumar, An insight into Judd-Ofelt analysis and non-contact optical thermometry of $LiCa_2Mg_2V_3O_{12}$:Dy^{3+} phosphors for multifunctional applications. *Opt. Mater.* 145 (2023) 114393, https://doi.org/10.1016/j.optmat.2023.114393.

121. S. Li, J. Guo, W. Shi, X. Hu, S. Chen, J. Luo, Y. Li, J. Kong, J. Che, H. Wang, B. Deng, R. Yu, Synthesis and characterization of wide band excited yellow phosphor $LaNb_2VO_9$: Dy^{3+} and application in potential fingerprint and lip print detection. *J. Lumin.* 244 (2022), 118681, https://doi.org/10.1016/j.jlumin.2021.118681.

122. A. R. Dhobale, M. Mohapatra, V. Natarajan, S. V. Godbole, Synthesis and photoluminescence investigations of the white light emitting phosphor, vanadate garnet, $Ca_2NaMg_2V_3O_{12}$ co-doped with Dy and Sm. *J. Lumin.* 132 (2012) 293–298, https://doi.org/10.1016/j.jlumin.2011.09.004.

123. Z. Liao, B. Cao, L. Li, Y. Cong, Y. He, B. Dong, Exploring the excitation spectrum behavior of Dy in $CaWO_4$ for a new excited-state-based ratiometric thermometry. *Appl. Mater. Today* 31 (2023) 101765, https://doi.org/10.1016/j.apmt.2023.101765.

124. Q. Liu, M. Wu, B. Chen, X. Huang, M. Liu, Y. Liu, K. Su, X. Min, R. Mi, Z. Huang, Optical thermometry based on fluorescence intensity ratio of Dy^{3+}-doped oxysilicate apatite warm white phosphor. *Ceram. Int.* 49 (2023) 4971–4978, https://doi.org/10.1016/j.ceramint.2022.10.012.

125. M. T. Abbas, S. A. Khan, J. Mao, N. Z. Khan, L. Qiu, J. Ahmed, X. Wei, Y. Chen, S. M. Alshehri, S. Agathopoulos, Optical thermometry based on the luminescence intensity ratio of Dy^{3+}-doped $GdPO_4$ phosphors. *J. Therm. Anal. Calorim.* 147 (21) (2022) 11769–11775, https://doi.org/10.1007/s10973-022-11415-3.

126. H. Zhang, B. Cao, Z. Liao, Y. Yang, J. Zhang, L. Li, Y. Cong, Y. He, Z. Zhang, Z. Feng, B. Dong, Energy transfer mechanism and new ratiometric thermometry strategy by the blue and yellow emissions of Dy. *Ceram. Int.* 48 (2022) 29838–29846, https://doi.org/10.1016/j.ceramint.2022.06.248.

127. Y. Li, X. T. Wei, H. M. Chen, G. Pang, Y. Pan, L. Gong, L. L. Zhu, G. Y. Zhu, Y. X. Ji, A new self-activated vanadate phosphor of $Na_2YMg_2(VO_4)_3$ and luminescence properties in Eu^{3+} doped $Na_2YMg_2(VO_4)_3$. *J. Lumin.* 168 (2015) 124–129, https://doi.org/10.1016/j.jlumin.2015.08.002.

128. P. Ranjith, S. Sreevalsa, J. Tyagi, K. Jayanthi, G. Jagannath, P. Patra, S. Ahmad, K. Annapurna, A. R. Allu, S. Das, Elucidating the structure and optimising the photoluminescence properties of $Sr_2Al_3O_6F{:}Eu^{3+}$ oxyfluorides for cool white-LEDs. *J. Alloys Compd.* 826 (2020) 154015, https://doi.org/10.1016/j.jallcom.2020.154015.

129. X. Qiao, S. Qi, Y. Lu, Y. Pu, Y. Huang, X. Wang, Synthesis, structure and red-emitting luminescence properties of Eu^{3+} activated perovskite-related tungstate $Ba_4Na_2W_2O_{11}$. *J. Alloys Compd.* 656 (2016) 189–195, https://doi.org/10.1016/j.jallcom.2015.09.182.

130. A. S. de Oliveira, B. H. S. T. da Silva, M. S. Góes, A. Cuin, H. de Souza, L. F. C. de Oliveira, G. P. de Souza, M. A. Schiavon, J. L. Ferrari, Photoluminescence, thermal stability and structural properties of Eu^{3+}, Dy^{3+} and Eu^{3+}/Dy^{3+} doped apatite-type silicates. *J. Lumin.* 227 (2020) 117500, https://doi.org/10.1016/j.jlumin.2020.117500.

131. C. L. Chen, J. Wang, S. Bi, Y. L. Huang, H. J. Seo, Energy transfer and luminescence quenching of Cr^{3+}-doped $LiGaW_2O_8$. *J. Alloys Compd.* 786 (2019) 1051–1059, https://doi.org/10.1016/j.jallcom.2019.01.277.

132. D. K. Xu, F. Y. Xie, L. Yao, Y. J. Li, H. Lin, A. M. Li, S. H. Yang, S. L. Zhong, Y. L. Zhang, Enhancing upconversion luminescence of highly doped lanthanide nanoparticles through phase transition delay. *J. Alloys Compd.* 815 (2020) 152622, https://doi.org/10.1016/j.jallcom.2019.152622.

133. H. Desirena, J. Molina, O. Meza, A. Benitez, J. Bujdud-Perez, J. Briones-Hernandez, Eu^{3+} heavily doped tellurite glass ceramic as an efficient red phosphor for white LED. *J. Lumin.* 250 (2022) 119080, https://doi.org/10.1016/j.jlumin.2022.119080.

134. J. Li, B. Liu, G. Liu, Q. Che, Y. Lu, Z. Liu, Red color $Sr_2NaMg_2V_3O_{12}{:}Eu^{3+}$ phosphor with high thermal stability for w-LEDs. *J. Rare Earths* 41 (2022) 1689–1695, https://doi.org/10.1016/j.jre.2022.08.022.

135. P. Rohilla, A. S. Rao, Synthesis optimisation and efficiency enhancement in Eu^{3+} doped barium molybdenum titanate phosphors for w-LED applications. *Mater. Res. Bull.* 150 (2022) 111753, https://doi.org/10.1016/j.materresbull.2022.111753.

136. F. M. Song, Synthesis and photoluminescence of new Eu^{3+}activated $Cs_2Ba(MoO_4)_2$ red-emitting phosphors with high color purity for white LEDs. *J. Lumin.* 239 (2021) 118324, https://doi.org/10.1016/j.jlumin.2021.118324.

137. P. Woźny, K. Soler-Carracedo, N. Stopikowska, I. R. Martín, M. Runowski, Structure-dependent luminescence of Eu^{3+}-doped strontium vanadates synthesized with different V : Sr ratios – Application in WLEDs and ultra-sensitive optical thermometry. *J. Mater. Chem.* C 11 (14) (2023) 4792–4807, https://doi.org/10.1039/D2TC05341A.

138. W. Wang, L. Li, J. Xie, Q. Cao, Y. Li, X. Shen, Y. Pan, Non-concentration quenching phosphor $Na_2YbMg_2(VO_4)_3{:}Eu^{3+}$ for WLEDs and optical thermometry applications. *J. Lumin.* 266 (2024) 120289, https://doi.org/10.1016/j.jlumin.2023.120289.

139. L. Peng, L. Li, F. Qin, C. Wang, Z. Zhang, A multi-mode self-referenced optical thermometer based on low-doped YVO_4:Eu^{3+} phosphor. *J. Lumin.* 263 (2023) 120168, https://doi.org/10.1016/j.jlumin.2023.120168.

140. J. Y. Chen, L. J. Li, J. Q. Chen, T. Pang, L. P. Chen, H. Guo, Eu^{3+} doped $Ca_3LiZnV_3O_{12}$ phosphors for four-mode optical thermometry. *J. Lumin.* 261 (2023) 119931, https://doi.org/10.1016/j.jlumin.2023.119931.

141. H. Xiao, Q. Meng, C. Wang, Optical thermometry based on fluorescence intensity ratio of doped ions and matrix in $CaWO_4$:Eu^{3+} phosphors. *J. Lumin.* 263 (2023) 119975, https://doi.org/10.1016/j.jlumin.2023.119975.

Chapter 7

Upconversion phosphors for luminescence-based optical thermometers

Abhilasha Jain, Yatish R. Parauha, and Sanjay J. Dhoble

7.1 BACKGROUND

The growing sensor market, especially in temperature sensors, is a direct response to the rising need for analyzing temperature variations across scientific disciplines and industries. Recent data indicates that the temperature measurement market is valued at 6412.5 million USD in 2020. Precise temperature information, a critical thermodynamic parameter, facilitates the diagnosis of medical conditions [1]. Temperature is a fundamental physical parameter deeply intertwined with our daily existence, exerting a profound influence across natural sciences, scientific research, industrial processes, and daily activities. Its significance is particularly pronounced in comprehending the intricate dynamics of biological systems, especially at the cellular level, necessitating precise temperature knowledge. Consequently, temperature monitoring has evolved into an indispensable facet of scientific research, finding crucial applications in various fields. In this context, rare-earth (RE) luminescent materials have emerged as valuable tools, leveraging their intrinsic connection between emission intensity or other features and temperature [2]. The advancement of microelectronics, optoelectronics, nanomedicine, and related disciplines has spurred increased interest in non-contact temperature sensors. These sensors, characterized by attributes such as high spatial resolution, sensitivity, wide temperature applicability, and measurement precision, are becoming pivotal in contemporary scientific and technological landscapes [2]. Thermometers fall into two primary types: contact and non-contact. Contact thermometers employ sensors that come into direct contact with the object to measure temperature, utilizing the heat transfer principle called "conduction [3]. These sensors need physical touch with the object to equilibrate their temperature with it. On the other hand, non-contact thermometers use infrared sensors to gauge temperature by detecting the infrared emissions from the object. This method eliminates the need for physical contact, offering a hands-free temperature measurement approach.

Non-contact thermometers have a big advantage over contact methods because they don't need physical contact with the object they're measuring.

DOI: 10.1201/9781032661537-7

This means they can be used safely from a distance. You can keep an eye on temperature continuously without slowing down or stopping the production line to put a thermometer on it. However, contact methods like thermocouples can be really accurate too. They're good for getting readings from the surface of things. But sometimes, other stuff can mess with the readings:

- The temperature of the air around it.
- How hot the thermocouple was to start with.
- How big and heavy the thermocouple is.
- If the surface is rusty.
- Losing heat through touching stuff.
- How hard the thermocouple is pressing on the surface.

All these things can make the readings way off. In fact, some thermocouples are too big and heavy, so they don't heat up enough from the surface they're measuring. Others are thin and bendy, so they don't press on the surface properly.

Traditional contact thermometers rely on the expansion and contraction of liquids to measure temperature, requiring direct contact with the object being measured. This method, known as thermal contact temperature detection, has limitations. With advancements in science and technology, there's a growing demand for temperature control and measurement in diverse environments, such as corrosive or electromagnetic fields. Additionally, in setups like diamond anvil cells (DACs) used for high-pressure studies, conventional thermometers struggle, especially in cases where sample sizes are very small. For applications like measuring temperature in microelectronic components or fast-moving objects, conventional sensors fall short due to their slow response and limited spatial resolution. To address these challenges, there's a need for non-contact temperature sensors offering rapid response, high spatial resolution, and the ability to operate without physical contact.

Optical temperature sensing emerges as a promising alternative, boasting advantages over traditional thermometers. Optical sensors deliver swift response times, precise spatial resolution, and operate without physical contact, making them suitable for complex and specialized applications where conventional methods are inadequate. Figure 7.1 shows the temperature sensor classification, fiber type, and sensor mechanism from the inside to the outside.

7.2 PHOSPHORS IN OPTICAL SENSORS

Optical sensors have received a lot of attention in a variety of fields because of their non-invasive nature, high sensitivity, and ability to detect multiple targets at once. Phosphors are important components of optical

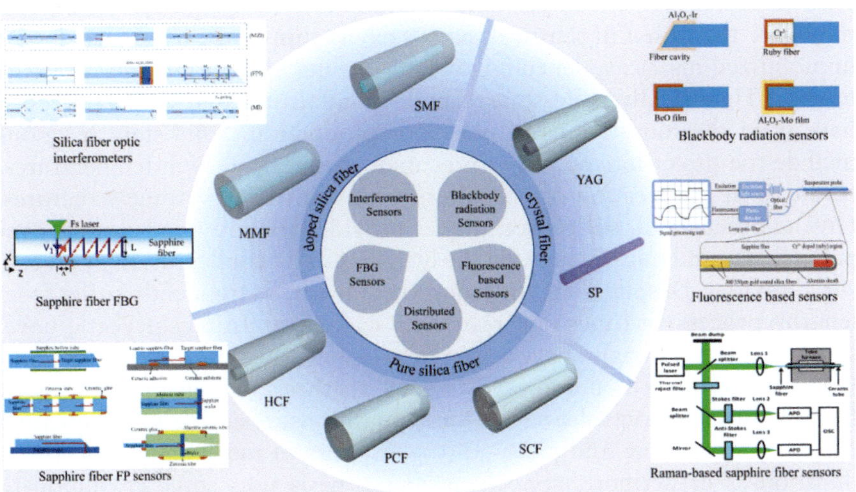

Figure 7.1 Classification of Temperature sensor "Open access article reference [4]"

sensors because they help convert incident photons into measurable signals [5]. Phosphors play an important role in optical sensing technology, allowing sensitive and selective detection of analytes in a variety of fields. Phosphor-based sensing is based on the interaction of incident light and phosphor luminescent properties. When exposed to light of a specific wavelength, phosphors absorb photons and undergo electronic transitions, resulting in the emission of light with a longer wavelength. This emitted light, which is typically in the visible or near-infrared spectrum, can be detected and quantified to determine the presence or concentration of analytes in the sample. The sensitivity and selectivity of phosphor-based sensors are determined by a variety of factors, including the phosphor material used, the excitation wavelength, and the detection technique.

Phosphors used in optical sensors include a wide range of materials such as inorganic compounds, organic dyes, and quantum dots. Inorganic phosphors, such as transition metal ions doped into host lattices or lanthanide-based compounds, are widely used due to their durability, photostability, and tunable luminescence [6–9]. Organic dyes have features such as high absorption coefficients and spectrum tunability, making them ideal for applications requiring narrow emission bands. Quantum dots, semiconductor nanocrystals with size-dependent optical characteristics, have high brightness and photostability, allowing for sensitive detection in biological and environmental sensing.

The synthesis of phosphor-based sensors entails the production of phosphor materials and their incorporation into sensor platforms. Various processes, such as solid-state reactions, sol-gel procedures, chemical vapor deposition, and hydrothermal synthesis, are used to produce phosphors with regulated shape, composition, and luminescence [10–13]. These phosphors are then used in sensor designs such as thin films, nanoparticles, or immobilized matrices, depending on the sensing application and detecting method. Historically, solid-state reaction, combustion, and sol-gel processes have been the foundations of phosphor production. Solid-state reactions include the direct interaction of precursor materials at high temperatures, resulting in the creation of phosphors with distinct crystalline structures. On the other hand, sol-gel techniques provide flexibility and control over particle size and morphology via the hydrolysis and condensation of precursor alkoxides. Despite their efficiency, these methods frequently suffer from lengthy processing times and restricted scalability. In recent years, novel synthesis methodologies have arisen to overcome the limits of old methods. One such strategy is microwave-assisted synthesis, which provides quick heating and fine control over reaction conditions, resulting in the creation of highly crystalline and phase-pure phosphors in much shorter synthesis durations. Furthermore, sonochemical synthesis uses sonic cavitation to generate localized high temperatures and pressures, allowing for the rapid nucleation and development of phosphor nanoparticles with improved characteristics.

The introduction of nanotechnology has transformed phosphor production, allowing the creation of tiny phosphors with distinct optical and sensing characteristics. Bottom-up techniques such as hydrothermal and solvothermal processes enable fine control of particle size, shape, and crystallinity, resulting in increased luminescence efficiency and sensor performance. Furthermore, incorporating quantum dots and other nanomaterials into phosphor matrices has created new opportunities for multifunctional sensors with higher sensitivity and selectivity.

Phosphor synthesis for optical sensors demands careful tweaking of material properties to satisfy specific application needs. Dopant ions can be used to change emission spectra and increase sensor sensitivity to target analytes. Furthermore, surface functionalization techniques enable the immobilization of recognition elements such as antibodies and aptamers onto phosphor nanoparticles, allowing for the creation of selective and label-free sensing platforms for biomedical and environmental monitoring.

7.3 PRINCIPLE OF OPTICAL THERMOMETRY SENSING

Optical thermometry sensing relies on the fundamental interaction between electromagnetic radiation and matter to estimate temperature properly. One common theory is based on Planck's equation of blackbody radiation (Figure 7.2), which defines the spectrum distribution of an object's thermal radiation. According to this equation, the intensity and spectral features of

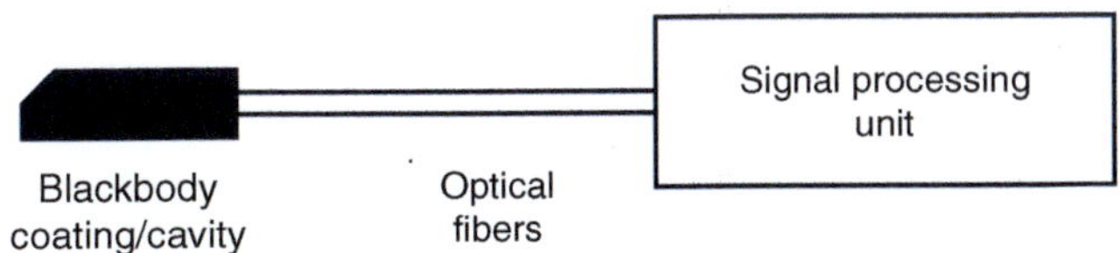

Figure 7.2 Optical fiber deployment for measurement using the blackbody principles. "Open access article reference [14]"

emitted radiation are completely determined by the object's temperature, making it a reliable basis for temperature measurement.

Generally, optical thermometry sensors use a variety of methods to detect and quantify thermal radiation. One method includes employing photodetectors to measure the intensity of emitted radiation at various wavelengths. The sensor can detect the temperature of an item by analyzing the spectral distribution of radiation using Planck's law. Another principle of optical thermometry sensing takes advantage of temperature-dependent features of materials, including organic fluorophores, metal complexes, and nanoparticles doped with lanthanide [15–18]. Lanthanide-doped nanoparticles stand out as current optical nanosensors for temperature sensing. For example, fluorescent materials alter in emission intensity or spectra as temperature varies, enabling for exact temperature readings. Similarly, materials experiencing phase transitions, such as liquid crystals or thermochromic dyes, can be used as sensitive temperature sensors. Fluorescence temperature sensors are classified into three types: intensity type, decay lifespan type, and fluorescence intensity ratio (FIR) type. Each method has advantages and limitations and is suitable to a given scope and application.

7.3.1 Fluorescence intensity method

The fluorescence intensity ratio (FIR) method calculates temperature by comparing the intensity of light emitted from two closely related energy levels. In comparison to other approaches, it has a higher sensitivity and is less impacted by issues such as power fluctuations or electromagnetic interference [19]. Essentially, it determines the ratio of light emitted from two energy levels linked by temperature. Furthermore, using frequency upconversion emission for temperature monitoring provides benefits over downshifting emission. Fluorescence intensity-based temperature sensing is based on the temperature-dependent emission properties of fluorescent probes. The fluorescence intensity approach relies on temperature-induced variations in the quantum yield and spectrum characteristics of fluorescent compounds. As the temperature rises, molecule vibrations and collisions promote non-radiative routes, resulting in decreased fluorescence intensity. Conversely, lowering the temperature lowers these non-radiative processes, resulting in increased fluorescence intensity. Fluorescence intensity temperature measurement is a direct approach where the emitted fluorescence intensity is studied in relation to temperature. The factors affecting the intensity can be summarized as follows [20]:

$$I = 2.3\, k\varphi I_0 \varepsilon l c \tag{7.1}$$

These include the excitation light intensity (I_0), a constant related to the measured system (k), energy transfer (ET) efficiency (φ), molar absorbance (ε), absorption path length (l), and concentration of the fluorescent substance (c). As long as I_0 and c are determined, I only depend on φ. ET efficiency involves interactions between matrix materials and RE ions, and electronic non-radiative transition probabilities of RE ions, both of which typically increase with temperature. This results in a non-monotonic relationship between I and temperature, with a turning point on the I–T curve. Below this turning point, the influence of temperature on ET efficiency dominates, causing I to increase with temperature. Above the turning point, increased phonon interactions lead to non-radiative transitions, causing I to weaken until fluorescence quenching occurs. Another method involves studying the relationship between thermal coupling energy levels (TCLs) of excited states and temperature, or the energy gap (ΔE) between the ground state and nearby excited states. This method is simple; however, the absolute luminescence intensity is influenced by fluctuations in laser excitation power, environmental factors, and thermal quenching effects on I.

7.3.2 Lifetime method

This method relies on the temperature-dependent collisional quenching of excited-state fluorophores, where higher temperatures accelerate the quenching process, resulting in shorter fluorescence lifetimes. By calibrating the relationship between fluorescence lifetime and temperature, this technique offers a non-invasive and highly sensitive approach for temperature measurement in various applications. When a laser stimulates a substance, its molecules absorb energy and transition from their normal state to an energized one known as the excited state. Later, they release the excess energy and revert to their former condition, producing radiation in the process. Once the laser light has stopped, the strength of the radiated radiation progressively decreases over time [20].

$$I_t = I_0 \exp\!\left(\frac{-t}{\tau}\right) \tag{7.2}$$

where I_0 represents the intensity at $t=0$, and τ is the lifetime, which is defined as $I(\tau)=I_0/e$. In semiconductor theory, the intensity of lattice vibrations increases with temperature. This allows more molecules to participate in light absorption, which accelerates the light-quenching process.

The temperature impacts the pace of this process, as represented by the lifetime τ. Thus, measuring τ allows us to determine the ambient temperature.

7.3.3 Spectral intensity ratio method

The spectral intensity ratio method for optical temperature monitoring compares the intensities of two distinct spectral lines emitted or absorbed by a material as a function of temperature. Using the temperature-dependent behavior of these spectral lines, which often originate from an atomic or molecular transition, the ratio of their intensities provides a direct indicator of the temperature of the material. This approach has many advantages, including excellent precision and dependability, making it useful for a variety of applications such as industrial process monitoring, environmental sensing, and biomedical diagnostics.

The FIR approach is based on changes in the intensity ratio of specific ions of emission bands as temperature varies. These ions, such as Pr^{3+}, Nd^{3+}, Ho^{3+}, Er^{3+}, and Tm^{3+}, have energy levels that are well-suited to this approach. When the energy gap between two levels is narrow, their emission bands overlap extensively, whereas a wider gap may result in insufficient population of the higher energy level within the temperature range under consideration. Thermalization happens when the higher energy level gets more occupied than the lower one as temperature increases. It can be understood by Boltzmann distribution [15]:

$$\text{FIR} = \text{LIR} = \frac{I_2}{I_1} = B_{\exp}\left(\frac{\Delta E}{k_B T}\right) \tag{7.3}$$

The Fluorescence Intensity Ratio (FIR) and Luminescence Intensity Ratio (LIR) are measures used in thermometry, directly linked to the temperature of a luminescent material through Boltzmann distribution (k_B). FIR is determined by the ratio of integrated intensities (I_2/I_1) of two transitions, where I_2 is from a higher energy transition and I_1 from a lower energy one. The energy difference between the centroids of these transitions (ΔE) is crucial, often obtained from emission spectra or fitting procedures. In the formula, ΔE is multiplied by the Boltzmann constant (k_B) and the absolute temperature (T), along with a constant (B) dependent on various factors such as degeneracies of states and transition branching ratios relative to the ground state.

7.4 UP CONVERSION PHOSPHORS FOR TEMPERATURE SENSING

The FIR approach has gained popularity for its non-contact temperature probing capabilities at the nanometer to submicron scale. Several fluorescence-based materials have been investigated for this purpose, including dye-sensitized polymeric particles, quantum dots, and luminous materials doped with rare earth (RE) ions [21,22]. RE ion-doped materials are particularly intriguing because of their narrow absorption and emission lines, which include specific thermally coupled energy levels (TCLs) that can be used to measure temperature. Among them, RE ion-doped materials have sparked widespread interest due to RE ions showing distinct features, such as narrow absorption and emission bands. These ions have specific energy levels, known as thermally coupled energy levels (TCLs), which can be used for a variety of purposes. For example, some transitions between TCLs, such as the $^2H_{11/2}$ and $^4S_{3/2}$ levels for Er^{3+}, $^5F_4/^5S_2$, $^5F_{2,3}/^3K_8$, $^5G_6/^4F_1$ for Ho^{3+}, $^3F_{2,3}$ and $^3H_4$ for Tm^{3+}, $^5D_0$ and $^5D_1$ for Eu^{3+}, are critical for producing certain emissions [23,24]. However, the energy gap between these two TCLs should be between 200 and 2000 cm^{-1} to avoid overlapping emission bands and guarantee efficient energy transfer at higher temperatures [25].

Recently, there has been an increase in research concentrating on optical thermometers that use the up-conversion (UC) emission phenomenon of RE ions. In this mechanism, RE ions absorb near-infrared (NIR) light and emit visible light, a well-known phenomenon [26]. The efficiency of UC emission is controlled by the host lattice's interaction with dopants. Host materials with lower phonon frequencies typically provide higher UC efficiency. While fluoride compounds are widely studied due to their low phonon energy, they are sensitive to surface contamination and provide environmental concerns during synthesis. In the past years, various UC luminescence materials have been investigated and reported. In this chapter, we discussed some highly potential UC luminescence materials, which may be applicable for temperature sensors.

Recently, Upadhayay and colleagues investigated the $GdNbO_4:Tm^{3+}/Yb^{3+}$ phosphor using a solid-state reaction technique to investigate its potential for upconversion emission, optical thermometry, latent fingerprint visualization, and anti-counterfeiting applications [27]. They studied temperature sensing capabilities by analyzing the fluorescence intensity ratio (FIR) of thermally and non-thermally coupled levels of the Tm^{3+} ion. Their findings showed that the non-thermally coupled level is approximately 45 times more sensitive than the thermally coupled level [27].

Zhang and colleagues developed a new type of chiral nanostructured material called D-BYEP, made of bismuth oxychloride co-doped with Er^{3+} and Yb^{3+} ions [28]. They grew these crystals hydrothermally and then subjected them to air annealing, using chiral sugar alcohols to induce chirality. When excited with a 980 nm near-infrared laser, D-BYEP emitted bright red light. The researchers tested its temperature-sensing abilities using two different energy levels of Er^{3+}: thermally coupled (TCLs) and non-thermally coupled (NTCLs). They found that at 298 K, D-BYEP exhibited maximum relative sensitivities of 0.84% K^{-1} and 1.30% K^{-1} for TCLs and NTCLs, respectively. These findings suggest that D-BYEP could serve as promising materials for optical temperature sensors.

Ni et al. [29] developed a series of $Ba_3Y_{2-x-y}Yb_xHo_yB_4O_{12}$ ($0.01 \leq x \leq 0.35$, $0.005 \leq y \leq 0.035$) upconversion phosphors, using a straightforward solid-state method. When these phosphors are excited by light at 980 nm, they emit strong green light at 551 nm, weaker red light at 668 nm, and near-infrared light at 755 nm. They found that when the composition is $x=0.15$ and $y=0.01$, the luminescent intensity is the highest. Using this finding, they investigated the temperature-sensing properties according to the FIR method at different wavelengths. They discovered that at 478 K, the maximum relative sensitivity (SR) reaches 1.53% K^{-1}. This suggests that $BYBO:Yb^{3+}/Ho^{3+}$ has promising potential for applications in temperature sensing.

Vega-Pallauta and colleagues [30] developed a $LaAlO_3$ phosphor using a Pechini sol-gel method. When excited with light at 975 nm, this material exhibits up-conversion, generating emissions at wavelengths of 533, 545, and 670 nm. They investigated its potential for optical temperature sensing between 299 and 327 K using luminescence intensity ratios. They found that at 327 K, the intensity ratio I_{670}/I_{545} showed the highest sensitivity at 5.21×10^{-3} K^{-1}, while at 299 K, the ratio $I_{533/545}$ exhibited a relative sensitivity of 0.87%. The temperature uncertainty was determined to be 0.2 K, suggesting that this material could serve as an effective temperature-sensing device based on both thermal and non-thermal energy levels.

Guo et al. [31] synthesized UC phosphors of La_2MgGeO_6 doped with Er^{3+} and Yb^{3+} using a solid-state reaction method. Their study explored a multi-mode optical thermometry technique using the FIR method between different energy levels and the fluorescence lifetime of specific states. Through these methods, they achieved maximum relative sensitivities of 1.130% K^{-1}, 0.623% K^{-1}, and 2.914% K^{-1}, respectively. The findings demonstrate the excellent temperature measurement capabilities of the produced samples, characterized by a wide temperature range, high sensitivity, and stability.

Duan and colleagues demonstrated the remarkable green emission characteristics of Ba^{2+}-doped $Ca_9Y(VO_4)_7:Yb^{3+}/Er^{3+}$ phosphors prepared via the solid-state reaction method [32]. When exposed to a 980 nm near-infrared laser, these phosphors exhibit strong green luminescence at 530 and 557 nm, along with weaker red luminescence at 664 nm. The ratio of integrated intensities of green to red luminescence is 32.1, indicating a significant enhancement. This enhancement is attributed to defect bands that aid in the absorption of excitation photons and efficient energy transfer from electrons to Yb^{3+} and Er^{3+} ions. Notably, the maximum sensitivity to temperature changes, as indicated by the sensitivity coefficients (S_A and S_R), is 1.44% K^{-1} (at 548 K) and 1.28% K^{-1} (at 298 K), respectively, making these phosphors promising candidates for temperature sensing applications.

Yun and colleagues synthesized phosphors by doping Erbium (Er), Ytterbium (Yb), and Neodymium (Nd) into $LiLn(WO_4)_2$ (where $Ln = Y$, La, or Gd) using a solid-state method [33]. They found that among the three host materials, the one with the most intense green emission under 785 nm excitation had the lowest phonon density and strong absorption properties. They also studied the effect of Nd^{3+} and Yb^{3+} concentrations on luminescent properties, as well as the energy transfer from Nd^{3+} to Yb^{3+} to Er^{3+} under 785 nm excitation. The ratio of integrated intensity from certain transitions of Er^{3+} was used for optical temperature sensing, known as fluorescence intensity ratio (FIR). LGW exhibited a high absolute sensitivity of about 0.0070 K^{-1} at 439 K and a wide temperature sensing range from 200 to 600 K under 785 nm excitation compared to other systems doped with Er^{3+}.

Xiao and colleagues synthesized a series of $La_2Mo_2O_9$ phosphors doped with Yb^{3+}, $Er^{3+}/Tm^{3+}/Ho^{3+}$ ions, which exhibit three-primary color up-conversion luminescence [34]. They evaluated the optical temperature sensing capabilities based on the typical levels of Yb^{3+}, $Er^{3+}/Tm^{3+}/Ho^{3+}$ co-doped systems. The $La_2Mo_2O_9$:Yb^{3+}, Er^{3+} phosphor showed a maximum relative sensitivity of 1.15% K^{-1} and an absolute sensitivity of 10.03×10^{-3} K^{-1}. For $La_2Mo_2O_9$:Yb^{3+}, Er^{3+}, the maximum relative sensitivity was 2.29% K^{-1} with an absolute sensitivity of 0.284×10^{-3} K^{-1}. $La_2Mo_2O_9$:Yb^{3+}, Ho^{3+} exhibited a maximum relative sensitivity of 0.30% K^{-1} and an absolute sensitivity of 20×10^{-3} K^{-1}. These multi-color up-conversion luminescence phosphors with high-temperature sensing sensitivities hold promising applications in anti-counterfeiting, three-dimensional displays, and non-contact temperature sensing.

Researchers have identified various other luminescent materials that could have potential applications in temperature sensing. Such as Yb^{3+}/Tm^{3+} codoped $KLa(MoO_4)_2$ [35], $Lu_6O_5F_8$: 1%Er^{3+}/10%Yb^{3+} nanoparticles [36], LuAG:Yb^{3+}/Tm^{3+} polycrystals [37], $LiGd(WO_4)_2$: Er^{3+}, Yb^{3+}, Nd^{3+} microparticles [38], $LiYF_4$:20Yb^{3+}, 1Ho^{3+} (mol%) phosphor [39], $La_{9.67}Si_6O_{26.5}$:Yb^{3+}–Er^{3+}/Ho^{3+} [40], $Yb_3Al_5O_{12}$:Er^{3+} [41], ScF_3:Yb^{3+}/Tm^{3+} [42], etc.

7.5 SUMMARY AND FUTURE PROSPECT

Upconversion phosphors are emerging as potential materials for luminescence-based optical thermometers due to their unique ability to convert low-energy photons into higher-energy emissions. These materials have various advantages, including high sensitivity, rapid response times, and non-contact temperature sensing. In recent years, great work has been made in understanding the fundamental mechanisms that regulate the up-conversion process, as well as modifying these phosphors' features for specific temperature sensing applications. Researchers have investigated a variety of measures to improve the efficiency and reliability of upconversion thermometry, including optimizing the composition and morphology of the phosphor particles and inventing innovative manufacturing techniques.

Looking ahead, upconversion phosphors for luminescence-based optical thermometers appear to have a promising future, with applications spanning from biomedical imaging to industrial process monitoring. Continued research efforts are likely to focus on increasing these materials' performance properties, such as temperature sensitivity, dynamic range, and stability. Furthermore, incorporating upconversion phosphors into modern sensing platforms, such as microfluidic devices or wearable sensors, may open up new avenues for real-time temperature monitoring in a variety of situations. Overall, current improvements in upconversion phosphor technology have the potential to significantly improve the capabilities of luminescence-based optical thermometers and drive innovation in temperature sensing approaches.

REFERENCES

1. L. Marciniak, K. Kniec, K. Elżbieciak-Piecka, K. Trejgis, J. Stefanska, M. Dramićanin, Luminescence thermometry with transition metal ions. A review, *Coord. Chem. Rev.* 469 (2022). https://doi.org/10.1016/j.ccr.2022.214671.

2. C.D.S. Brites, S. Balabhadra, L.D. Carlos, Lanthanide-based thermometers: At the cutting-edge of luminescence thermometry, Adv. Opt. Mater. 7 (2019) 1801239. https://doi.org/10.1002/adom.201801239.

3. R.S. Diarah, C. Osueke, A. Adekunle, S. Adebayo, A.B. Aaron, O.O. Joshua, Types of temperature sensors, in: *Wirel. Sens. Networks - Des. Appl. Challenges*, 2023: pp. 1–18. https://doi.org/10.5772/intechopen.110648.

4. S. Ma, Y. Xu, Y. Pang, X. Zhao, Y. Li, Z. Qin, Z. Liu, P. Lu, X. Bao, Optical fiber sensors for high-temperature monitoring: a review, *Sensors* 22 (2022). https://doi.org/10.3390/s22155722.

5. A. Solaimuthu, A.N. Vijayan, P. Murali, P.S. Korrapati, Nano-biosensors and their relevance in tissue engineering, *Curr. Opin. Biomed. Eng.* 13 (2020) 84–93. https://doi.org/10.1016/j.cobme.2019.12.005.

6. L.D. Carlos, R.A.S. Ferreira, V. de Zea Bermudez, Progress on lanthanide-based organic–inorganic hybrid phosphors, *Chem. Sci. Trans.* 40 (2011) 536–549.

7. P. Singh, S. Kachhap, P. Singh, S.K. Singh, Lanthanide-based hybrid nanostructures: Classification, synthesis, optical properties, and multifunctional applications, *Coord. Chem. Rev.* 472 (2022) 2147958. https://doi.org/10.1016/j.ccr.2022.214795.

8. S.K. Gupta, K. Sudarshan, R.M. Kadam, Optical nanomaterials with focus on rare earth doped oxide : A Review, *Mater. Today Commun.* 27 (2021) 102277. https://doi.org/10.1016/j.mtcomm.2021.102277.

9. P. Singh, S. Kachhap, M. Sharma, Lanthanide-doped materials for optical applications, in: *Handb. Mater. Sci.*, Springer, 2023: pp. 99–127. https://doi.org/10.1007/978-981-99-7145-9_4.

10. K.K. Markose, R. Anjana, A. Antony, M.K. Jayaraj, Synthesis of Yb^{3+}/Er^{3+} co-doped Y_2O_3, YOF and YF_3 UC phosphors and their application in solar cell for sub-bandgap photon harvesting, *J. Lumin.* 204 (2018) 448–456. https://doi.org/10.1016/j.jlumin.2018.08.005.

11. Z. Ma, J. Gou, Y. Zhang, Y. Man, G. Li, C. Li, J. Tang, Yb^{3+}/Er^{3+} co-doped Lu_2TeO_6 nanophosphors: Hydrothermal synthesis, upconversion luminescence and highly sensitive temperature sensing performance, *J. Alloys Compd.* (2018). https://doi.org/10.1016/j.jallcom.2018.09.175.

12. P. Du, L. Luo, H. Park, J. Su, highly-efficient infrared-to-visible upconversion material for optical temperature sensors and optical heaters, *Chem. Eng. J.* 306 (2016) 840–848. https://doi.org/10.1016/j.cej.2016.08.007.

13. L. Li, S. Fu, Y. Zheng, C. Li, P. Chen, G. Xiang, S. Jiang, X. Zhou, Near-ultraviolet and blue light excited Sm^{3+} doped Lu_2MoO_6 phosphor for potential solid state lighting and temperature sensing, *J. Alloys Compd.* 738 (2018) 473–483. https://doi.org/10.1016/j.jallcom.2017.12.169.

14. M. Mikolajek, R. Martinek, J. Koziorek, S. Hejduk, J. Vitasek, A. Vanderka, R. Poboril, V. Vasinek, R. Hercik, Temperature measurement using optical fiber methods: Overview and evaluation, *J. Sensors.* 2020 (2020). https://doi.org/10.1155/2020/8831332.

15. M. Runowski, Chapter: 10. Pressure and temperature optical sensors: Luminescence of lanthanide-doped nanomaterials for contactless nano-manometry and nanothermometry, Elsevier Inc., 2020. https://doi.org/10.1016/B978-0-12-816699-4.00010-4.

16. D. Jaque, F. Vetrone, Luminescence nanothermometry, *Nanoscale* 4 (2012) 4301–4326. https://doi.org/10.1039/c2nr30764b.

17. C.D.S. Brites, P.P. Lima, N.J.O. Silva, A. Mill, V.S. Amaral, D. Carlos, Nanoscale thermometry at the nanoscale, *Nanoscale View* 4 (2012) 4799–4829. https://doi.org/10.1039/c2nr30663h.

18. C.D.S. Brites, A. Milla, L.D. Carlos, Lanthanides in luminescent thermometry, in: *Handb. Phys. Chem. Rare Earths*, Elsevier, 2016: pp. 339–427. https://doi.org/10.1016/bs.hpcre.2016.03.005.

19. M.M. Upadhyay, K. Kumar, Gd^{3+} ion induced UV upconversion emission and temperature sensing in $Tm^{3+}/Yb^{3+}:Y_2O_3$ phosphor, *J. Rare Earths* 41 (2023) 1295–1301. https://doi.org/10.1016/j.jre.2022.07.005.

20. Y. Zhao, X. Wang, Y. Zhang, Y. Li, X. Yao, Optical temperature sensing of up-conversion luminescent materials: Fundamentals and progress, *J. Alloys Compd.* (2019) 152691. https://doi.org/10.1016/j.jallcom.2019.152691.

21. J. Yao, M. Yang, Y. Duan, Chemistry, biology, and medicine of fluorescent nanomaterials and related systems: New insights into biosensing, bioimaging, genomics, diagnostics, and therapy, *Chem. Rev.* 114 (2014) 6130–6178. https://doi.org/10.1021/cr200359p.

22. L.J. Mohammed, K.M. Omer, Carbon dots as new generation materials for nanothermometer: Review, *Nanoscale Res. Lett.* 15 (2020) 182.

23. W. Chen, G. Zhang, Y. Wang, D. Trans, D.O.I. Cdtg, Temperature sensor based on the enhanced upconversion luminescence of Li^+ doped $NaLuF_4$: Yb^{3+}, Tm^{3+}/Er^{3+} nano/microcrystals, *Dalt. Trans.* 47 (2018) 8656–8662. https://doi.org/10.1039/C8DT00928G.

24. G. Bao, P.A. Tanner, A stoichiometric terbium-europium dyad molecular thermometer: Energy transfer properties, *Light Sci. Appl.* 7, 96, (2018). https://doi.org/10.1038/s41377-018-0097-7.

25. K. Li, D. Zhu, H. Lian, Up-conversion luminescence and optical temperature sensing properties in novel KBaY(MoO$_4$)$_3$: Yb3+, Er3+ materials for temperature sensors, *J. Alloys Compd.* (2019) 152554. https://doi.org/10.1016/j.jallcom.2019.152554.

26. N. Rakov, F. Matias, G.S. Maciel, Temperature sensing performance of Er^{3+}: Yb^{3+} co-doped CaF$_2$ ceramic powders using near-infrared light, *J. Rare Earths* (2024) 3–7.

27. M.M. Padhyay, N.K. Mishra, K. Kumar, Spectrochimica acta part A: Molecular and upconversion luminescence based temperature sensing properties and anti-counterfeiting applications of, *Spectrochim. Acta A Mol. Biomol. Spectrosc.* 304 (2024) 123333.

28. H. Zhang, H. Chen, L. Xu, Y. Li, Z. Song, D. Zhou, Q. Wang, Z. Long, Y. Yang, Y. Wen, J. Han, Y. Gao, J. Qiu, Er^{3+}/Yb^{3+} ions doped chiral inorganic nanostructured BiOCl phosphor as a novel temperature sensing material, *J. Am. Ceram. Soc.* (2024) 1–7. https://doi.org/10.1111/jace.19695.

29. Z. Ni, Y. Zhu, H. Zhang, Y. Su, Yb^{3+}/Ho^{3+} doped Ba$_3$Y$_2$B$_4$O$_{12}$ phosphors: Upconversion luminescence and temperature sensing properties, *J. Lumin.* 268 (2024) 120427. https://doi.org/10.1016/j.jlumin.2023.120427.

30. M.A. Vega-Pallauta, R. Castillo, K. Soler-Carracedo, I.R. Martín, Near-infrared to visible up-conversion in Er^{3+} doped LaAlO$_3$ phosphors and their assessment as an optical temperature sensor, J. Lumin. 267 (2024) 3–7.

31. J. Guo, Y. Shen, L. Chen, D. Deng, S. Xu, A multi-mode optical thermometry based on the up-conversion LaMgGeO: Er, Yb phosphor, *J. Lumin.* 266 (2024) 120331. https://doi.org/10.1016/j.jlumin.2023.120331.

32. B. Duan, C. Ding, Y. Wu, Y. Li, W. Jin, F. Wang, J. Hu, Infrared physics & technology defective band enhance the single green upconversion luminescence of Ca$_9$Y(VO$_4$)$_7$: Yb^{3+}/Er^{3+} phosphors by doping with Ba^{2+} ions, *Infrared Phys. Technol.* 136 (2024) 105100.

33. R. Yun, J. He, L. Luo, X. Liu, Z. Nie, W. Zhao, H. Wen, Upconversion luminescence and temperature sensing properties in LiGd(WO): Er, Yb, Nd microparticles under 785 nm excitation, *Ceram. Int.* 47 (2021) 16062–16069.

34. Q. Xiao, X. Dong, X. Yin, H. Wang, H. Zhong, K. Liu, B. Dong, X. Luo, Promising Yb^{3+}-sensitized La$_2$Mo$_2$O$_9$ phosphors for multi-color up-conversion luminescence and optical temperature sensing, *J. Alloys Compd.* 895 (2022) 162686. https://doi.org/10.1016/j.jallcom.2021.162686.

35. Y. Zhang, B. Wang, Y. Liu, G. Bai, Z. Fu, H. Liu, Upconversion luminescence and temperature sensing characteristics of Yb^{3+}/Tm^{3+}:KLa(MoO$_4$)$_2$ phosphors, *Dalt. Trans.* 50 (2021) 1239–1245. https://doi.org/10.1039/d0dt03979a.

36. Y. Li, L. Guo, Up-conversion luminescence of Lu$_6$O$_5$F$_8$:1%Er^{3+}/10%Yb^{3+} nanoparticles for temperature sensing and Cu^{2+} detection, *Opt. Mater. (Amst).* 115 (2021) 111031.

37. H. Zhou, N. An, K. Zhu, J. Qiu, L. Yue, L. Wang, L. Ye, Optical temperature sensing properties of Tm^{3+}/Yb^{3+} co-doped LuAG polycrystalline phosphor based on up-conversion luminescence, *J. Lumin.* 229 (2021) 117656.

38. R. Yun, J. He, L. Luo, X. Liu, Z. Nie, W. Zhao, H. Wen, Upconversion luminescence and temperature sensing under 785 nm excitation, *Ceram. Int.* 47 (2021) 16062–16069.

39. W. Li, L. Hu, W. Chen, S. Sun, M. Guzik, G. Boulon, The effect of temperature on green and red upconversion emissions of $LiYF_4$:20Yb^{3+}, 1Ho^{3+} and its application for temperature sensing, *J. Alloys Compd.* 866 (2021) 158813.

40. J. Zhanga, J. Chena, Y. Zhang, Temperature-sensing luminescent materials $La_{9.67}Si_6O_{26.5}$:Yb^{3+}-Er^{3+}/Ho^{3+} based on pump-power-dependent upconversion luminescence, *Inorg. Chem. Front.* 7 (2020) 4892–4901. https://doi.org/10.1039/D0QI01058H.

41. X. Li, Z. Guan, Z. Liu, R. Shen, Y. Li, R. Zhou, H. Yu, L. Li, B. Chen, Impact of nitridation on the up-conversion luminescence property of $Yb_3Al_5O_{12}$: Er^{3+} phosphors upon 980 nm excitation, *Inorg. Chem. Commun.* 160 (2024) 111905. https://doi.org/10.1016/j.inoche.2023.111905.

42. M. Dai, Y. Li, Z. Wang, A. Li, Thermally boosted upconversion luminescence and high-performance thermometry in ScF_3:Yb^{3+}/Tm^{3+} nanorods with negative thermal expansion, *J. Lumin.* 265 (2024) 120219.

Carbon-based materials for luminescence thermometers

S. Satyanarayana Reddy and Nunna Guru Prakash

8.1 INTRODUCTION

Temperature is a key thermodynamic physical quantity; accurate temperature measurement is of great significance and has a wide range of applications in science and engineering [1,2]. In particular, thermometry is employed in medical science for identifying various diseases such as stroke, cancer, or inflammation, one of whose early signs is the emergence of localized temperature anomalies [3]. Conventional thermometers are classified into four categories based on their sensing mechanism techniques. They are (i) liquid-filled glass thermometers based on the thermal expansion of the liquid [4]; (ii) thermocouples based on the thermionic effect [5]; (iii) thermistors based on changing resistance [6]; and (iv) optical temperature sensors [7,8]. The first three types of thermometer including the liquid-filled glass thermometers, thermocouples, and thermistors are classified as contact thermometers. Direct contact between the thermometer and the substrate is necessary for contact-based thermometers. Hence, these types of sensing are not suitable for sub-micron scale material temperature measurement in the medical field due to lack of sensitivity and accuracy. Luminescent thermometry often employs pellets that create a thin film or coating when mixed with a binder. The prevailing type of luminous temperature probes is commonly employed either as compressed pellets, with or without calcination, or as coatings after blending with binders.

The most recent breakthrough is non-contact nanomaterial/molecular thermometry. The fluorescence light from the material at different temperatures measures the temperature based on their variation in emission intensity, band shape, Stokes shift, and decay lifetime concerning temperature with proper calibration [9–11]. The medical field could greatly benefit from the use of molecular thermometers for disease diagnosis. Nanomaterials have recently captured the attention of scientists, particularly semiconductors [12], polymers [13], nanodiamonds [14], lanthanide, and transition metal nanoparticles [15], and have been employed as thermal sensors with sub-micron thermal resolution. The material with highly sensitive and high-resolution thermometers, which can be integrated into living

DOI: 10.1201/9781032661537-8

systems and measure a broad range of temperatures with high accuracy, could be a game-changer in many fields of chemical, biological, and physical study. However, these nanomaterials have limitations in use along with their advantages. Such nanomaterials have a thermosensitive characteristic within the physiological range, while polymers exhibit poor photostability and pronounced cross-sensitivity to oxygen, and significant pH dependence on the fluorophore's lifetime [16].

Another class of thermometers is semiconductor quantum dots (SQDs) have been reported by considering their properties. They have a high quantum yield, a long lifetime before photobleaching, and sufficient biocompatibility following proper surface modification. They can also be easily coupled to proteins and DNA for sensing and imaging [17–19]. Additionally, SQD thermal sensors are immune to pH shifts and other environmental variations expected in a cell [20]. The quantity of luminescence quenching and spectral shift is greatly affected by the size and composition of the SQDs [21,22]. SQDs have favorable properties such as photostability, quantum efficiency, and adjustable fluorescence. However, due to their inherent blinking, SQDs are not suitable for long-term single-molecule tracking. Furthermore, SQD's primary drawback is their toxicity, which is brought on by the presence of heavy toxic metals (cadmium). As a result, they have limited biological and environmental applications, and SQDs are expensive due to the rare availability of precursors in nature [23].

As a result of the environmental risk of biological uses of such toxic compounds, the development of carbon-based materials is more fascinating and necessary. Because of their unique features, carbon-based materials may be used to overcome the drawbacks of nanomaterials and SQDs for medical applications. When compared to earlier nanomaterials, carbon-based nano and quantum dot materials exhibit high sensitivity, sub-degree accuracy, low toxicity, simple synthesis, low-cost precursor, intrinsic biocompatibility, photostability, and biocompatibility [24–29]. The fluorescent nanodiamonds (NDs) are the first carbon-based nanoparticles to be identified as a novel member of fluorescence thermometry [28]. The fluorescence of NDs is triggered by nitrogen-vacancy color centers [21,22]; the precise mechanism of nitrogen-vacancy defect fluctuations with temperature is explained in Session 8.2. Accurate measurement of these color centers is crucial to the concept of nitrogen-vacancy-based thermometry, and they may be optically detected with great spatial resolution [30,31]. Nevertheless, the limited fluorescence efficiency and lack of controllability significantly restrict the usability of fluorescent NDs [32].

A relatively new category of carbon-based nanomaterials is the exceptionally luminescent carbon dots (CDs). Recently, there has been a notable surge in the interest of scientists toward using luminescent CDs as temperature sensors due to their unique properties, such as low toxicity, highly luminescent, bright photoluminescence, facile preparation, water-solubility, photochemical stability, low-cost precursor, great biocompatibility, and photostability [33,34]. CDs are zero-dimensional and spherical in structure with the particle less than 10 nm particle size [35–37]. Furthermore, CDs have a wide range of applications in the fields including targeted drug delivery [38], sensing including bio and chemical [39,40], biological imaging [41,42], pharmaceutical analysis [43], and catalysis [44–46]. In essence, a specific circumstance is required for a precise temperature measurement. That is, within the specified temperature range, CDs must exhibit a discernible shift in their photoluminescence [47,48]. The present challenge and the work are progressing on practical application requisites of photostability, pH stability, and shelf life [49–51]. Moreover, simple procedures can change and functionalize CDs' surface state, enabling the customization of solubility, stability, physicochemical properties, and quantum yields for experimental needs.

8.2 CLASSIFICATION OF CARBON-BASED MATERIALS FOR LUMINESCENCE THERMOMETRY

The importance of carbon-based luminescence thermometry in the environmental hazard for biological applications and the toxicity of earlier compounds have piqued the interest of researchers. Nanodiamonds (NDs) are the first material reported as a luminescence nanothermometer in the family of carbon-based nanomaterials [27]. Further, highly luminous carbon dots (CDs) are more recent and more important members of the carbon-based nanomaterials family as luminescence thermometers due to their remarkable characteristics, strong photoluminescence, photochemical stability, water solubility, excellent biocompatibility, and lack of toxicity.

8.2.1 Luminescence centers of nanodiamonds

Luminous imperfections, like impurities and color centers, may be present in the crystal lattice of the nanodiamonds [52]. The most famous of them are the nitrogen-vacancy (NV) color centers, which have a persistent fluorescence spectrum extending from the visible to near-infrared (NRI) regions. The particle size of the nanodiamond plays a major role in the variation of bandwidth and lifetime of the fluorescent spectra [53]. The characteristic emission of ND particularly at 575 nm and 637 nm corresponds to the neutral nitrogen-vacancy center (NV^0) and negatively charged nitrogen-vacancy centers (NV^-) are also known as allocate 5 and 6 electrons, respectively [54–56]. Each charge state population can be regulated to match a maximum luminous intensity by considering the wavelength and power of an excitation [57]. Moreover, the NV centers jump between the two charge states continually under 450–575 nm excitation [58] and 532 nm excitation NV^0 and NV^- configurations exist at about 30% and 70%, respectively [59]. Hence, fluorescent nanodiamonds are a good probe for biomedical thermal imaging applications [14,60]. Furthermore, the NV center is highly attractive due to photostability [61–63]. These nanodiamonds' temperature sensitivity comes from nitrogen-vacancy color centers, point defects with a nitrogen atom replacing a lattice carbon atom, and an adjacent empty site.

8.2.2 Luminescence centers of carbon dots

As of yet, no reliable mechanism has been established to account for luminescence emission and behavior concerning the temperature of CDs. Wang et al. demonstrated the CD's emission mechanism that several carbon atoms with carbonyl or carboxyl groups near the edge of the carbon backbone could be the source of the CD's green fluorescence emission [64].

According to earlier research reports, the mechanism involved in the variation of fluorescence emission concerning the temperature is due to the thermal activation/deactivation of surface non-radiative trap/defect channels [65,66]. Two possible mechanisms of CD fluorescence emissions with increasing temperature are shown in Figure 8.1. Blue line transition represents transition at low temperature and red belongs to high-temperature transition. In general, the intensity of the emission increases with increasing temperature due to the radiative emission of photons, which are not triggered by the non-radiative channels at low temperatures [8,67]. Conversely, for specific defects or trap states, the fluorescence intensity decreases with increasing temperature. This reduction is attributed to the activation of additional non-radiative channels, facilitating the transfer of excited electrons through non-radiative transitions[3,68–71].

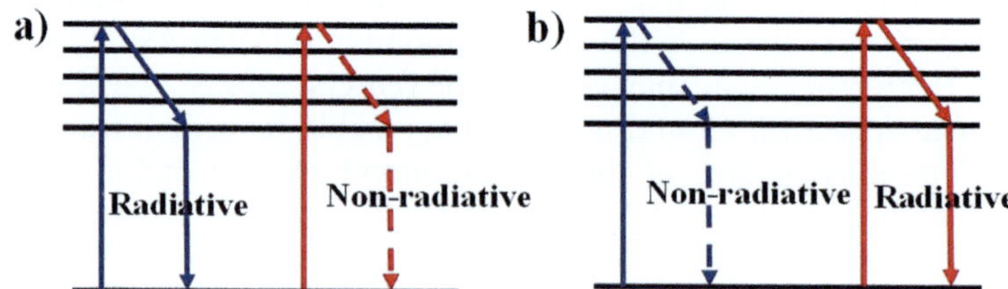

Figure 8.1 Schematic illustration of CD fluorescence emission mechanisms, (a) increasing the non-radiative transition and (b) increasing the radiative transition with increasing temperature.

8.3 SYNTHESIS OF CARBON-BASED MATERIALS

The discovery of inventive methods to develop any nanomaterial and quantum dots is a critical and challenging task for luminescence thermometers. Chemical techniques to enhance the properties of carbon-based materials have gained popularity in recent years. Significant emphasis has already been paid to the synthesis of NDs and CDs for high efficiency, high quality, high stability, low cost, and large-scale productivity applications. Some of the unique techniques of persistent luminescence NDs and CDs synthesis are discussed below.

8.3.1 Nanodiamonds

The size of the principal particles in ND particles can be readily classified into three groups: nanocrystalline particles (few tens of nm), ultra-nanocrystalline particles (several nm) [72], and diamondoids (1–2 nm) [73]. The margin of 10 nm between two classes of NDs may seem arbitrarily chosen, but it represents a variation in the properties of the final products made using various-sized diamond nanoparticles [74].

8.3.1.1 Laser ablation

Pulsed laser ablation is the simplest technique and has become an attractive method for the synthesis of the liquid phase of ND through graphene dispersions [75] or on a carbon target made of pyrolytic graphite [76,77]. About 0.5 mg/mL of graphene powder is disseminated in solvent (e.g., isopropyl alcohol) by sonicating the solution for an hour. The resultant graphene powder dispersed solution will be black in color. Following that, the graphene dispersed sample underwent the Nd-YAG laser irradiation (wavelength 532 nm, power 2 W, pulse duration 5 ns, energy 80 mJ/pulse, and pulse rate 10 Hz) focused by a lens with about 100–200 mm focal length for about 30 minutes or until the entire solution was turned into a plane yellow-colored solution. The size of the laser spot on the target surface can be approximated to be between 0.1 and 0.5 mm. As a result, the fluence of each pulse ranges between 40 and 1000 J/cm^2. In order to prevent gravitational settling at the solution's base during the laser irradiation, it is magnetically stirred throughout the laser irradiation process. The laser-ablated solution is then centrifuged or filtered to collect the precipitation of the ND's sample and dried overnight at 70°C.

8.3.1.2 High-energy ball milling

The monocrystalline ND particles produced from high-pressure and high-temperature (HPHT) synthetic diamond and natural diamond powders are commercially accessible at Pureon (https://pureon.com/products/micron-diamond-powders/), with the smallest typical particle size being a few tens of nanometers [78]. The commercially available microcrystalline diamond powder with high impact strength, toughness, and high thermal stability is the basic material used. The raw microcrystalline diamond powder is milled nominally at room temperature using an oscillatory mill. The complex movements of the grinding vial in conjunction with back-and-forth oscillation and lateral movement of its bottom are used to optimize the collisions between the powder, ball, and inner wall of the vial in order to guarantee that the as-milled powder pattern is uniform. Owing to the intense oscillation, which had an amplitude of 5 cm and a speed of 750 rpm, the ball velocities were calculated to be 6 m/s, resulting in significant collision milling energy [79]. Furthermore, the hardened chrome steel alloy is used to make the grinding vial and balls, and isopropanol is a surface-active agent (2 mL/g) which allows for the operation to be performed under extreme conditions. An aliquot is ball milled in a planetary ball mill under argon for 24 hours of effective grinding time. This process yields approximately about 15% of the microdiamond bulk into 10 nm ND particles. The resultant nanoparticles of every 1 g of powder are then subjected to purification with 50 mL of a sulfuric and hydrochloric mixture (1:2 v/v) in a glass autoclave at 200°C for three hours until the liquid's characteristic dark yellowish color – which is associated with iron salts – completely vanished. Furthermore, the diamond sample must be washed with distilled water using ultrasonication and thoroughly suspended in ethyl alcohol using a surfactant. Finally, it was centrifuged to collect the fine diamond particles.

8.3.1.3 Chemical vapor deposition

Figure 8.2 shows the flow chart of ND growing on Si substrate by a low-pressure microwave-plasma chemical vapor deposition (CVD) reactor. Double-polished silicon wafers of (100) orientation and 200–600 μm thickness can be used as a substrate. The wafers are undergoing a cleaning and seeding process. The cleaning was composed ultrasonic bath in acetone and 2-propanol for 10 minutes each and then in an $HF:H_2O$ (1:9) solution followed by drying after each step [80,81]. After cleaning the substrate is seeded with 5–7 nm nanodiamond suspension by immersed in the suspension or drop cast around 2000 rpm and rinsed with methanol and dried under compressed air flow. An average seeding density ranging from 3×10^{11} to 8×10^{11} cm^{-2} may be obtained. Then the substrate was transferred in a molybdenum holder into the CVD chamber with the MW power of 2 kW keeping a constant 0.5 mbar process pressure to grow the crystal. The $H_2/CH_4/CO_2$ gas mixture combination can be used at about 78 sccm m^{-1} of total gas flow rate for 1–2 hours based on the required size of the nanodiamond.

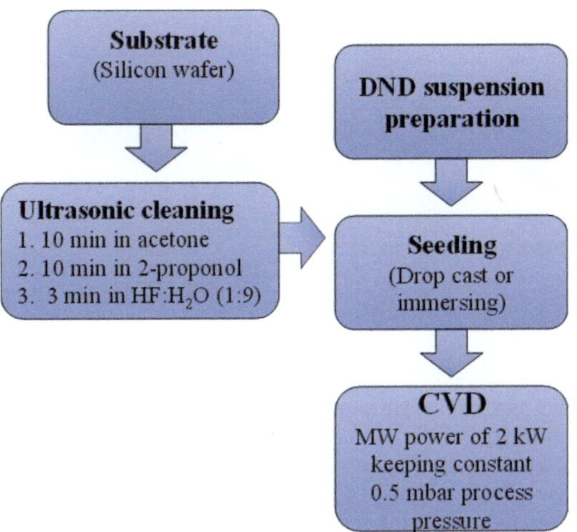

Figure 8.2 Flow chart of ND growing through the CVD method.

The substrate holder is not necessary to heat in addition to the diffusive plasma. The grown particles are crystalline in shape, with an average particle size of roughly 50 nm [78].

8.3.2 Carbon dots

The approaches to CD preparation can be divided into two categories: top-down and bottom-up methods. Several synthesis processes for CD production have been developed over the years. The bottom-up approach of hydrothermal, solvothermal, and microwave-assisted synthesis techniques has recently attracted a great deal of researcher's attention. Similarly, laser ablation in solution has recently received a lot of attention as a single-step top-down technique. These syntheses of CD's methods are briefly explained here.

8.3.2.1 *Hydrothermal/solvothermal*

The hydrothermal/solvothermal [1,3,82–84] synthesis process is one of the most widely used methods that has drawn the attention of scientists and professionals working in the field of material science. As the name implies, heat is involved in the synthesis process, and the word "thermal" refers to heat and hydro/solvo refers to either water as a solvent or any other appropriate solvent or combination of solvents. The schematic representation of the synthesis process of the hydrothermal/solvothermal method is shown in Figure 8.3. The pure and doped CDs are

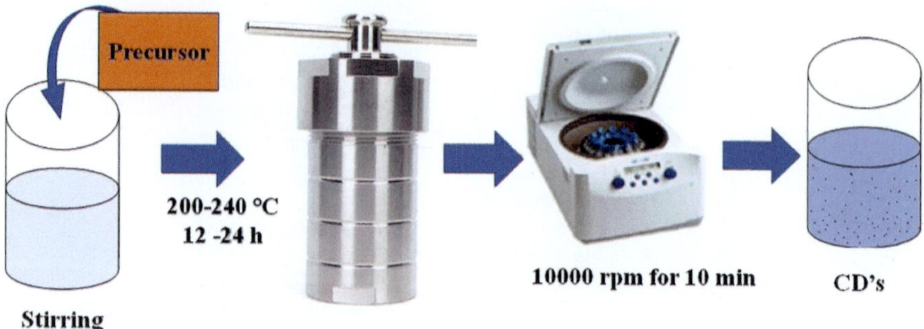

Figure 8.3 Schematic representation of the synthesis of CDs by hydrothermal/solvo-thermal method.

synthesized using a two-step hydrothermal method. Typically, carbon and dopant precursors are dissolved in distilled water/solvent under vigorous stirring to form a homogeneous solution. The homogeneous solution is transferred into a Teflon-lined stainless-steel autoclave and maintained in the range from 200°C to 240°C for 12–20 hours. Then, the reaction system was allowed to cool down to room temperature naturally. In order to remove any remaining impurities, the sample is washed numerous times with ethanol and acetone at a sample-to-solvent ratio of 1:10. After each wash, the precipitate is collected by centrifugation at 10,000–15,000 rpm for 10 minutes. The resulting suspensions containing CDs were filtered through 0.22 mm filter membranes and then subjected to dialysis for about 72 hours.

8.3.2.2 Laser ablation

Laser ablation is the process of removing/trimming material from a solid substance/microparticle. In this technique, the bulk carbon solution is prepared by dispersing the 20 mg of micro carbon powder into 50 mL of solvent by ultrasonication. The dispersed 10 mL of solution is transferred into a 50 mL glass beaker for laser irradiation. For irradiation, a Ti: sapphire femtosecond laser system with a central wavelength of 800 nm, pulse duration of 150 fs, and repetition rate of 1 kHz can be used [85]. The schematic representation of the laser ablation technique is shown in Figure 8.4. The laser beam can be focused into the suspension with different focus spot sizes of about 8, 16, and 32 μm using lenses with focal lengths of about 50, 100, and 150 mm, respectively. A magnetic stirrer has been used during laser irradiation to prevent the initial powders from settling due to gravity. Then the solution is centrifuged at less than 4000 rpm for 10 minutes to separate larger carbon particles and CDs can be obtained in the superna-tant. The mixture is then dialyzed using a dialysis membrane to remove

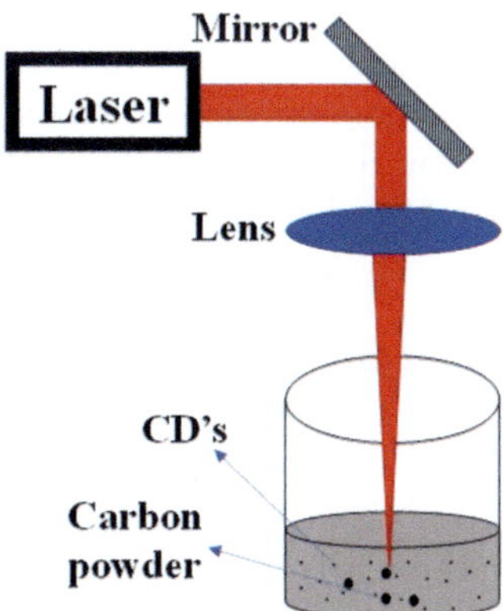

Figure 8.4 Schematic representation of CDs using laser ablation technique.

unreacted materials and solvents. The CD's precipitate can be collected by centrifugation at 10,000–15,000 rpm for 10 minutes. CDs were dispersed in water for further characterization.

8.3.2.3 Microwave-assisted synthesis

Microwave-assisted synthesis has revolutionized the way of producing QDs and nanomaterials. Many researchers are moving away from conventional heating synthesis methods and moving toward microwave-assisted synthesis because it offers clean synthesis with the added advantages of increased reaction rates, greater selectivity, economy, and higher yields for the synthesis of a large number of QDs and nanomaterials [86]. Dipolar polarization and conduction are the fundamental mechanisms identified in microwave-assisted synthesis. As a result, microwave radiation is frequently employed in chemical synthesis as a heating source.

The temperature-controlled or modified microwave-assisted synthesis method is the quickest method to synthesize the QDs and nanoparticles [87]. The synthesis of CDs using a temperature-controlled microwave is a one-step reaction with the carbon and dopant sources as a precursor dissolved in the solvent [67]. Further, a solution of 0.1 M carbon source in the solvent is prepared. The 20 mL of the mixed solution is stirred and sonicated until it gets a clear homogeneous solution. The clear solution is then transferred into the 50 mL glass beaker or 35 mL microwave reaction vial and heated to 180°C for different time intervals ranging from 5 to 15 minutes in steps of 15 minutes gap between 5 minutes. The microwave-reacted solution is cooled down to room temperature naturally, and then the obtained mixture is dialyzed for over five days with the water changed twice a day using a dialysis membrane to remove unreacted materials and solvent. The material is subsequently passed through a 0.2 mm nylon filter to eliminate any aggregates. In order to remove any leftover contaminants, the sample is washed several times with ethanol and acetone with a 1:10 sample-to-solvent ratio. The precipitate is collected after each wash by centrifugation at 10,000 rpm for 10 minutes. The resultant material was dried overnight in an oven at 70°C and resuspended in water.

8.4 TEMPERATURE SENSING USING VARIOUS LUMINESCENCE FUTURES OF NANODIAMONDS

A few published articles on NDs that exhibit temperature-dependent fluorescence are tabulated in Table 8.1. New temperature measuring techniques are needed to meet the challenge of accurately sensing temperature with high precision and high spatial resolution. Nitrogen-vacancy (NV) centers, which are the basis of an atomic-scale thermal sensor, have recently been constructed and successfully used to demonstrate temperature sensing at the mK level in a few tens of nanometers of diamond.

Choe et al. [88] reported a temperature-sensing mechanism based on the NV center in both experimental and theoretical aspects. Figure 8.5a shows the schematic representation of the continuous wave electron spin resonance (CW-ESR) measurement. This has a resistive heater and a TEC cooler thermally linked to a diamond plate with NV centers. To excite the NV center and detect the PL signal, a confocal optics setup consists of a 532 nm laser and an APD single photon detector. A microwave wire is positioned near the diamond to control the NV's spin transitions at 2.87 GHz. The CW-ESR spectra show a single NV center at two different temperatures (297 and 385 K) shown in Figure 8.5c. At each temperature, there are two resonance dips in the CW-ESR spectrum, which correspond to the spin transitions between the $m_s = 0$ state and the $m_s = \pm 1$ states, respectively, and the corresponding energy level diagram is shown in Figure 8.5b. The measurement over a wide range of temperatures from 297 to 333 K shows a linear decay response with the amount of shift in the central frequency of NV centers with a precision of 0.2 K, as shown in Figure 8.5d. Furthermore, hypothetically analyze three sensing techniques and analyze temperature sensitivity to determine the ideal measurement duration and NV concentration for the ensemble measurement. This technique can be applicable to a variety of nanotechnology sectors that require precise temperature monitoring with great spatial resolution, including miniaturized electronics, integrated nano-photonics, and molecular biomedicines.

Wang et al. [89] reported a high-sensitivity temperature detection using an implanted single NV center array in the diamond. The implanted single NV center in a diamond was subjected to the high-order thermal Carr-Purcell-Meiboom-Gill (TCPMG) method in a static magnetic field. It was observed that under minor detunings for the two driving microwave

Table 8.1 Temperature-dependent fluorescent of a few carbon-based materials recorded so far

Carbon-based material	Synthesis method	Scheme of temperature sensing	The linear range of the temperature ($^{\circ}C$ or K)	Sensitivity ($^{\circ}C^{-1}$)	References
NDs	Purchased	FWHM, intensity decay, and band shift	11–150 K 200–296 K	–	[32]
ND-PMA-Eu/Tb	ND-rare earth hybrid materials	The ratio of Tb to Eu luminescence intensity	77–277 K	–	[88]
CDs	Self-promoted and self-controlled	Time decay, intensity decay, and band shift	77–300 K	–	[65]
	Microwave-assisted	FL Intensity variation	5–60	3.71%	[67]
	Hydrothermal	FL Intensity decay	25–95	–	[89]
	Laser ablation		5–85	1.48%	[90]
Passivated CDs	Hydrothermal		15–60	–	[1]
Nitrogen-doped CDs	Reflux		20–80	0.85%	[91]
	Hydrothermal		25–95	–	[92]
Nitrogen and sulfur co-doped CDs	Reflux		20–80	–	[93]
	Hydrothermal		25–75	–	[94]
			10–90	–	[68]
			5–75	0.41% $^{\circ}C^{-1}$	[3]
			5–75	–	[95]
			283–343 K	0.64%	[70]
		Particle size variation	20–80	–	[82]
Nitrogen and chlorine co-doped CDs	Hydrothermal	FL Intensity decay	10–60	–	[71]

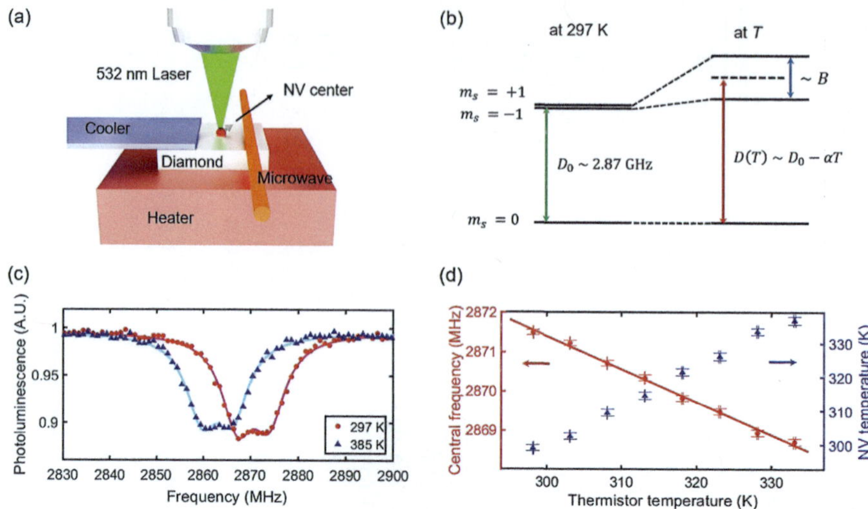

Figure 8.5 (a) Schematics of an experimental setup for CW-ESR measurement, (b) energy level diagram of the NV center's ground spin states, (c) CW-ESR measurements at two distinct temperatures, 297 and 385 K, on a single NV center, and (d) plotting the temperature and central frequency from the NV center against the heater's thermistor temperature. (Reproduced from Ref. [88] with permission from Elsevier.)

frequencies, the oscillation frequency of the induced fluorescence of the NV center is almost equivalent to the average of the two driving fields' detunings. The zero-field splitting D for the NV center and the associated temperature might be ascertained from the result. The results of the experiment demonstrated that the high-order TCPMG's coherence time was successfully increased, namely for TCPMG-8, up to 108 μs, or around 14 times the value of 7.7 μs according to the thermal Ramsey method. The thermal sensitivity for this coherence time was found to be 10.1 mK/Hz$^{1/2}$. The result suggests that it may be possible to use high-quality diamonds with implanted NV centers to measure temperatures with high sensitivity and nanoscale resolution in the fields of materials science, biology, chemistry, and microelectronics [96–99].

Su et al. [32] reported the temperature-dependent blue fluorescence (400 nm) of NDs purchased from Zhongnan Jete Superabrasives. The purchased NDs were annealed in air at 420°C for 30 minutes to remove the amorphous carbon, then deposited on silicon substrate by spin coating for further characterization. The photoluminescence (PL) intensity of the nanoparticles (NDs) varies with temperature, as depicted in Figure 8.6a. As the temperature increases, there is a noticeable decrease in the strength of PL. The non-radiative traps have a thermal response, becoming activated as the temperature rises. This is evident from the observable widening of

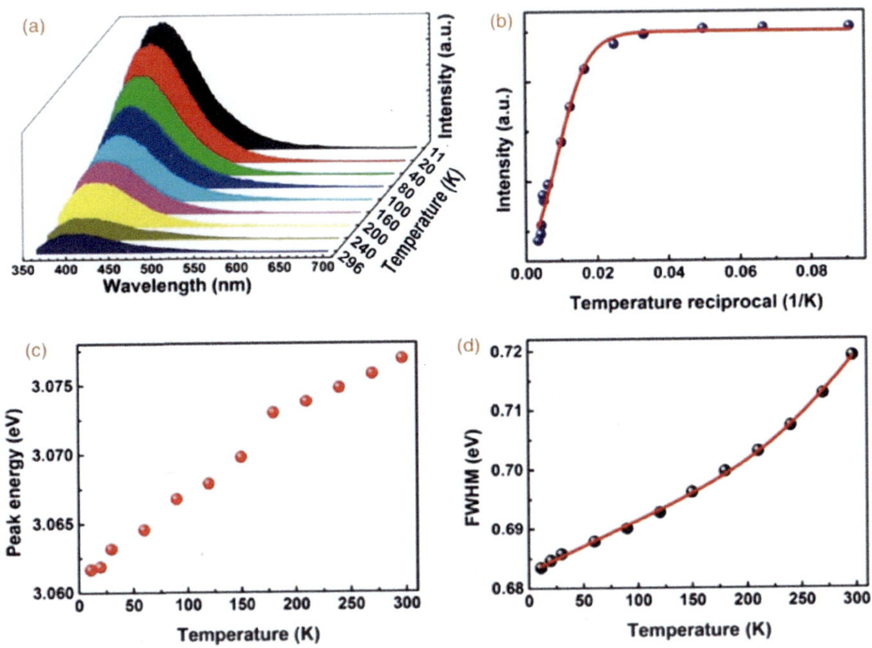

Figure 8.6 (a) NDs' recorded fluorescence spectra from 11 to 296 K. (b) The NDs' fluorescence intensity plotted using the Arrhenius equation. (c) The relationship between temperature and the peak energy and (d) the FHWM of the NDs as a function of temperature. (Reproduced from Ref. [32] with permission from IOP Publishing Ltd.)

their full width at half maximum (FHWM) with increasing temperature. The temperature-dependent quenching of PL was mathematically described by the following formula [90,91]:

$$I(T) = \frac{I(0)}{1 + C * \exp\left(-\dfrac{E_A}{K_B T}\right)} \tag{8.1}$$

where $I(T)$ and $I(0)$ are the PL intensity at temperature T and 0 K, respectively; E_A is the PL thermal quenching activation energy, and K_B is the Boltzmann constant; C is a constant. Figure 8.6b shows the variation of PL intensity with respect to the reciprocal of the temperature. The relationship between the PL intensity and temperature may be accurately fitted by Eq. (8.1), and determined activation energy E_A (21±1.7 meV). Excited electron relaxation in NDs can be caused by various methods, including radiative relaxation, heat escape, and surface activation. While the non-radiative channel remains inactive at low temperatures, electrons in an excited state can still interact with holes, resulting in the emission of photons. Elevated temperature induces the activation of non-radiative pathways, which confine carriers within NDs employing surface or defect states. Consequently, the PL intensity decreases as the temperature rises. The OH and C=O in NDs, suggest that surface defect states associated with these oxygen-containing groups are expected to have a substantial impact on the non-radiative process.

Figure 8.6c displays the peak energy of fluorescence concerning temperature. This energy shift toward shorter wavelengths, known as a blueshift, occurs as the temperature increases. This phenomenon is attributed to the presence of excited excitons in the higher-energy regions of functional group molecules. Figure 8.6d exhibits the variation of FWHM of the PL spectrum with respect to the temperature. It was observed that the FWHM increased with increasing temperature. It may be due to the scattering of ionized impurities and interactions between excitons and phonons also had a role in a slight expansion of the fluorescence bandwidth. The experimental results suggest that the fluorescence of the NDs may be attributed to radiative recombination, which happens via intrinsic transitions between highly localized π states. Moreover, the oxidation defect states play a crucial role in causing the shift in peak energy and the broadening of the FWHM which is characteristic of the NDs. Studying the temperature changes in luminescence can provide a more comprehensive understanding of the luminous process of NDs. This discovery also paves the path for improving the luminous properties of NDs and broadening their range of applications as a luminescence thermometer.

A self-calibrating luminous thermometer constructed from ND-rare earth hybrid materials ND-PMA-Eu, ND-PMA-Tb, and ND-PMA-Eu/Tb was described by Wu et al. [92]. ND-PMA-RE hybrid materials have been effectively produced through the covalent modification of NDs with pyromellitic acid (PMA), which has the ability to coordinate with Eu^{3+} and Tb^{3+} ions. Figure 8.7a–c shows the temperature-dependent PL emission spectra of (a) ND-PMA-Eu, (b) ND-PMA-Tb, and (c) ND-PMA-Eu/Tb recorded from 77 to 277 K under the 260 nm excitation. Variations of prominent peaks normalized intensity with temperature of the $^5D_0 \rightarrow {}^7F_2$ transition of ND-PMA-Eu and $^5D_4 \rightarrow {}^7F_5$ transition of ND-PMA-Tb are shown in Figure 8.7d. Changes in the normalized intensity of prominent peaks as a function of temperature of $^5D0 \rightarrow {}^7F_2$ and $^5D_4 \rightarrow {}^7F_5$ transitions for ND-PMA-Eu/Tb are shown in Figure 8.7e, and the normalized intensity ratio of Tb^{3+} to Eu^{3+} for NDPMA-Eu/Tb is shown in Figure 8.7f. The ND-PMA-Eu/Tb co-doped hybrid composites show a remarkable luminous behavior that changes with temperature and a strong linear relationship between the ratio of Tb to Eu luminescence intensity from 77 to 277 K. By utilizing its temperature sensitivity and self-calibration mechanism, this device has the potential to function as a self-calibrating ratiometric luminous thermometer.

8.5 TEMPERATURE SENSING USING VARIOUS LUMINESCENCE FUTURES OF CARBON DOTS

Table 8.1 shows the few published articles on carbon dots that exhibit temperature-dependent fluorescence. Macarian et al. [67] produced microwave-synthesized dual-fluorescing carbon dots (dCDs) for fluorescence (FL)-based ratiometric temperature sensing. FL spectrum of dCDs with varying temperatures from 5°C to 60°C under excitation at 640 and 405 nm are shown in Figure 8.8a and b respectively. With increasing temperatures, FL intensity increased 3.5-fold under the excitation wavelength at 640 nm, and the integral area under the curve showed a linear response from 5°C to 60°C (Figure 8.8b). Also, FL spectra of dCD with excitation wavelength at 405 nm show two well-resolved blue and red bands. The blue band integral intensity ratio decreased by 1.3-fold, whereas the red band increased by threefold (Figure 8.8d). The ratio of the integrated areas of the red and blue FL components as a function of temperature is linear over 5°C–60°C has 0.048 per K thermal resolution and 1.97% per C thermal sensitivity.

The application of FL of CD in bio-imaging and sensing has a huge positive significance in the field of biomedicine. Mua et al. [70] prepared nitrogen, sulfur, and iodine-doped CD (N,S,I-CDs) by a facile hydrothermal reaction. Synthesized samples have exhibited a strong blue fluorescence

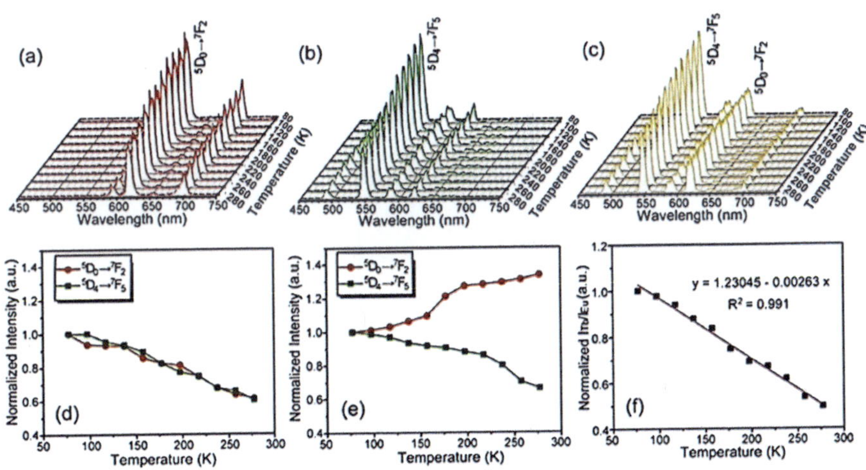

Figure 8.7 Temperature-dependent emission spectra of (a) ND-PMA-Eu, (b) ND-PMA-Tb, and (c) ND-PMA-Eu/Tb recorded between 77 and 277 K excited at 260 nm. Variation of prominent peaks normalized intensity with temperature of (d) the $^5D_0 \rightarrow {}^7F_2$ transition of ND-PMA-Eu and $^5D_4 \rightarrow {}^7F_5$ transition of ND-PMA-Tb; (f) $^5D_0 \rightarrow {}^7F_2$ and $^5D_4 \rightarrow {}^7F_5$ transitions for ND-PMA-Eu/Tb; and (e) the normalized intensity ratio of Tb^{3+} to Eu^{3+} for NDPMA-Eu/Tb. (Reproduced from Ref. [92] with permission from IOP Publishing Ltd.)

with 32.4% fluorescence quantum yield. It also interacted with folic acid (FA) with good selectivity, developed a sensitive method for FA detection from 0.1 to 175 mM wide linear range with a detection limit of 84 nM, and was used for U-2 OS cell imaging. Further, FL intensity variation in the form of I/I_0 ratio of N,S,I-CDS was decreased linearly with respect to the increasing temperature ranging from 10°C to 80°C. Increasing the temperature from 10°C to 80°C resulted in a 68% drop in I/I_0. Figure 8.9 shows the confocal microscopy images of N,S,I-CDs-colon cancer cell HT-29 with corresponding fluorescence field at 15°C, 25°C, 35°C, and 15°C. Hence,

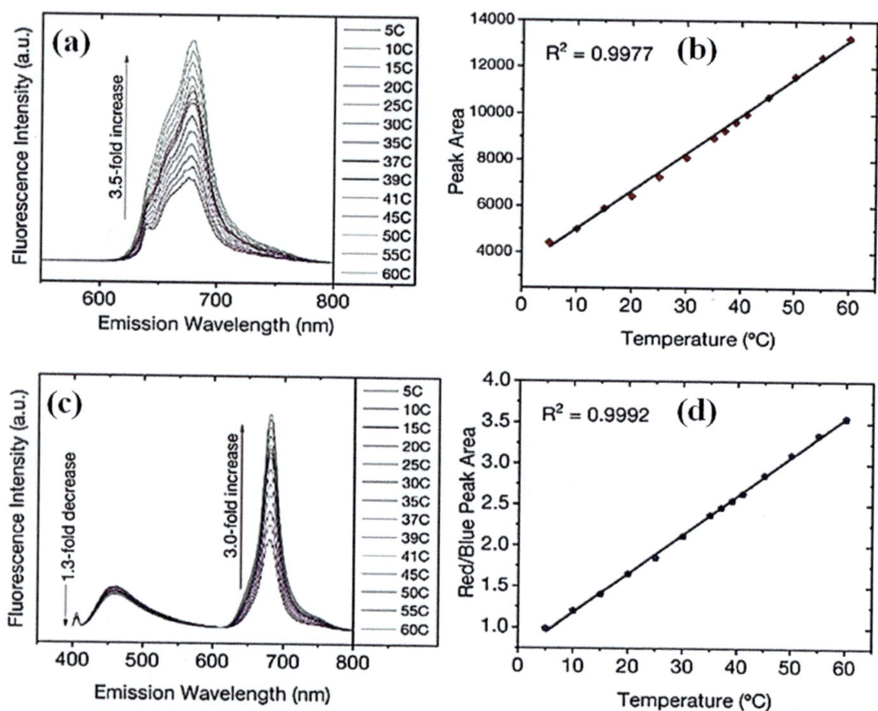

Figure 8.8 (a) Fluorescence spectrum of dCDs with temperature under excitation at 640 nm (b) FL integrated area over the range of 5°C–60°C; (c) fluorescence spectra of the dCDs as a function of temperature under excitation 405 nm (d) the ratio of the integrated areas of the red and blue fluorescence as a function of temperature. (Reproduced from Ref. [67] with permission from the Royal Society of Chemistry.)

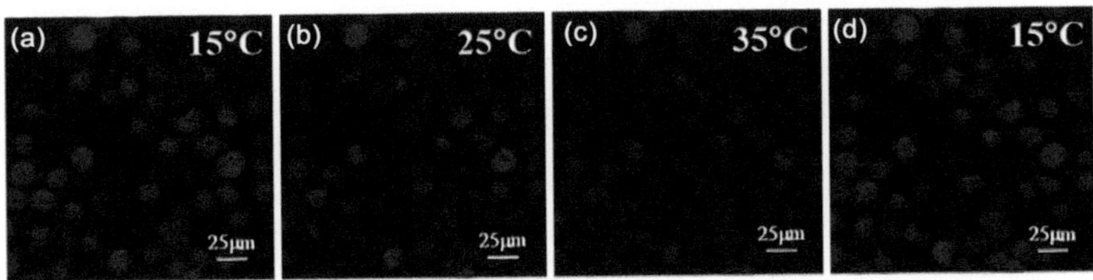

Figure 8.9 Confocal microscopy images of N,S,I-CDs-colon cancer cell HT-29 with corresponding fluorescence field at 15°C, 25°C, 35°C and 15°C, respectively. (Reproduced from Ref. [70] with permission from Elsevier).

as per the results of the N,S,I-CDs sensor, it demonstrated high sensitivity, linearity, and reversibility in the 10°C–80°C temperature range, successfully sensing cell HT-29 samples. The study suggests that the N,S,I-CDs probe could be a promising material for detecting FA in cells and accurately recognizing cell temperature.

He et al. [82] reported the development of a label-free fluorescent probe for hematin and temperature sensing using multifunctional N, Cl co-doped CDs (NCl-CDs), which were synthesized using a one-step hydrothermal process. Figure 8.10a and b illuminate the hydrated particle size of NCl-CDs at different temperatures. Research demonstrates that as the temperature rises, the hydration particle size of NCl-CDs increases, leading to aggregation formation and fluorescence suppression. As the temperature decreases, the hydration particle size of NCl-CDs decreases, suggesting that chilling may cause depolymerization. Figure 8.10c shows that NCl-CDs as a temperature probe exhibit a reversible temperature response and a linear association. A cyclic operation assessed the stability of fluorescence severity in NCl-CDs. Figure 8.10d shows that the fluorescence severity of NCl-CDs remained stable after five cycles of repetition, demonstrating a high stability of the temperature-dependent fluorescence severity. The NCl-CDs exhibit temperature-dependent fluorescence with notable linear feedback between 20°C and 80°C. Additionally, hematin effectively reduces fluorescence intensity in NCl-CDs through the inner filter effect (IFE). The sensing technology showed high selectivity for hematin, with a detection limit of 0.16 μM, making it suitable for human serum samples (Figure 8.10).

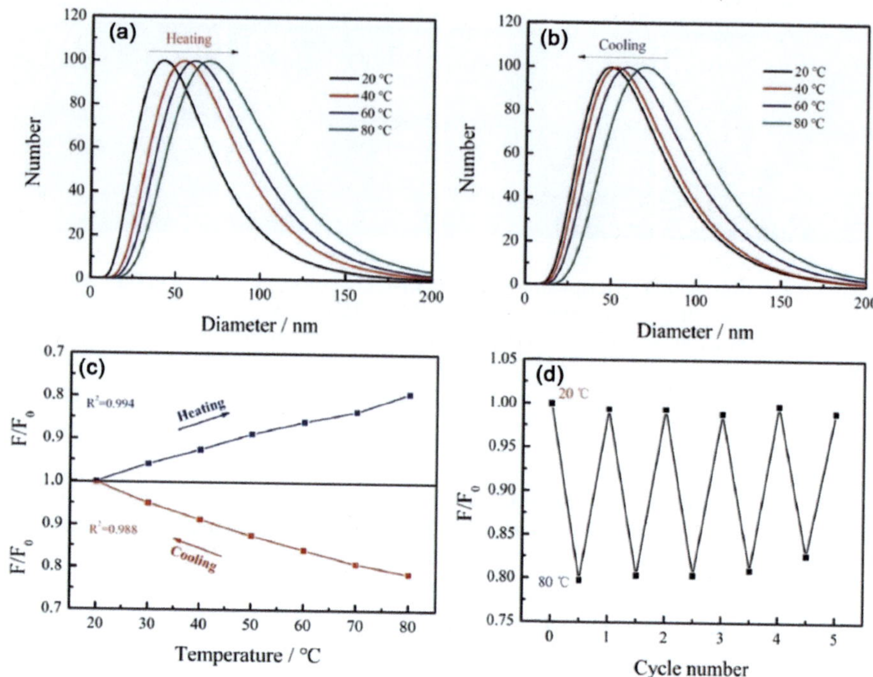

Figure 8.10 (a) Heating and (b) cooling change CDs hydrated particle size. (c) The F/F_0-temperature graphs of NCl-CDs after heating and cooling (F_0 is FL at 20°C). (d) Temperature response cycle experiment over NCl-CDs. (Reproduced from Ref. [82] with permission from Elsevier.)

ACKNOWLEDGMENT

The author has devoted this chapter to the memory of Dr. B.N. Lakshminarasappa, a retired Physics Professor from Bangalore University in Bengaluru, Karnataka, India, who departed on January 02, 2024. He was a very exceptional experimenter who served as an inspiration to many scholars. Prof. BNL was a sincere and hardworking physicist who was always eager to explore new horizons in physics in general and luminescence in particular.

REFERENCES

1. C. Wang, Z. Xu, H. Cheng, H. Lin, M.G. Humphrey, C. Zhang, A hydrothermal route to water-stable luminescent carbon dots as nanosensors for pH and temperature, *Carbon N. Y.* 82 (2015) 87–95. https://doi.org/10.1016/j.carbon.2014.10.035.

2. F. Amon, C. Pearson, Thermal Imaging in Firefighting and Thermography Applications, in: 2010: pp. 279–331. https://doi.org/10.1016/S1079-4042(09)04305-7.

3. J. Zhang, D. Nan, S. Pan, H. Liu, H. Yang, X. Hu, N,S co-doped carbon dots as a dual-functional fluorescent sensor for sensitive detection of baicalein and temperature, spectrochim. *Acta Part A Mol. Biomol. Spectrosc.* 221 (2019) 117161. https://doi.org/10.1016/j.saa.2019.117161.

4. A. Ostadfar, Real Time Measurement Techniques of Biofluids, in: *Biofluid Mech.*, Elsevier, 2016: pp. 295–322. https://doi.org/10.1016/B978-0-12-802408-9.00008-9.

5. A.Z. Paydar, S.K.M. Balgehshiri, B. Zohuri, Heated Junction Thermocouple System, in: *Adv. React. Concepts*, Elsevier, 2023: pp. 341–380. https://doi.org/10.1016/B978-0-443-18989-0.00007-7.

6. K. Siren, G. Rosén, J. Vad, P.V. Nielsen, Experimental Techniques, in: *Ind. Vent. Des. Guideb.*, Elsevier, 2001: pp. 1105–1195. https://doi.org/10.1016/B978-012289676-7/50015-1.

7. Z.P. Tshabalala, D.N. Oosthuizen, H.C. Swart, D.E. Motaung, Tools and Techniques for Characterization and Evaluation of Nanosensors, in: *Nanosensors for Smart Cities*, Elsevier, 2020: pp. 85–110. https://doi.org/10.1016/B978-0-12-819870-4.00005-0.

8. L.J. Mohammed, K.M. Omer, Carbon dots as new generation materials for nanothermometer: Review, *Nanoscale Res. Lett.* 15 (2020) 182. https://doi.org/10.1186/s11671-020-03413-x.

9. S. Wang, S. Westcott, W. Chen, Nanoparticle luminescence thermometry, *J. Phys. Chem. B.* 106 (2002) 11203–11209. https://doi.org/10.1021/jp026445m.

10. S. Uchiyama, A. Prasanna de Silva, K. Iwai, Luminescent molecular thermometers, *J. Chem. Educ.* 83 (2006) 720. https://doi.org/10.1021/ed083p720.

11. E.J. McLaurin, L.R. Bradshaw, D.R. Gamelin, Dual-emitting nanoscale temperature sensors, *Chem. Mater.* 25 (2013) 1283–1292. https://doi.org/10.1021/cm304034s.

12. Ž. Antić, K. Prashanthi, S. Kuzman, J. Periša, Z. Ristić, V.R. Palkar, M.D. Dramićanin, Ratiometric temperature measurement using negative thermal quenching of intrinsic BiFeO3 semiconductor nanoparticles, *RSC Adv.* 10 (2020) 16982–16986. https://doi.org/10.1039/D0RA01896A.

13. J. Qiao, Y.-H. Hwang, C.-F. Chen, L. Qi, P. Dong, X.-Y. Mu, D.-P. Kim, Ratiometric fluorescent polymeric thermometer for thermogenesis investigation in living cells, *Anal. Chem.* 87 (2015) 10535–10541. https://doi.org/10.1021/acs.analchem.5b02791.

14. S. Claveau, J.-R. Bertrand, F. Treussart, Fluorescent nanodiamond applications for cellular process sensing and cell tracking, *Micromachines* 9 (2018) 247. https://doi.org/10.3390/mi9050247.

15. L. Đačanin Far, M. Dramićanin, Luminescence thermometry with nanoparticles: A review, *Nanomaterials* 13 (2023) 2904. https://doi.org/10.3390/nano13212904.

16. M.D. Dramićanin, Trends in luminescence thermometry, *J. Appl. Phys.* 128 (2020). https://doi.org/10.1063/5.0014825.

17. J. Lee, N.A. Kotov, Thermometer design at the nanoscale, *Nano Today* 2 (2007) 48–51. https://doi.org/10.1016/S1748-0132(07)70019-1.

18. C.D.S. Brites, P.P. Lima, N.J.O. Silva, A. Millán, V.S. Amaral, F. Palacio, L.D. Carlos, Thermometry at the nanoscale, *Nanoscale* 4 (2012) 4799. https://doi.org/10.1039/c2nr30663h.

19. P. Kumbhakar, A. Roy Karmakar, G.P. Das, J. Chakraborty, C.S. Tiwary, P. Kumbhakar, Reversible temperature-dependent photoluminescence in semiconductor quantum dots for the development of a smartphone-based optical thermometer, *Nanoscale* 13 (2021) 2946–2954. https://doi.org/10.1039/D0NR07874C.

20. J.-M. Yang, H. Yang, L. Lin, Quantum dot nano thermometers reveal heterogeneous local thermogenesis in living cells, *ACS Nano* 5 (2011) 5067–5071. https://doi.org/10.1021/nn201142f.

21. K. Okabe, R. Sakaguchi, B. Shi, S. Kiyonaka, Intracellular thermometry with fluorescent sensors for thermal biology, *Pflügers Arch. Eur. J. Physiol.* 470 (2018) 717–731. https://doi.org/10.1007/s00424-018-2113-4.

22. D. Jaque, B. del Rosal, E.M. Rodríguez, L.M. Maestro, P. Haro-González, J.G. Solé, Fluorescent nanothermometers for intracellular thermal sensing, *Nanomedicine* 9 (2014) 1047–1062. https://doi.org/10.2217/nnm.14.59.

23. A. Rakovich, T. Rakovich, Semiconductor versus graphene quantum dots as fluorescent probes for cancer diagnosis and therapy applications, *J. Mater. Chem. B.* 6 (2018) 2690–2712. https://doi.org/10.1039/C8TB00153G.

24. H.-G. Ye, X. Lu, R. Cheng, J. Guo, H. Li, C.-F. Wang, S. Chen, Mild bottom-up synthesis of carbon dots with temperature-dependent fluorescence, *J. Lumin.* 238 (2021) 118311. https://doi.org/10.1016/j.jlumin.2021.118311.

25. V.D.N. Bezzon, T.L.A. Montanheiro, B.R.C. de Menezes, R.G. Ribas, V.A.N. Righetti, K.F. Rodrigues, G.P. Thim, Carbon nanostructure-based sensors: A brief review on recent advances, *Adv. Mater. Sci. Eng.* 2019 (2019) 1–21. https://doi.org/10.1155/2019/4293073.

26. C. Li, Y. Wang, H. Jiang, X. Wang, Review—intracellular sensors based on carbonaceous nanomaterials: A review, *J. Electrochem. Soc.* 167 (2020) 037540. https://doi.org/10.1149/1945-7111/ab67a3.

27. T. Plakhotnik, D. Gruber, Luminescence of nitrogen-vacancy centers in nanodiamonds at temperatures between 300 and 700 K: Perspectives on nanothermometry, *Phys. Chem. Chem. Phys.* 12 (2010) 9751. https://doi.org/10.1039/c001132k.

28. T. Plakhotnik, H. Aman, NV-centers in nanodiamonds: How good they are, *Diam. Relat. Mater.* 82 (2018) 87–95. https://doi.org/10.1016/j.diamond.2017.12.004.

29. Q. Sun, S. Li, T. Plakhotnik, A.D. Greentree, Fluorescence profile of a nitrogen-vacancy center in a nanodiamond, *Phys. Rev. A.* 107 (2023) 013720. https://doi.org/10.1103/PhysRevA.107.013720.

30. R. Piñol, C.D.S. Brites, N.J. Silva, L.D. Carlos, A. Millán, Nanoscale Thermometry for Hyperthermia Applications, in: *Nanomater. Magn. Opt. Hyperth. Appl.*, Elsevier, 2019: pp. 139–172. https://doi.org/10.1016/B978-0-12-813928-8.00006-5.

31. G. Kucsko, P.C. Maurer, N.Y. Yao, M. Kubo, H.J. Noh, P.K. Lo, H. Park, M.D. Lukin, Nanometre-scale thermometry in a living cell, *Nature* 500 (2013) 54–58. https://doi.org/10.1038/nature12373.

32. L.-X. Su, Q. Lou, J.-H. Zang, C.-X. Shan, Y.-F. Gao, Temperature-dependent fluorescence in nanodiamonds, *Appl. Phys. Express.* 10 (2017) 025102. https://doi.org/10.7567/APEX.10.025102.

33. M. Liu, Optical properties of carbon dots: A review, *Nanoarchitectonics* 1 (2020) 1–12. https://doi.org/10.37256/nat.112020124.1-12.

34. W. Ghann, V. Sharma, H. Kang, F. Karim, B. Richards, S.M. Mobin, J. Uddin, M.M. Rahman, F. Hossain, H. Kabir, N. Uddin, The synthesis and characterization of carbon dots and their application in dye sensitized solar cell, *Int. J. Hydrogen Energy.* 44 (2019) 14580–14587. https://doi.org/10.1016/j.ijhydene.2019.04.072.

35. X. Wang, Y. Feng, P. Dong, J. Huang, A mini review on carbon quantum dots: Preparation, properties, and electrocatalytic application, *Front. Chem.* 7 (2019). https://doi.org/10.3389/fchem.2019.00671.

36. L. Wang, Y. Yin, A. Jain, H.S. Zhou, Aqueous phase synthesis of highly luminescent, nitrogen-doped carbon dots and their application as bioimaging agents, *Langmuir* 30 (2014) 14270–14275. https://doi.org/10.1021/la5031813.

37. Z. Liu, M. Chen, Y. Guo, J. Zhou, Q. Shi, R. Sun, Oxidized nanocellulose facilitates preparing photoluminescent nitrogen-doped fluorescent carbon dots for Fe^{3+} ions detection and bioimaging, *Chem. Eng. J.* 384 (2020) 123260. https://doi.org/10.1016/j.cej.2019.123260.

38. Q. Duan, L. Ma, B. Zhang, Y. Zhang, X. Li, T. Wang, W. Zhang, Y. Li, S. Sang, Construction and application of targeted drug delivery system based on hyaluronic acid and heparin functionalised carbon dots, *Colloids Surf. B Biointerfaces* 188 (2020) 110768. https://doi.org/10.1016/j.colsurfb.2019.110768.

39. K.M. Omer, A.Q. Hassan, Chelation-enhanced fluorescence of phosphorus doped carbon nanodots for multi-ion detection, *Microchim. Acta* 184 (2017) 2063–2071. https://doi.org/10.1007/s00604-017-2196-1.

40. Y. Coşkun, F.Y. Ünlü, T. Yılmaz, Y. Türker, A. Aydogan, M. Kuş, C. Ünlü, Development of highly luminescent water-insoluble carbon dots by using calix[4]pyrrole as the carbon precursor and their potential application in organic solar cells, *ACS Omega* 7 (2022) 18840–18851. https://doi.org/10.1021/acsomega.2c01795.

41. Y. Song, S. Zhu, B. Yang, Bioimaging based on fluorescent carbon dots, *RSC Adv.* 4 (2014) 27184. https://doi.org/10.1039/c3ra47994c.

42. H. Zhang, G. Wang, Z. Zhang, J.H. Lei, T.-M. Liu, G. Xing, C.-X. Deng, Z. Tang, S. Qu, One step synthesis of efficient red emissive carbon dots and their bovine serum albumin composites with enhanced multi-photon fluorescence for in vivo bioimaging, *Light Sci. Appl.* 11 (2022) 113. https://doi.org/10.1038/s41377-022-00798-5.

43. S. Dugam, S. Nangare, P. Patil, N. Jadhav, Carbon dots: A novel trend in pharmaceutical applications, *Ann. Pharm. Françaises* 79 (2021) 335–345. https://doi.org/10.1016/j.pharma.2020.12.002.

44. A. Paul, M. Kurian, Catalytic Applications of Carbon Dots, in: *Carbon Dots Anal. Chem.*, Elsevier, 2023: pp. 337–344. https://doi.org/10.1016/B978-0-323-98350-1.00010-4.

45. C. Rosso, G. Filippini, M. Prato, Carbon dots as nano-organocatalysts for synthetic applications, *ACS Catal.* 10 (2020) 8090–8105. https://doi.org/10.1021/acscatal.0c01989.

46. B.B. Chen, M.L. Liu, C.Z. Huang, Carbon dot-based composites for catalytic applications, *Green Chem.* 22 (2020) 4034–4054. https://doi.org/10.1039/D0GC01014F.

47. H. Yi, J. Liu, J. Yao, R. Wang, W. Shi, C. Lu, Photoluminescence mechanism of carbon dots: Triggering multiple color emissions through controlling the degree of protonation, *Molecules* 27 (2022) 6517. https://doi.org/10.3390/molecules27196517.

48. H. Bao, Y. Liu, H. Li, W. Qi, K. Sun, Luminescence of carbon quantum dots and their application in biochemistry, *Heliyon* 9 (2023) e20317. https://doi.org/10.1016/j.heliyon.2023.e20317.

49. R. Yadav, Vikas, V. Lahariya, M. Tanwar, R. Kumar, A. Das, K. Sadhana, A study on the photophysical properties of strong green-fluorescent N-doped carbon dots and application for pH sensing, *Diam. Relat. Mater.* 139 (2023) 110411. https://doi.org/10.1016/j.diamond.2023.110411.

50. C. Liu, F. Zhang, J. Hu, W. Gao, M. Zhang, A mini review on pH-sensitive photoluminescence in carbon nanodots, *Front. Chem.* 8 (2021). https://doi.org/10.3389/fchem.2020.605028.

51. Y. Chen, G. Xiong, L. Zhu, J. Huang, X. Chen, Y. Chen, M. Cao, Enhanced fluorescence and environmental stability of red-emissive carbon dots via chemical bonding with cellulose films, *ACS Omega* 7 (2022) 6834–6842. https://doi.org/10.1021/acsomega.1c06426.

52. F. Jelezko, J. Wrachtrup, Single defect centres in diamond: A review, *Phys. Status Solidi.* 203 (2006) 3207–3225. https://doi.org/10.1002/pssa.200671403.

53. E.R. Wilson, L.M. Parker, A. Orth, N. Nunn, M. Torelli, O. Shenderova, B.C. Gibson, P. Reineck, The effect of particle size on nanodiamond fluorescence and colloidal properties in biological media, *Nanotechnology* 30 (2019) 385704. https://doi.org/10.1088/1361-6528/ab283d.

54. D.Y. Fedyanin, M. Agio, Ultrabright single-photon source on diamond with electrical pumping at room and high temperatures, *New J. Phys.* 18 (2016) 073012. https://doi.org/10.1088/1367-2630/18/7/073012.

55. Z. Su, Z. Ren, Y. Bao, X. Lao, J. Zhang, J. Zhang, D. Zhu, Y. Lu, Y. Hao, S. Xu, Luminescence landscapes of nitrogen-vacancy centers in diamond: Quasi-localized vibrational resonances and selective coupling, *J. Mater. Chem. C* 7 (2019) 8086–8091. https://doi.org/10.1039/C9TC01954E.

56. A. Vervald, S. Burikov, N. Borisova, I. Vlasov, K. Laptinskiy, T. Laptinskaya, O. Shenderova, T. Dolenko, Fluorescence properties of nanodiamonds with NV centers in water suspensions, *Phys. Status Solidi.* 213 (2016) 2601–2607. https://doi.org/10.1002/pssa.201600215.

57. N. Aslam, G. Waldherr, P. Neumann, F. Jelezko, J. Wrachtrup, Photo-induced ionization dynamics of the nitrogen vacancy defect in diamond investigated by single-shot charge state detection, *New J. Phys.* 15 (2013) 013064. https://doi.org/10.1088/1367-2630/15/1/013064.

58. K. Beha, A. Batalov, N.B. Manson, R. Bratschitsch, A. Leitenstorfer, Optimum photoluminescence excitation and recharging cycle of single nitrogen-vacancy centers in ultrapure diamond, *Phys. Rev. Lett.* 109 (2012) 097404. https://doi.org/10.1103/PhysRevLett.109.097404.

59. G. Waldherr, J. Beck, M. Steiner, P. Neumann, A. Gali, T. Frauenheim, F. Jelezko, J. Wrachtrup, Dark states of single nitrogen-vacancy centers in diamond unraveled by single shot NMR, *Phys. Rev. Lett.* 106 (2011) 157601. https://doi.org/10.1103/PhysRevLett.106.157601.

60. G. Petrini, E. Moreva, E. Bernardi, P. Traina, G. Tomagra, V. Carabelli, I. Pietro Degiovanni, M. Genovese, Is a quantum biosensing revolution approaching? Perspectives in NV-assisted current and thermal biosensing in living cells, *Adv. Quantum Technol.* 3 (2020). https://doi.org/10.1002/qute.202000066.

61. S.-J. Yu, M.-W. Kang, H.-C. Chang, K.-M. Chen, Y.-C. Yu, Bright fluorescent nanodiamonds: No photobleaching and low cytotoxicity, *J. Am. Chem. Soc.* 127 (2005) 17604–17605. https://doi.org/10.1021/ja0567081.

62. Y.Y. Hui, C.-L. Cheng, H.-C. Chang, Nanodiamonds for optical bioimaging, *J. Phys. D Appl. Phys.* 43 (2010) 374021. https://doi.org/10.1088/0022-3727/43/37/374021.

63. F. Pedroza-Montero, K. Santacruz-Gómez, M. Acosta-Elías, E. Silva-Campa, D. Meza-Figueroa, D. Soto-Puebla, B. Castaneda, E. Urrutia-Bañuelos, O. Álvarez-Bajo, S. Navarro-Espinoza, R. Riera, M. Pedroza-Montero, Thermometric characterization of fluorescent nanodiamonds suitable for biomedical applications, *Appl. Sci.* 11 (2021) 4065. https://doi.org/10.3390/app11094065.

64. L. Wang, S.-J. Zhu, H.-Y. Wang, S.-N. Qu, Y.-L. Zhang, J.-H. Zhang, Q.-D. Chen, H.-L. Xu, W. Han, B. Yang, H.-B. Sun, Common origin of green luminescence in carbon nanodots and graphene quantum dots, *ACS Nano.* 8 (2014) 2541–2547. https://doi.org/10.1021/nn500368m.

65. P. Yu, X. Wen, Y.-R. Toh, J. Tang, Temperature-dependent fluorescence in carbon dots, *J. Phys. Chem. C* 116 (2012) 25552–25557. https://doi.org/10.1021/jp307308z.

66. F. Yuan, L. Ding, Y. Li, X. Li, L. Fan, S. Zhou, D. Fang, S. Yang, Multicolor fluorescent graphene quantum dots colorimetrically responsive to all-pH and a wide temperature range, *Nanoscale* 7 (2015) 11727–11733. https://doi.org/10.1039/C5NR02007G.

67. J.-R. Macairan, D.B. Jaunky, A. Piekny, R. Naccache, Intracellular ratiometric temperature sensing using fluorescent carbon dots, *Nanoscale Adv.* 1 (2019) 105–113. https://doi.org/10.1039/C8NA00255J.

68. Z. Song, F. Quan, Y. Xu, M. Liu, L. Cui, J. Liu, Multifunctional N,S co-doped carbon quantum dots with pH- and thermo-dependent switchable fluorescent properties and highly selective detection of glutathione, *Carbon N. Y.* 104 (2016) 169–178. https://doi.org/10.1016/j.carbon.2016.04.003.

69. Y. Hu, J. Yang, L. Jia, J.-S. Yu, Ethanol in aqueous hydrogen peroxide solution: Hydrothermal synthesis of highly photoluminescent carbon dots as multifunctional nanosensors, *Carbon N. Y.* 93 (2015) 999–1007. https://doi.org/10.1016/j.carbon.2015.06.018.

70. Z. Mua, J. Huaa, Y. Yanga, N,S,I co-doped carbon dots for folic acid and temperature sensing and applied to cellular imaging, *Spectrochim. Acta Part A Mol. Biomol. Spectrosc.* 224 (2020) 117444. https://doi.org/10.1016/j.saa.2019.117444.

71. C. Li, Z. Qin, M. Wang, W. Liu, H. Jiang, X. Wang, Manganese oxide doped carbon dots for temperature-responsive biosensing and target bioimaging, *Anal. Chim. Acta* 1104 (2020) 125–131. https://doi.org/10.1016/j.aca.2020.01.001.

72. O.A. Williams, Ultrananocrystalline diamond for electronic applications, *Semicond. Sci. Technol.* 21 (2006) R49–R56. https://doi.org/10.1088/0268-1242/21/8/R01.

73. J.E. Dahl, S.G. Liu, R.M.K. Carlson, Isolation and structure of higher diamondoids, nanometer-sized diamond molecules, *Science* 299 (2003) 96–99. https://doi.org/10.1126/science.1078239.

74. D. Zupančič, P. Veranič, Nanodiamonds as possible tools for improved management of bladder cancer and bacterial cystitis, *Int. J. Mol. Sci.* 23 (2022) 8183. https://doi.org/10.3390/ijms23158183.

75. A. Thomas, M.S. Parvathy, K.B. Jinesh, Synthesis of nanodiamonds using liquid-phase laser ablation of graphene and its application in resistive random access memory, *Carbon Trends.* 3 (2021) 100023. https://doi.org/10.1016/j.cartre.2020.100023.

76. D. Amans, A.-C. Chenus, G. Ledoux, C. Dujardin, C. Reynaud, O. Sublemontier, K. Masenelli-Varlot, O. Guillois, Nanodiamond synthesis by pulsed laser ablation in liquids, *Diam. Relat. Mater.* 18 (2009) 177–180. https://doi.org/10.1016/j.diamond.2008.10.035.

77. M. Cazzanelli, L. Basso, C. Cestari, N. Bazzanella, E. Moser, M. Orlandi, A. Piccoli, A. Miotello, Fluorescent nanodiamonds synthesized in one-step by pulsed laser ablation of graphite in liquid-nitrogen, C. 7 (2021) 49. https://doi.org/10.3390/c7020049.

78. V. Danilenko, O.A. Shenderova, Advances in Synthesis of Nanodiamond Particles, in: *Ultananocrystalline Diam.*, Elsevier, 2012: pp. 133–164. https://doi.org/10.1016/B978-1-4377-3465-2.00005-0.

79. C.R. Lin, D.H. Wei, M.K. Ben Dao, R.J. Chung, M.H. Chang, Nanocrystalline diamond particles prepared by high-energy ball milling method, *Appl. Mech. Mater.* 284–287 (2013) 168–172. https://doi.org/10.4028/www.scientific.net/AMM.284-287.168.

80. J. Zalieckas, P. Pobedinskas, M.M. Greve, K. Eikehaug, K. Haenen, B. Holst, Large area microwave plasma CVD of diamond using composite right/left-handed materials, *Diam. Relat. Mater.* 116 (2021) 108394. https://doi.org/10.1016/j.diamond.2021.108394.

81. R. Salerno, B. Pede, M. Mastellone, V. Serpente, V. Valentini, A. Bellucci, D.M. Trucchi, F. Domenici, M. Tomellini, R. Polini, Etching kinetics of nanodiamond seeds in the early stages of CVD diamond growth, *ACS Omega* 8 (2023) 25496–25505. https://doi.org/10.1021/acsomega.3c03080.

82. W. He, Z. Huo, X. Sun, J. Shen, Facile and green synthesis of N, Cl-dual-doped carbon dots as a label-free fluorescent probe for hematin and temperature sensing, *Microchem. J.* 153 (2020) 104528. https://doi.org/10.1016/j.microc.2019.104528.

83. Z. Zhu, Z. Sun, Z. Guo, X. Zhang, Z.C. Wu, A high-sensitive ratiometric luminescent thermometer based on dual-emission of carbon dots/Rhodamine B nanocomposite, *J. Colloid Interface Sci.* 552 (2019) 572–582. https://doi.org/10.1016/j.jcis.2019.05.088.

84. J. Liu, X. Kang, H. Zhang, Y. Liu, C. Wang, X. Gao, Y. Li, Carbon dot-based nanocomposite: Long-lived thermally activated delayed fluorescence for lifetime thermal sensing, *Dye. Pigment.* 181 (2020) 108576. https://doi.org/10.1016/j.dyepig.2020.108576.

85. V. Nguyen, L. Yan, J. Si, X. Hou, Femtosecond laser-induced size reduction of carbon nanodots in solution: Effect of laser fluence, spot size, and irradiation time, *J. Appl. Phys.* 117 (2015). https://doi.org/10.1063/1.4909506.

86. A.S. Grewal, K. Kumar, S. Redhu, S. Bhardwaj, J. Nayak, L. Memorial, Microwave Assisted Synthesis: A Green Chemistry Approach, in: 2013. https://api.semanticscholar.org/CorpusID:59941493.

87. S.S. Reddy, N. Chauhan, K.R. Nagabhushana, F. Singh, TL/OSL properties of beta irradiated Al_2O_3:Tm^{3+} phosphor synthesized by microwave combustion method, *Mater. Res. Bull.* 104 (2018) 236–243. https://doi.org/10.1016/j.materresbull.2018.04.019.

88. S. Choe, J. Yoon, M. Lee, J. Oh, D. Lee, H. Kang, C.-H. Lee, D. Lee, Precise temperature sensing with nanoscale thermal sensors based on diamond NV centers, *Curr. Appl. Phys.* 18 (2018) 1066–1070. https://doi.org/10.1016/j.cap.2018.06.002.

89. J. Wang, F. Feng, J. Zhang, J. Chen, Z. Zheng, L. Guo, W. Zhang, X. Song, G. Guo, L. Fan, C. Zou, L. Lou, W. Zhu, G. Wang, High-sensitivity temperature sensing using an implanted single nitrogen-vacancy center array in diamond, *Phys. Rev. B* 91 (2015) 155404. https://doi.org/10.1103/PhysRevB.91.155404.

90. A. Al Salman, A. Tortschanoff, M.B. Mohamed, D. Tonti, F. van Mourik, M. Chergui, Temperature effects on the spectral properties of colloidal CdSe nanodots, nanorods, and tetrapods, *Appl. Phys. Lett.* 90 (2007). https://doi.org/10.1063/1.2696687.

91. P. Yu, X. Wen, Y.-R. Toh, J. Tang, Temperature-dependent fluorescence in Au 10 nanoclusters, *J. Phys. Chem. C* 116 (2012) 6567–6571. https://doi.org/10.1021/jp2120077.

92. X.-Y. Wu, Q. Zhao, D.-X. Zhang, Y.-C. Liang, K.-K. Zhang, Q. Liu, L. Dong, C.-X. Shan, A self-calibrated luminescent thermometer based on nanodiamond-Eu/Tb hybrid materials, *Dalt. Trans.* 48 (2019) 7910–7917. https://doi.org/10.1039/C9DT00850K.

93. H. Zhang, J. You, J. Wang, X. Dong, R. Guan, D. Cao, Highly luminescent carbon dots as temperature sensors and "off-on" sensing of Hg^{2+} and biothiols, *Dye. Pigment.* 173 (2020) 107950. https://doi.org/10.1016/j.dyepig.2019.107950.

94. V. Nguyen, L. Yan, H. Xu, M. Yue, One-step synthesis of multi-emission carbon nanodots for ratiometric temperature sensing, *Appl. Surf. Sci.* 427 (2018) 1118–1123. https://doi.org/10.1016/j.apsusc.2017.08.133.

95. Y. Yang, W. Kong, H. Li, J. Liu, M. Yang, H. Huang, Y. Liu, Z. Wang, Z. Wang, T.-K. Sham, J. Zhong, C. Wang, Z. Liu, S.-T. Lee, Z. Kang, Fluorescent N-doped carbon dots as in vitro and in vivo nanothermometer, *ACS Appl. Mater. Interfaces* 7 (2015) 27324–27330. https://doi.org/10.1021/acsami.5b08782.

96. T.S. Atabaev, S. Sayatova, A. Molkenova, I. Taniguchi, Nitrogen-doped carbon nanoparticles for potential temperature sensing applications, *Sens. Bio-Sensing Res.* 22 (2019) 100253. https://doi.org/10.1016/j.sbsr.2018.100253.

97. W. Shi, F. Guo, M. Han, S. Yuan, W. Guan, H. Li, H. Huang, Y. Liu, Z. Kang, N,S co-doped carbon dots as a stable bio-imaging probe for detection of intracellular temperature and tetracycline, *J. Mater. Chem. B* 5 (2017) 3293–3299. https://doi.org/10.1039/C7TB00810D.

98. X. Cui, Y. Wang, J. Liu, Q. Yang, B. Zhang, Y. Gao, Y. Wang, G. Lu, Dual functional N- and S-co-doped carbon dots as the sensor for temperature and Fe^{3+} ions, *Sens. Actuat. B Chem.* 242 (2017) 1272–1280. https://doi.org/10.1016/j.snb.2016.09.032.

99. Z. Zhang, Y. Liu, Z. Yan, J. Chen, Simultaneous determination of temperature and erlotinib by novel carbon-based sensitive nanoparticles, *Sens. Actuat. B Chem.* 255 (2018) 986–994. https://doi.org/10.1016/j.snb.2017.08.153.

Chapter 9

Nanocomposites for luminescence thermometers

Vaishali Raikwar and Nikhilesh Bajaj

9.1 INTRODUCTION

Nanotechnology has opened many new areas and horizons due to technological advances. The tremendous growth in nanoscale research in almost every field of life is proof of this fact. The miniaturization and effectiveness in the improvement of the properties is the ultimate effect of nanoscience and nanotechnology. There is a great demand for thermometry in micro-optics, microelectronics, nanomedicines, photonics, and microfluidics. The spatial resolution is an important parameter of a given system to resolve temperature changes occurring in two points separated by a certain distance and is defined as the minimum distance between points presenting a temperature difference higher than δT [1]

$$\delta x = \frac{\delta T}{|\overline{\nabla T}|\,\mathrm{max}}$$

where $|\overline{\nabla T}|\,\mathrm{max}$ denotes the maximum temperature gradient of the mapping. The temperature gradient for a one-dimensional temperature profile can be stated as

$$|\overline{\nabla T}|\,\mathrm{max} = \left|\frac{dT}{dx}\right|\mathrm{max}$$

Like the spatial resolution, the temporal resolution of the measurement (δt) can be defined as the minimum time interval between measurements presenting a temperature difference higher than δT. Both temporal and spatial resolutions are important to evaluate the applicability of a thermometer for dynamic temperature measurements [2].

In various situations like intracellular temperature fluctuations or temperature mapping of nanoscale circuits and microfluidics, conventional thermometers are not useful. The surrounding parameters can drastically affect the spatial resolution, thereby affecting the performance of the

DOI: 10.1201/9781032661537-9

sensing device. Hence, it is essentially required that at nano scale level the accurate measurement of temperature in numerous electronic and optical devices be done with precision.

9.2 OPTICAL THERMOMETERS – MICRO AND NANOSCALE

Temperature, which is a fundamental physical quantity, has been generally used in day-to-day life activities, industries, research laboratories, and various scientific research. Many types of thermometers with various working principles have been designed and used by various stakeholders of society. Among these, the use of mercury-based thermometers, thermocouples, and thermistors that sense temperature are the common ones. But in many extreme and harsh environmental conditions, fast-moving objects, or very small-sized objects, these bulky, traditional thermal sensors cannot be used. Here came the need of developing highly sensitive micro and nanoscale optical thermometers. The temperature distributions inside a living cell greatly affect the thermodynamics and functions of cellular components. Extraordinary heat is produced in many cellular pathogeneses of diseases; hence, it is very necessary to use intracellular temperature mapping of living cells so that better diagnosis and therapies can save many precious lives [3–6].

Typically, the thermometers are categorized as primary and secondary thermometers. Primary thermometers are those that give the temperatures directly without calibration, and they are often inconvenient from a practical standpoint due to low reproducibility, slow response time, and lack of ease of operation [7–9]. For secondary thermometers, the information related to the measured quantity is not sufficient to derive directly the temperature, but it requires calibration against a reference system. Temperature change is proportional to the measured parameter, and it must be calibrated using known reference values.

9.2.1 Micro thermometers

Micro thermometers are miniature temperature sensors that are designed to measure and monitor temperature in small-scale applications. They are designed to be miniature in size, often on the scale of micrometers or millimeters, allowing them to be integrated into tiny devices or systems. They are typically built using microelectromechanical systems (MEMS) technology, which allows for the fabrication of tiny sensors with high precision and accuracy. These thermometers can be used in various industries and fields, such as biomedical, aerospace, automotive, and electronics, where precise temperature measurements are required. Micro thermometers are highly compact temperature sensors that utilize advanced technologies to measure and monitor temperature in small-scale environments. In this current era of nanotechnological advancements, areas such as micro-optics, microfluidics, microelectronics, and photonics have reached such a height that it is impossible to use conventional thermometry to get measurements with high resolution. It is indeed a matter of concern if the temperature mapping of microcircuits or micro fluids is not done with great accuracy [10–15].

9.2.2 Nanothermometers

When the size of systems in which the temperature must be measured reduces to the nanoscale, the accuracy and resolution in the measurement of temperature becomes important. At such a small size, the minimal length scales for the existence of temperature have been well described by Hartmann et al. [16]. Numerous limitations of contact thermometers to work at submicron scale led to the development of non-contact thermometry techniques such as optical thermometry, luminescence, IR thermography, thermoreflectance, etc.

9.3 FUNDAMENTAL MECHANISMS AND TYPES

There are four dominant mechanisms for optical thermometry. They are the luminescence intensity ratio (LIR) of thermally coupled levels (TCLs), temperature sensing via luminescence lifetime, temperature sensing via spectral shift, and temperature detection using low-lying thermally populated levels. Figure 9.1 shows these mechanisms.

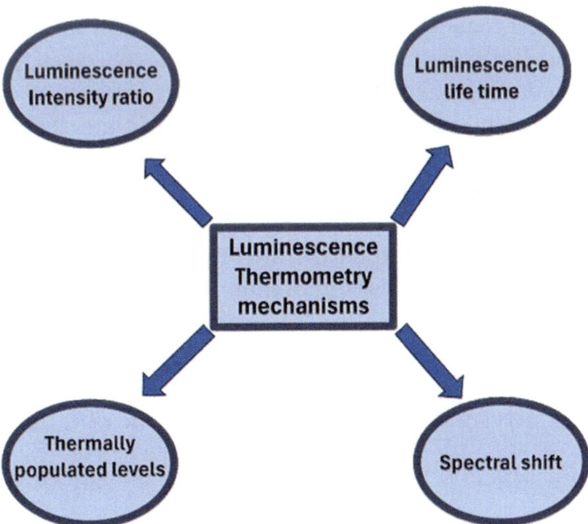

Figure 9.1 Fundamental mechanisms of optical thermometry.

9.3.1 Luminescence intensity ratio

This is the most common technique for sensing the temperature. In this technique, the temperature measurement is based on the changing of the emission intensity ratio of two emitting levels with temperature change. If these excited emitting levels are thermally coupled, then the population redistribution will follow Bolzman Law. According to the reports, the relative population of the TCL is shown as a Boltzman-type population distribution [17]. The parameter fluorescence emission intensity ratio (FIR) which is the result of TCL of active ions can be expressed as $\mathrm{FIR} = \dfrac{I_U}{I_L} = a^* e^{\frac{\Delta E}{kT}} + b$ where a is a constant, b is a modifying factor, T is the absolute temperature, k is the Boltzmann constant, and ΔE is the energy gap between the two excited states. The intensity ratios of the near infrared-red FIR1 (I750/I650) and green-red FIR2 (I545/I650) were calculated and plotted as a function of temperature. From the graph, it is observed that the fittings match well with the experimental data. The fitting coefficient, a, b and ΔE, matched with those calculated from the emission spectra. Thus, it can be inferred that the temperature can be accurately measured from FIR1 and FIR2 at the temperature range from 300 to 500 K [18].

Also, the variation of sensitivity with temperature is an important parameter for optical temperature sensor. The absolute sensitivity S_a can be defined as:

$$S_a = \frac{1}{\mathrm{FIR}} \frac{d\mathrm{FIR}}{dT} = \frac{\Delta E}{kT^2}$$

It was observed that with the increase of temperature, the absolute sensitivity S_a keeps decreasing and reaches to maximum value about 1.53% K^{-1} for FIR1 and 0.9% K^{-1} for FIR2 at 300 K, respectively. Hence, it was demonstrated that the near infrared-red absolute sensitivity is more sensitive for use as a thermometer at room temperature than the green-red [19]. The optical sensors are thus based on the fact that any change in the temperature inside the micro-cavity leads to a shift in the wavelength of the resonances [20].

Apart from sensitivities, the temperature determination or temperature ambiguity (T or δT) is another important parameter that can be defined as the lowest measurable variation of heating value in a given temperature measurement. It is defined as $\delta T = \dfrac{1}{S_r} \times 100\% \times \dfrac{\partial \mathrm{FIR}}{\mathrm{FIR}}$ where δFIR is the ambiguity in determining FIR values and can be obtained by using the equation: $\delta\mathrm{FIR} = \mathrm{FIR} + \sqrt{\left(\dfrac{\delta I_1}{I_1}\right)^2 + \left(\dfrac{\delta I_2}{I_2}\right)^2}$ where δI_1 and δI_2 are uncertainties in intensities of intensities of corresponding emissions [21].

9.3.2 Temperature sensing via luminescence lifetime

When a light is incident on a material, the energy of the light is absorbed by the material, and the electrons are transferred to excited state from where they decay back to the ground state. The difference in energy level is in the form of luminescence. This emission intensity can be expressed as $I_t = I_0 e^{\frac{-t}{\tau}}$ where I_t and I_0 are the emission intensities at the time t and $t=0$, respectively. τ represents the luminescence life time. When a temperature rises, the rate of lattice vibrations will be enhanced, which resulted in increase in number of molecules that are absorbed that quenches the emission light. As τ defines the atmospheric temperature at a particular time, it has many benefits of estimation of temperature that can be then related to other emission characteristics such as blackbody emission characteristics, flame emission and absorption, etc. The value of τ is not affected by surrounding parameters such as the shift in the power of irradiation source and the efficiency of communication channel such as optical fiber or the degree of pairing. It is also observed that the under high temperature the effect of thermally quenched intensity can be efficient in context with τ. That means if the thermal quenching is high and the ratio of signal to noise is small, then the decaying life time would be exact value. The d-block elements such as lanthanides are an excellent thermal probe property as their τ value changes considerably with temperature. Hence, it is concluded that high value of τ is ideal for temperature detection [22].

9.3.3 Temperature sensing via spectral shift

Unlike intensity-based nanothermometry, the spectral shift in the emission spectra provides thermal readings from the temperature-induced spectral shift of luminescence lines. It solely depends on the determination of the spectral position of the luminescence lines. It also has the advantage of non-dependence on the luminescence intensity. The temperature recording is not affected by fluctuations in luminescence intensity resulting from variations in the local concentration of emitting centers. Consequently, the small temperature variations can cause a significant shift in the luminescence peaks. A high-resolution thermometer needs this characteristic. Since this kind of mechanism is observed and applied in many luminescent systems, it has been successfully implemented in quantum dots-based optical nanothermometry. The composites made by using quantum dots show an excellent quantum size effect with good resolution. As the temperature increases, quenching in the emission intensity and the spectral shift in the emission spectra has been reported in many quantum dots-based systems [23–25]. The temperature-induced spectral shift occurs because of many phenomena. It is the resultant effect of these combined phenomena. The important parameter in this regard is the thermal coefficient of spectral shift $\dfrac{d\lambda}{dT}$ where λ is the spectral position of emission spectrum. It depends on the significant change in the energy gap of the confined energy (E_{conf}) within the lattice and that of electron-phonon coupling energy (E_{el-ph}) and it is given by:

$$\frac{d\lambda}{dT} \propto \frac{dE_g^o}{dT} + \frac{dE_{conf}}{dT} + \frac{dE_{el-ph}}{dT} \tag{9.1}$$

wherein the first term represents the thermal coefficient of the band gap energy of a bulk material. This term depends only on the constituent elements of quantum dots material or its composite formed with another matrix. The second term depends on the spectral thermal coefficient of confined energy levels of the quantum dots as well as the size of the quantum dots. As the size of the quantum dots decreases, there is an increase in $\dfrac{dE_{conf}}{dT}$ [25]. The third term represents the thermal coefficient of electron-phonon interaction energy. This is caused by the temperature dependence of density of states and the phonon energy of the lattice vibrations [26]. Thus, the quantum dots have the ability of using them as optical probes for thermal detection. They can be easily incorporated in any matrix to form a nanocomposite or a biological system to explore more possibility of applications in biological field.

9.3.4 Temperature detection using low-lying thermally populated levels

The non-contact nano-thermometers based on rare earth-doped luminescent nanocomposites have been extensively studied for their excellent temperature sensitivity. These studies are mainly focused on the fluorescent intensity ratio (FIR) of two thermally coupled excited energy levels (TCEEL) of trivalent rare earth ions such as Er^{3+}, Tm^{3+}, Yb^{3+}, Pr^{3+}, Dy^{3+}, or Gd^{3+} [27–30]. The FIR techniques have the advantage of no effectiveness of susceptibility of fluctuation in excitation source intensity. But at low temperature, it loses the thermal coupling, resulting in a deviation from the theoretical value calculated by the Boltzmann distribution law. The comparative study of both FIR method of thermally coupled electron energy levels and another method based on two thermally coupled ground energy levels was done which shows the effectiveness of later in sensing of temperature with great precision. According to first mechanism, the rare earth ions occupying the ground state get excited to a high-energy state as an effect of temperature. These excited ions relax to the two energy levels that are thermally coupled and reach to a transient thermal equilibrium state. Then they get decayed to other energy level. If γ_{12} is the excitation rate of the ions from energy level 1 to 2 and γ_{21} is the relaxation rate of ions from energy level 2 to 1, then the ratio of these two rates is given by

$$\frac{\gamma_2}{\gamma_1} = \frac{g_2}{g_1} e^{-\frac{\Delta E}{K_B T}} \tag{9.2}$$

where g_2 and g_1 are degeneracies of the two energy levels. As compared to relaxation rate γ_{21}, the excitation rate γ_{12} decreases drastically with decreasing temperature. Hence, when the temperature becomes sufficiently low, the excitation rate is so slow that the thermo-equilibrium can no longer be achieved before the decaying process to some low-energy level.

The earlier mechanisms of detection of temperature have disadvantage that a very small range of spectral region is available in which the emission light can be used. Hence, another method for detection was proposed in which the thermal population of low-lying levels just above ground level has a key role. The thermal energy will increase the population of higher level, and the population of luminescent level will then again be increased after the excitation laser pulse is applied. The coupling between the lower energy levels can be maintained even at low temperature level. They used Eu^{3+} doped $Ca_3Sc_2Si_3O_{12}$ (CSSO:Eu^{3+}) as a model system [31].

9.4 LUMINESCENT NANOCOMPOSITE MATERIALS FOR OPTICAL THERMOMETRY

9.4.1 Lanthanide-doped nanocomposites

Luminescent thermography using lanthanide ions became very popular due to the stability and narrow emission band profiles of the ions that covers the entire electromagnetic spectrum giving additional advantage of high emission quantum yields. Rare earth-doped nanophosphor materials can be incorporated in various matrices like glass, acrylic, or polymer to produce a nanocomposite that can be used as an optical thermometer to sense temperature in a wide range of temperatures. $Bi_4Ti_3O_{12}$ doped with Yb^{3+} and Er^{3+} nanocrystalline composite was developed to enable online temperature measurement under less power density to reduce the thermal effect of the laser. It shows an excellent precision of temperature measurement in the range of 293–573 K [32]. $Ba_{3-x}Sr_xLu_4O_9$ ($B_{3-x}S_xLO$) phosphor was doped with trivalent rare earth ions Er and Yb to establish energy transfer between Yb^{3+} to Er^{3+} and increase the luminescence of Er^{3+} when it is excited by 980 nm laser light [33]. Numerous efforts have been made to improve the material's optical thermometry property till date by using different host matrix and dopant combinations [34–38]. The host matrix thus is expected to have high chemical and thermal stability and low vibrational energy. To address these characteristics, many new host matrices such as oxide, fluoride, tungstates, phosphates, perovskites, garnets, etc. are tried that are doped with lanthanide ions to generate luminescence [39–55]. The real-time thermogenesis in a single HeLa cell was discussed. The temperature variation of a very small value can be detected [56]. The change in emitter colors with temperature change was reported. In the polymer matrix, lanthanide ions Eu and Tb with linkers are incorporated for temperature sensing [57]. The researchers worldwide have identified rare earth-doped luminescent materials as one of the most sensitive and versatile systems for contactless thermometers in a myriad field, including nanomedicine. The efforts are focused on the design and development of multifunctional nanothermometers with many new spectral ranges, enhanced brightness, and sensitivities. However, there is an urgent need to address certain issues in the assessment of the measurements provided by rare earth-doped luminescent nanothermometers.

9.4.2 Transition metal-doped nanocomposites

The transition metal-doped luminescent materials having high sensitivity toward temperature has been an attractive research area nowadays with lots of attention given to the synthesis of nanocomposites for a plausible application in nano thermometry [58]. The lanthanide-based phosphors have physicochemical and photochemical stability. There are numerous sensitizer and activator combinations available as inorganic luminescent phosphors to be feasible for various applications in myriad fields. When the inorganic host matrix is doped with lanthanide ions along with transition metal ions, the temperature sensitivity increases. In addition to excellent thermoluminescence, they can be synthesized in any form as nano powder, nanofilm, composites, glasses, glass-ceramics, coatings, etc. Among all the materials available for nanothermometry, they offer good repeatability, high sensitivity toward temperature, and a broader range of temperature measurements [59–61]. When these transition metals are introduced in the nanomaterials matrix and some excitation source is used, the population of the excited state of the ions causes energy transfer to the lanthanide ions that is assisted by phonons of vibrational state. These vibrational states are temperature-dependent. The obvious reason is the energy mismatch between the excited states of transition metal ions and lanthanide cations leading to strong thermal energy-based sensing action. This concept is very useful in the development of transition metal co-doped luminescent nanocomposites. The spectral range can be enhanced using suitable inorganic host and co-dopant lanthanide trivalent ions. Many transition metal ions show excellent temperature sensitivity, and three of them viz. Cr^{3+}, Mn^{4+}, and Ti^{3+}/Ti^{4+} are the potential candidates that show prominent results [62–66]. The Cr^{3+} ions are characterized by emission in red to near infrared spectral region, and the emission spectral shape depends on crystal field strength. The other transition metal ion Mn^{4+} has higher crystal field strength than Cr^{3+}. This factor affects the thermal quenching of luminescence. The Ti^{3+} ions have $3d^1$ configuration that leads to restriction in the transition between 2_E and 4_{T_2} states. These ions have two broad emission bands [67–70].

9.4.3 Luminescent dyes-based nanocomposites

Organic dyes are luminescent materials whose photoluminescence intensity and lifetime are dependent on temperature. Therefore, it can be used as a thermosensor. They are vast in numbers and can be tailor-made. Their properties such as absorption and emission wavelength, solubility, spectral wavelength range, and their ability to be functionalized to target specific analytes can be set to be suitable for a particular application. They are termed as "Temperature indicators," and in biological systems, RhB [71], cyanine dyes [72], and fluorescein isothiocyanates [73] are the popular choice in temperature detection. In the literature, we can find the use of dyes to label antibodies and antigens [74,75] or to improve the chemical and mechanical stability of a polymer matrix they can be impregnated in it [76,77]. There are numerous examples of dyes being used as temperature sensors, the most recent being the dialkylaminostyrylhetarene dyes [78] and Rhodamine 6G Fluorophore [79]. Figure 9.2(A) shows the nanocomposite film made by incorporating a Rhodamine 6G Fluorophore in aluminum nanoparticles. The laser heating in the experimental setup evaluates the temperature-dependent fluorescence and collects fluorescence image sequences during laser heating that focus on specific aluminum nanoparticles within the nanocomposites (Figure 9.2B).

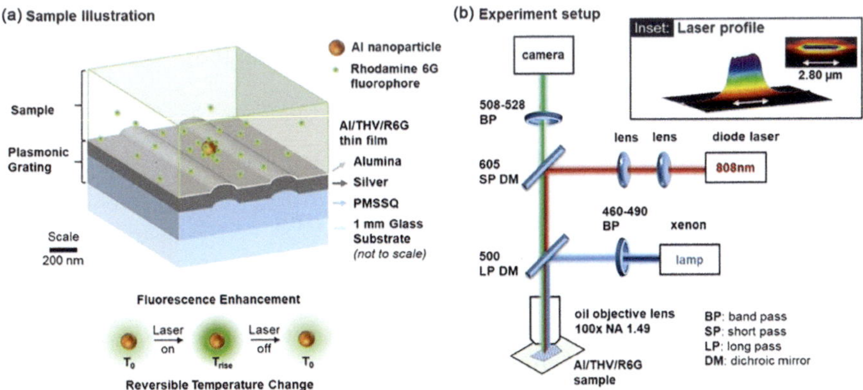

Figure 9.2 (a) A nanocomposite thin film deposited on a plasmonic grating substrate. The use of Rhodamine 6G fluorophore has been depicted here. (b) The experimental setup for temperature measurement. (Reprinted from Ref. [79] (https://creativecommons.org/licenses/by/4.0/).)

The metal oxide framework using lanthanide was a common approach due to lanthanide's wide spectral range and chemical and thermal stability [80,81]. To get a multiparameter probe design, fluorescence dyes become the popular choice due to their wide spectral range from UV region to near IR region [82]. This new approach helps the workers in this field to develop a multiparameter sensing system along with a reliable temperature system using optical mode [83].

9.4.4 Semiconductor quantum dot nanocomposites

Semiconductor quantum dots are very small in size, have tuneable luminescence properties, and are favorite candidates in display and optical devices. These are also potential candidates for nanoscale thermometry as they exhibit temperature-dependent photoluminescence. They have photostability and high quantum yield. Common materials in these areas are cadmium-based binary or ternary semiconductor materials. Many times, core–shell structure is preferred for achieving some outstanding properties [84,85]. It was shown that ZnS-coated CdSe quantum dots are capable of sensing temperature changes with characteristics [86]. Quantum dots tend to aggregate easily, hence to avoid agglomeration, some kind of coating or passivation is done. For example, ZnS-coated CdS quantum dots not only show excellent nanoscale homogeneous particles but also there is enhancement in optical property in the temperature range of 100–315 K [87]. Recently, the CdS/ZnS core/shell QDs were studied for their plausible application as a temperature sensor by observing their PL spectra in the temperature range of 77–297 K. The researchers found that the band edge emission intensity (BEE) decreases with increase in temperature, and the surface state emission (SEE) intensity first increases and then decreases. They also observed that there is a shift of emission intensity with broadening of the PL spectrum [88].

9.5 APPLICATION AREAS

9.5.1 Biological and medical applications

The biological system temperature measurement is an important issue in medical field because without destroying the biological tissue or biomaterial to be monitored and to enable accurate temperature measurement, a technologically important class of contactless thermometers are developed that uses IR rays as a source. But they cannot penetrate biological tissue or any other materials of interest without using invasive tools. On the contrary, methods that are based on nuclear magnetic resonance (NMR) phenomena resulted in a most suitable method for detailed thermal examination of 3D objects containing large amount of water like body tissue. MR-guided high-intensity focused ultrasound (MR-HIFU) is

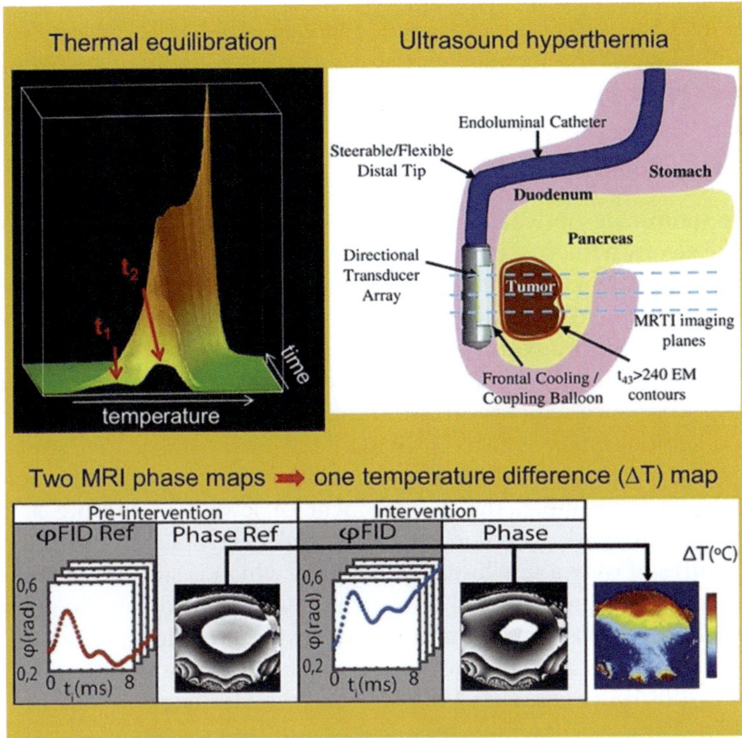

Figure 9.3 Phase-based temperature difference mapping using multiple coil elements for magnetic resonance thermometry (MRT). (Reproduced from Ref. [95]. This is an open access article under the CC BY-NC-ND license (http://creative-commons.org/licenses/by-nc-nd/4.0/).)

the latest technology for noninvasive thermal treatment of many malignant tumors [89–94] (Figure 9.3).

9.5.2 Nanoelectronics and photonics

With the advancement in semiconductor technology using nano-sized materials, the need to monitor the temperature in high-power nanoelectronics and photonics applications arises. The crucial parameter of controlling the temperature at this level has been a matter of discussion for many years. Some materials like two-dimensional materials and diamond color centers are potential candidates in this regard as both materials can withstand high temperatures up to several hundred degrees. These two-dimensional materials can be used to prepare photonic or electronic devices that are in the form of stacks [96–98]. Remote optical thermometry applications can be realized using thin films of two-dimensional materials or single-walled carbon tubes that act as thin film transistors and as temperature sensors. They can also be feasible to use as a temperature measuring system when there is a phase change in the materials or a magnetic material system. The main challenge in these systems is the accuracy of temperature measurement and simultaneously recording electrical, magnetic, or photoluminescent properties.

9.5.3 Microscopy

Remote nanoscale optical thermometry could be useful in other applications like electron microscope imaging. During imaging, the nanostructures are known to warm up by electron beams used by the microscopy devices, and it is difficult to quantify unequivocally their temperature. However, recent technological advancements demonstrated that emitters within the imaged material itself or fluorescent particles dispersed on the sample can be used to measure temperatures in addition to putting optical fibers and other collection optics in situ. Drawing from this, comparative procedures can be adjusted to gauge the temperature of developing semiconductors or materials – for example, the development of jewels containing a variety of focuses that work as color centers. Jewel development can happen at temperatures of only a few hundred degrees, which is still beneath the base temperature a business optical pyrometer can gauge. Using the embedded color centers, remote optical thermometry could provide a precise temperature at the crystal facet and help comprehend the nanodiamond's growth dynamics [99,100]. Figure 9.4 shows that the Germanium vacancy color centers in diamonds can emit high fraction of fluorescence in zero phonon lines, which can be collected by some optical system like fiber optic and thereby reducing heating due to conversion of pump light into phonons.

9.5.4 Microfluidic and surface study

Many application areas need to measure temperature that contains microfluidic channels and understand the mechanism of solid-liquid interfaces. Organic dyes and upconverted luminescent nanoparticles are two potential candidates in this regard. Knowing about the temperature inside

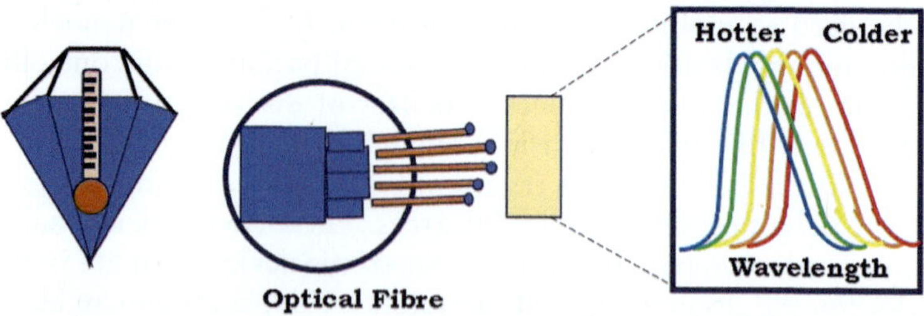

Figure 9.4 Germanium vacancy color centers in diamonds acting as temperature sensors.

microfluidic systems accurately is crucial for many devices and applications, such as biofluidic devices and chemical microreactors and biofluidic devices [101,102]. There are application areas like the control of microprocessor temperature [103], biophysical processes [104], microfluid motion [105,106], and lab-on-chip technologies [107–109] where the need for non-invasive micro-scale temperature sensing is very necessary. However, the thermal equilibrium between the probe and the sample is the key factor, which is a must step that exchanges heat facilitating the measurement of temperature on the microliter to nanoliter scale. Thus, this is a very challenging step [110].

9.5.5 Temperature sensors in extreme temperature regions

The advancement in science and technology makes it possible to measure the temperature in harsh environments such as nuclear energy production, aerospace, metallurgy, and underground wells. Optical fiber high-temperature sensing technology evolved in recent years as high-temperature detection has been a constant challenge in optical fiber sensor research [111]. Lanthanide-based luminescent optical nanothermometers have impacted the temperature determination. However, the deterioration of sensitivity and resolution with thermal quenching of the luminescence at high temperatures is a matter of concern. In extreme temperature regions, luminescent nanocomposites offer several advantages over traditional ones. They have excellent temperature stability that makes them potential candidates for applications at very high or low temperatures. Additionally, they have high sensitivity to temperature sensitivity to temperature changes making them ideal for temperature measurements. They can be engineered to record certain temperature ranges [112]. Some other studies showed temperature sensing above 1000 K. The first report discussed the synthesis of the Gd_2O_3:Yb^{3+}–Er^{3+} nanorods decorated with gold nanoparticles, which act as luminescent-plasmonic nanomaterials. They showed that this material has the potential to work as an optical heater/thermometer. They used two thermionic parameters, viz. luminescent intensity ratio (LIR) of Er^{3+} cations in the temperature range of 300–1050 K and the black body emission of the nanomaterial in the temperature range of 1200–2000 K [113]. The phenomenon of converting two or more low-energy photons into a high-energy photon emission is known as up-conversion (UC). Such materials show high luminescent efficiency, better light output, and better thermal stability. Yb^{3+}/Er^{3+} co-doped UC luminescent phosphor materials have been frequently reported as promising candidates for ratiometric thermometers [114–118]. Nd^{3+} is a new doping cation for its excellent properties. It possesses two sets of TCLs of $^4F_{7/2}$/$^4F_{5/2}$ and $^4F_{5/2}$/$^4F_{3/2}$, both of which possess an energy gap of around $1000\,cm^{-1}$, making them highly suitable as ratiometric thermometers [119–121].

9.6 LIMITATIONS AND CHALLENGES

Despite recent advance research in the development of sophisticated optical thermometers at nanoscale levels, there are many issues to be addressed till date. This field of research can gain momentum only when these issues are resolved at optimum level. These issues include the thorough understanding of heat transfer and energy transfer mechanisms, the optimization of the temperature sensing and recording system, the development of cost effective and high sensitivity sensors, the synthesizing techniques for efficient material for sensing and the expansion of various application areas. The challenges include the lack of collaborative efforts among chemists, physicists, biologists, medical field researchers, and engineers who are the stake holders in the field of optical thermometry. The exchange of ideas and innovative practices in the respective fields should be happened among the stake holders.

9.7 FUTURE SCOPE AND OPPORTUNITIES

Luminescence thermometry has been proven to be superior to conventional thermometry due to its excellent properties such as high sensitivity toward temperature, spectral shift, tuneable emission bandwidth, fluorescence lifetime, absolute intensity, and LIR. This thermometry has a wide range of applications such as in electrical and mechanical engineering, nanotechnology, microfluidics, and biomedicine to name a few. The temperature is one of the most often measured quantities and is an integral part of our day-to-day life; its application area will be widened in the future also. With advancements in Science and Technology, the need for accuracy in temperature measurement is increasing by many folds. Hence, it is high time that more research avenues should be opened for this area of research. With the availability of many types of materials that can be used in the preparation of the temperature sensing devices, optical thermometry research will be a blooming area in which many new techniques and devices have a tremendous scope.

REFERENCES

1. Kim, K., Jeong, W., Lee, W., & Reddy, P. (2012). Ultra-high vacuum scanning thermal microscopy for nanometer resolution quantitative thermometry. *Acs Nano*, 6(5), 4248–4257. https://doi.org/10.1016/B978-0-12–813928-8.00006-5.
2. *Nanomaterials for Magnetic and Optical Hyperthermia Applications*, edited by Raluca Maria Fratila and Jesus Martinez De La Fuente (2018). https://doi.org/10.1016/C2017-0-00855-8.

3. Okabe, K., Inada, N., Gota, C., Harada, Y., Funatsu, T., & Uchiyama, S. (2012). Intracellular temperature mapping with a fluorescent polymeric thermometer and fluorescence lifetime imaging microscopy. *Nature Communications*, *3*(1), 1–9. https://www.nature.com/articles/ncomms1714.

4. Lowell, B. B., & Spiegelman, B. M. (2000). Towards a molecular understanding of adaptive thermogenesis. *Nature*, *404*(6778), 652–660. https://doi.org/10.1038/35007527.

5. Monti, M., Brandt, L., Ikomi-Kumm, J., & Olsson, H. (1986). Microcalorimetric investigation of cell metabolism in tumour cells from patients with non-Hodgkin lymphoma (NHL). *Scandinavian Journal of Haematology*, *36*(4), 353–357. https://doi.org/10.1111/j.1600-0609.1986.tb01749.x.

6. McCabe, K. M., & Hernandez, M. (2010). Molecular thermometry. *Pediatric Research*, *67*(5), 469–475. https://doi.org/10.1203/PDR.0b013e3181d68cef.

7. Kauppinen, J. P., Loberg, K. T., Manninen, A. J., Pekola, J. P., & Voutilainen, R. A. (1998). Coulomb blockade thermometer: Tests and instrumentation. *Review of Scientific Instruments*, *69*, 4166–4175. https://pubs.aip.org/aip/rsi/article-abstract/69/12/4166/448300.

8. Rubin, L. G. (1997). Cryogenic thermometry: A review of progress since 1982. *Cryogenics*, *37*, 341–356. https://doi.org/10.1016/S0011-2275(97)00009-X.

9. Preston-Thomas, H. (1990). The international temperature scale of 1990 (ITS–90). *Metrologia*, *27*, 3–10. https://doi.org/10.1088/0026–1394/27/1/002.

10. Liu, W., & Yang, B. (2007). Thermography techniques for integrated circuits and semiconductor devices. *Sensor Review*, *27*(4), 298–309. https://doi.org/10.1108/02602280710821434

11. Mao, H., Yang, T., & Cremer, P. S. (2002). A microfluidic device with a linear temperature gradient for parallel and combinatorial measurements. *Journal of the American Chemical Society*, *124*(16), 4432–4435. https://doi.org/10.1021/ja017625x.

12. Gosse, C., Bergaud, C., & Löw, P. (2009). Molecular probes for thermometry in microfluidic devices. In *Thermal Nanosystems and Nanomaterials* (pp. 301–341). Berlin, Heidelberg: Springer Berlin Heidelberg. https://doi.org/10.1007/978-3-642-04258-4_10.

13. Barilero, T., Le Saux, T., Gosse, C., & Jullien, L. (2009). Fluorescent thermometers for dual-emission-wavelength measurements: Molecular engineering and application to thermal imaging in a microsystem. *Analytical Chemistry*, *81*(19), 7988–8000. https://doi.org/10.1021/ac901027f.

14. Feng, J., Tian, K., Hu, D., Wang, S., Li, S., Zeng, Y., ... & Yang, G. (2011). A triarylboron-based fluorescent thermometer: Sensitive over a wide temperature range. *Angewandte Chemie International Edition*, *50*(35), 8072–8076. https://doi.org/10.1002/anie.201102390.

15. Graham, E. M., Iwai, K., Uchiyama, S., De Silva, A. P., Magennis, S. W., & Jones, A. C. (2010). Quantitative mapping of aqueous microfluidic temperature with sub-degree resolution using fluorescence lifetime imaging microscopy. *Lab on a Chip*, *10*(10), 1267–1273. https://doi.org/10.1039/B924151E.

16. Hartmann, M. (2006). Minimal length scales for the existence of local temperature. *Contemporary Physics*, *47*(2), 89–102. https://doi.org/10.1080/00107510600581136.

17. Berthou, H., & Jörgensen, C. K. (1990). Optical-fiber temperature sensor based on upconversion-excited fluorescence. *Optics Letters, 15*(19), 1100–1102. https://doi.org/10.1364/OL.15.001100.

18. Li, H., Zhang, Y., Shao, L., Htwe, Z., & Yuan, P. (2017). Luminescence probe for temperature sensor based on fluorescence intensity ratio. *Optical Materials Express, 7*(3), 1077–1083. https://doi.org/10.1364/OME.7.001077.

19. Rai, V. K. (2007). Temperature sensors and optical sensors. *Applied Physics B, 88*, 297–303. https://doi.org/10.1007/s00340-007-2717-4.

20. Soler-Carracedo, K., Martín, I. R., Lahoz, F., Vasconcelos, H. C., Lozano-Gorrín, A. D., Martín, L. L., & Paz-Buclatin, F. (2020). Er^{3+}/Ho^{3+} codoped nanogarnet as an optical FIR based thermometer for a wide range of high and low temperatures. *Journal of Alloys and Compounds, 847*, 156541. https://doi.org/10.1016/j.jallcom.2020.156541.

21. Baker, S. N., McCleskey, T. M., & Baker, G. A. (2005). *An Ionic Liquid-Based Optical Thermometer*, ACS Publications. ACS Symposium Series Vol. 902 Chapter 14, pp 171–181. https://pubs.acs.org/doi/abs/10.1021/bk-2005-0902.ch014.

22. Labrador-Paez, L., Pedroni, M., Speghini, A., Garcia-Sole, J., Haro-Gonzalez, P., & Jaque, D. (2018). Reliability of rare-earth-doped infrared luminescent nanothermometers. *Nanoscale, 10*(47), 22319–22328. https://doi.org/10.1039/C8NR07566B.

23. Maestro, L. M., Jacinto, C., Silva, U. R., Vetrone, F., Capobianco, J. A., Jaque, D., & Solé, J. G. (2011). CdTe quantum dots as nanothermometers: Towards highly sensitive thermal imaging. *Small, 7*(13), 1774–1778. https://doi.org/10.1002/smll.201002377.

24. Narayanaswamy, A., Feiner, L. F., Meijerink, A., & Van der Zaag, P. J. (2009). The effect of temperature and dot size on the spectral properties of colloidal InP/ZnS core–shell quantum dots. *ACS Nano, 3*(9), 2539–2546. https://doi.org/10.1021/nn9004507.

25. Dai, Q., Zhang, Y., Wang, Y., Hu, M. Z., Zou, B., Wang, Y., & Yu, W. W. (2010). Size-dependent temperature effects on PbSe nanocrystals. *Langmuir, 26*(13), 11435–11440. https://pubs.acs.org/doi/abs/10.1021/la101545w.

26. Olkhovets, A., Hsu, R. C., Lipovskii, A., & Wise, F. W. (1998). Size-dependent temperature variation of the energy gap in lead-salt quantum dots. *Physical Review Letters, 81*(16), 3539. https://doi.org/10.1103/PhysRevLett.81.3539.

27. Tang, W., Wang, S., Li, Z., Sun, Y., Zheng, L., Zhang, R., … & Yu, M. (2016). Ultrahigh-sensitive optical temperature sensing based on ferroelectric Pr^{3+}-doped $(K_{0.5}Na_{0.5})\,NbO_3$. *Applied Physics Letters, 108*(6). https://doi.org/10.1063/1.4941669.

28. Zheng, K., Liu, Z., Lv, C., & Qin, W. (2013). Temperature sensor based on the UV upconversion luminescence of Gd^{3+} in Yb^{3+}–Tm^{3+}–Gd^{3+} codoped $NaLuF_4$ microcrystals. *Journal of Materials Chemistry C, 1*(35), 5502–5507. https://doi.org/10.1039/C3TC30763H.

29. Wang, X., Liu, Q., Cai, P., Wang, J., Qin, L., Vu, T., & Seo, H. J. (2016). Excitation powder dependent optical temperature behavior of Er^{3+} doped transparent $Sr_{0.69}La_{0.31}F_{2.31}$ glass ceramics. *Optics Express, 24*(16), 17792–17804. https://doi.org/10.1364/OE.24.017792.

30. Xing, L., Xu, Y., Wang, R., Xu, W., & Zhang, Z. (2014). Highly sensitive optical thermometry based on upconversion emissions in Tm^{3+}/Yb^{3+} codoped $LiNbO_3$ single crystal. *Optics Letters*, *39*(3), 454–457. https://doi.org/10.1364/OL.39.000454.

31. Zhao, L., Cai, J., Hu, F., Li, X., Cao, Z., Wei, X., ... & Duan, C. K. (2017). Optical thermometry based on thermal population of low-lying levels of Eu^{3+} in $Ca_{2.94}Eu_{0.04}Sc_2Si_3O_{12}$. *RSC Advances*, *7*(12), 7198–7202. https://doi.org/10.1039/C6RA28431K.

32. Pan, E., Bai, G., Wang, L., Lei, L., Chen, L., & Xu, S. (2019). Lanthanide ion-doped bismuth titanate nanocomposites for ratiometric thermometry with low pump power density. *ACS Applied Nano Materials*, *2*(11), 7144–7151. https://doi.org/10.1021/acsanm.9b01631.

33. Hu, J., Zhang, X., Zheng, H., Lu, F., Peng, X., Wei, R., ... & Guo, H. (2022). Improved photoluminescence and multi-mode optical thermometry of Er^{3+}/Yb^{3+} co-doped $(Ba, Sr)_3Lu_4O_9$ phosphors. *Ceramics International*, *48*(3), 3051–3058. https://doi.org/10.1016/j.ceramint.2021.10.080.

34. Lv, H., Du, P., Li, W., & Luo, L. (2022). Tailoring of upconversion emission in Tm^{3+}/Yb^{3+}-codoped $Y_2Mo_3O_{12}$ submicron particles via thermal stimulation engineering for non-invasive thermometry. *ACS Sustainable Chemistry & Engineering*, *10*(7), 2450–2460. https://doi.org/10.1021/acssuschemeng.1c07323.

35. Yang, X., Zhu, Y., Li, T., Long, S., & Wang, B. (2023). High-accuracy dual-mode optical thermometry based on up-conversion luminescence in $Er^{3+}/Ho^{3+}-Yb^{3+}$ doped $LaNbO_4$ phosphors. *Ceramics International*, *49*(13), 21932–21940. https://doi.org/10.1016/j.ceramint.2023.04.017.

36. Goderski, S., Runowski, M., Woźny, P., Lavín, V., & Lis, S. (2020). Lanthanide upconverted luminescence for simultaneous contactless optical thermometry and manometry–sensing under extreme conditions of pressure and temperature. *ACS Applied Materials & Interfaces*, *12*(36), 40475–40485. https://doi.org/10.1021/acsami.0c09882.

37. Lei, Z., Liu, R., Sun, L., Wang, X., Hu, C., Zou, Y., ... & Zhong, D. (2023). An up-conversion $Ba_3In(PO_4)_3{:}Er^{3+}/Yb^{3+}$ phosphor that enables multi-mode temperature measurements and wide-gamut 'temperature mapping'. *Dalton Transactions*, *11*. https://doi.org/10.1039/D3DT01575K.

38. Zhang, H., Liang, Y., Yang, H., Liu, S., Li, H., Gong, Y., ... & Li, G. (2020). Highly sensitive dual-mode optical thermometry in double-perovskite oxides via Pr^{3+}/Dy^{3+} energy transfer. *Inorganic Chemistry*, *59*(19), 14337–14346. https://doi.org/10.1021/acs.inorgchem.0c02118.

39. Fan, Y., Xiao, Q., Yin, X., Lv, L., Wu, X., Dong, X., ... & Luo, X. (2022). Upconversion luminescence and optical temperature sensing of Er^{3+}-doped $La_2Mo_2O_9$ phosphors under 980 and 1550 nm excitation. *Solid State Sciences*, *132*, 106966. https://doi.org/10.1016/j.solidstatesciences.2022.106966.

40. Zhang, Z., Meng, Q., Bai, L., & Sun, W. (2022). Temperature-sensitive properties of $Y_2Ti_2O_7$: Pr^{3+}, Tb^{3+} phosphors. *Journal of Luminescence*, *251*, 119229. https://doi.org/10.1016/j.jlumin.2022.119229.

41. Yuan, M., Han, K., Wang, L., Yang, X., Yang, Z., Wang, H., & Xu, X. (2022). Highly thermally stable upconversion in copper (II)-doped LiYF4: Yb, Er microcrystals toward ultrahigh temperature (micro) thermometers. *Journal of Alloys and Compounds, 919*, 165844. https://doi.org/10.1016/j.jallcom.2022.165844.

42. Laia, A. S., Maciel, G. S., Rodrigues Jr, J. J., Dos Santos, M. A., Machado, R., Dantas, N. O., ... & Alencar, M. A. (2022). Lithium-boron-aluminum glasses and glass-ceramics doped with Eu^{3+}: A potential optical thermometer for operation over a wide range of temperatures with uniform sensitivity. *Journal of Alloys and Compounds, 907*, 164402. https://doi.org/10.1016/j.jallcom.2022.164402.

43. Cheng, H., Jiang, Z., Lai, F., Wang, H., Xiao, Z., Sun, J., & You, W. (2022). Simultaneous achievement of sensitivity enhancement and dual-mode luminescence through co-doping Eu^{3+} in Ca_2MgWO_6: Er^{3+}/Yb^{3+} phosphor. *Journal of Luminescence, 246*, 118804. https://doi.org/10.1016/j.jlumin.2022.118804.

44. Xu, W., Zhao, H., Zhang, Z., & Cao, W. (2013). Highly sensitive optical thermometry through thermally enhanced near-infrared emissions from Nd^{3+}/Yb^{3+} codoped oxyfluoride glass ceramic. *Sensors and Actuators B: Chemical, 178*, 520–524. https://doi.org/10.1016/j.snb.2012.12.050.

45. Tang, J., Du, P., Li, W., & Luo, L. (2020). Boosted thermometric performance in $NaGdF_4$: Er^{3+}/Yb^{3+} upconverting nanorods by Fe^{3+} ions doping for contactless nanothermometer based on thermally and non-thermally coupled levels. *Journal of Luminescence, 224*, 117296. https://doi.org/10.1016/j.jlumin.2020.117296.

46. Liao, J., Kong, L., Wang, M., Sun, Y., & Gong, G. (2019). Tunable upconversion luminescence and optical temperature sensing based on non-thermal coupled levels of Lu_3NbO_7: Yb^{3+}/Ho^{3+} phosphors. *Optical Materials, 98*, 109452. https://doi.org/10.1016/j.optmat.2019.109452.

47. Xing, L., Yang, W., Ma, D., & Wang, R. (2015). Effect of crystallinity on the optical thermometry sensitivity of Tm^{3+}/Yb^{3+} codoped $LiNbO_3$ crystal. *Sensors and Actuators B: Chemical, 221*, 458–462. https://doi.org/10.1016/j.snb.2015.06.132.

48. Sinha, S., & Kumar, K. (2018). Studies on up/down-conversion emission of Yb^{3+} sensitized Er^{3+} doped $MLa_2(MoO_4)_4$ (M= Ba, Sr and Ca) phosphors for thermometry and optical heating. *Optical Materials, 75*, 770–780. https://doi.org/10.1016/j.optmat.2017.10.036.

49. An, S., Chen, J., Zhang, J., Zhao, J., & Li, X. (2021). $NaGd_9$ $(SiO_4)6O_2$: Yb^{3+}-Er^{3+}/Tm^{3+}: Optical thermometric materials of high-sensitivity by using different strategies. *Journal of Luminescence, 239*, 118388. https://doi.org/10.1016/j.jlumin.2021.118388.

50. Kolesnikov, I. E., Kalinichev, A. A., Kurochkin, M. A., Mamonova, D. V., Kolesnikov, E. Y., Kurochkin, A. V., ... & Mikhailov, M. D. (2018). Y_2O_3: Nd^{3+} nanocrystals as ratiometric luminescence thermal sensors operating in the optical windows of biological tissues. *Journal of Luminescence, 204*, 506–512. https://doi.org/10.1016/j.jlumin.2018.08.050.

51. Jiang, G., Wei, X., Chen, Y., Duan, C., Yin, M., Yang, B., & Cao, W. (2015). Luminescent La_2O_2S: Eu^{3+} nanoparticles as non-contact optical temperature sensor in physiological temperature range. *Materials Letters, 143*, 98–100. https://doi.org/10.1016/j.matlet.2014.12.057.

52. Fu, L., Fu, Z., Yu, Y., Wu, Z., & Jeong, J. H. (2015). An Eu/Tb-codoped inorganic apatite $Ca_5(PO_4)_3F$ luminescent thermometer. *Ceramics International, 41*(5), 7010–7016. https://doi.org/10.1016/j.ceramint.2015.02.004.

53. Wu, Y., Zhang, Z., Suo, H., Zhao, X., & Guo, C. (2019). 808 nm light triggered up-conversion optical nano-thermometer YPO_4: $Nd^{3+}/Yb^{3+}/Er^{3+}$ based on FIR technology. *Journal of Luminescence, 214*, 116578. https://doi.org/10.1016/j.jlumin.2019.116578.

54. Chen, D., Wang, Z., Zhou, Y., Huang, P., & Ji, Z. (2015). Tb^{3+}/Eu^{3+}: YF_3 nanophase embedded glass ceramics: Structural characterization, tunable luminescence and temperature sensing behavior. *Journal of Alloys and Compounds, 646*, 339–344. https://doi.org/10.1016/j.jallcom.2015.06.030.

55. Wang, C., Jin, Y., Zhang, R., Yao, Q., & Hu, Y. (2022). A review and outlook of ratiometric optical thermometer based on thermally coupled levels and non-thermally coupled levels. *Journal of Alloys and Compounds, 894*, 162494. https://doi.org/10.1016/j.jallcom.2021.162494.

56. Suzuki, M., Tseeb, V., Oyama, K., & Ishiwata, S. I. (2007). Microscopic detection of thermogenesis in a single HeLa cell. *Biophysical Journal, 92*(6), L46–L48. https://doi.org/10.1529/biophysj.106.098673.

57. Hatanaka, M., Hirai, Y., Kitagawa, Y., Nakanishi, T., Hasegawa, Y., & Morokuma, K. (2017). Organic linkers control the thermosensitivity of the emission intensities from Tb (III) and Eu (III) in a chameleon polymer. *Chemical Science, 8*(1), 423–429. https://doi.org/10.1039/C6SC03006H.

58. Piotrowski, W., Kniec, K., & Marciniak, L. (2021). Enhancement of the Ln^{3+} ratiometric nanothermometers by sensitization with transition metal ions. *Journal of Alloys and Compounds, 870*, 159386. https://doi.org/10.1016/j.jallcom.2021.159386.

59. Duan, Y., Liu, Y., Zhang, G., Yao, L., & Shao, Q. (2021). Broadband Cr^{3+}-sensitized upconversion luminescence of $LiScSi_2O_6$: Cr^{3+}/Er^{3+}. *Journal of Rare Earths, 39*(10), 1181–1186. https://doi.org/10.1016/j.jre.2020.08.007.

60. Prorok, K., & Bednarkiewicz, A. (2016). Energy migration up-conversion of Tb^{3+} in Yb^{3+} and Nd^{3+} codoped active-core/active-shell colloidal nanoparticles. *ACS Chemisry of Materials, 28*, 2295–2300. https://doi.org/10.1021/acs.chemmater.6b00353

61. Liao, J., Kong, L., Wang, Q., Li, J., Peng, G., Sun, Y., & Wen, H. (2018). Sol-gel preparation and near-infrared emission properties of Yb^{3+} sensitized by Mn^{4+} in double-perovskite La_2ZnTiO_6. *Optical Materials, 84*, 82–88. https://doi.org/10.1016/j.optmat.2018.06.055.

62. Marciniak, L., Bednarkiewicz, A., Drabik, J., Trejgis, K., & Strek, W. (2017). Optimization of highly sensitive $YAG:Cr^{3+},Nd^{3+}$ nanocrystal-based luminescent thermometer operating in an optical window of biological tissues. *Physical Chemistry Chemical Physics, 19*, 7343–7351. https://doi.org/10.1039/c6cp07213e.

63. Chen, D., Zhou, Y., Xu, W., Zhong, J., Ji, Z., & Xiang, W. (2016). Enhanced luminescence of $Mn^{4+}:Y_3Al_5O_{12}$ red phosphor via impurity doping. *Journal of Materials Chemistry C, 4*, 1704–1712. https://doi.org/10.1039/c5tc04133c.

64. Trejgis, K., & Marciniak, L. (2018). The influence of manganese concentration on the sensitivity of bandshape and lifetime luminescent thermometers based on $Y_3Al_5O_{12}:Mn^{3+}$, Mn^{4+}, Nd^{3+} nanocrystals. *Physical Chemistry Chemical Physics, 20*, 9574–9581. https://doi.org/10.1039/c8cp00558c.

65. Drabik, J., Marciniak, Ł., Cichy, B. B., & Marciniak, L. (2018). New type of nanocrystalline luminescent thermometers based on Ti^{3+}/Ti^{4+} and Ti^{4+}/Ln^{3+} ($Ln^{3+}=Nd^{3+}$, Eu^{3+}, Dy^{3+}) luminescence intensity ratio. *The Journal of Physical Chemistry C, 122*, 14928–14936. https://doi.org/10.1021/acs.jpcc.8b02328.

66. Chaika, M., Tomala, R., Vovk, O., Nizhankovskyi, S., Mancardi, G., & Strek, W. (2020). Upconversion luminescence in $Cr^{3+}:YAG$ single crystal under infrared excitation. *Journal of Luminescence, 226*, 117467. https://doi.org/10.1016/j.jlumin.2020.117467.

67. Mares, J. A., Nie, W., & Boulon, G. (1991). Energy transfer processes between various Cr^{3+} and Nd^{3+} multisites in YAG:Nd, Cr. *Journal of Luminescence, 48–49*, 227–231. https://doi.org/10.1016/0022-2313(91)90110-H.

68. Avram, N. M., & Brik, M. G. (2013). *Optical Properties of 3d-Ions in Crystals: Spectroscopy and Crystal Field Analysis*. Jointly published with Tsinghua University Press, 8(268), 63. https://doi.org/10.1007/978-3-642-30838-3.

69. Pan, G. H., Zhang, L. L., Wu, H. H., Qu, X., Wu, H. H., Hao, Z., … & Zhang, J. (2020). On the luminescence of Ti^{4+} and Eu^{3+} in monoclinic ZrO_2: High performance optical thermometry derived from energy transfer. *Journal of Materials Chemistry C 8*, 4518–4533. https://doi.org/10.1039/c9tc06992e.

70. Elzbieciak, K., Bednarkiewicz, A., & Marciniak, L. (2018). Temperature sensitivity modulation through crystal field engineering in Ga^{3+} co-doped $Gd_3Al_5\text{-}xGaxO_{12}:Cr^{3+}$, Nd^{3+} nanothermometers. *Sensors and Actuators B: Chemical, 269*, 96–102. https://doi.org/10.1016/j.snb.2018.04.157

71. Ross, D., Gaitan, M., & Locascio, L. E. (2001). Temperature measurement in microfluidic systems using a temperature-dependent fluorescent dye. *Analytical Chemistry, 73*(17), 4117–4123. https://doi.org/10.1021/ac010370l.

72. Mikkelsen, R. B., & Wallach, D. F. (1977). Temperature sensitivity of the erythrocyte membrane potential as determined by cyanine dye fluorescence. *Cell Biology International Reports, 1*(1), 51–55. https://doi.org/10.1016/0309-1651(77)90009-1.

73. Guan, X., Liu, X., & Su, Z. (2007). Preparation and photophysical behaviors of fluorescent chitosan bearing fluorescein: Potential biomaterial as temperature/pH probes. *Journal of Applied Polymer Science, 104*(6), 3960–3966. https://onlinelibrary.wiley.com/doi/abs/10.1002/app.26200.

74. He, X., Duan, J., Wang, K., Tan, W., Lin, X., & He, C. (2004). A novel fluorescent label based on organic dye-doped silica nanoparticles for HepG liver cancer cell recognition. *Journal of Nanoscience and Nanotechnology, 4*(6), 585–589. https://www.ingentaconnect.com/contentone/asp/jnn/2004/00000004/00000006/art00007.

75. Wang, F., Tan, W. B., Zhang, Y., Fan, X., & Wang, M. (2005). Luminescent nanomaterials for biological labelling. *Nanotechnology, 17*(1), R1–R13. https://iopscience.iop.org/article/10.1088/0957-4484/17/1/R01/meta.

76. Köse, M. E., Carroll, B. F., & Schanze, K. S. (2005). Preparation and spectroscopic properties of multiluminophore luminescent oxygen and temperature sensor films. *Langmuir, 21*(20), 9121–9129. https://pubs.acs.org/doi/abs/10.1021/la050997p.

77. Köse, M. E., Omar, A., Virgin, C. A., Carroll, B. F., & Schanze, K. S. (2005). Principal component analysis calibration method for dual-luminophore oxygen and temperature sensor films: Application to luminescence imaging. *Langmuir, 21*(20), 9110–9120. https://pubs.acs.org/doi/full/10.1021/la050999+.

78. Akhmadeev, B. S., Gerasimova, T. P., Gilfanova, A. R., Katsyuba, S. A., Islamova, L. N., Fazleeva, G. M., ... & Mustafina, A. R. (2022). Temperature-sensitive emission of dialkylaminostyrylhetarene dyes and their incorporation into phospholipid aggregates: Applicability for thermal sensing and cellular uptake behavior. *Spectrochimica Acta Part A: Molecular and Biomolecular Spectroscopy, 268,* 120647. https://www.sciencedirect.com/science/article/pii/S1386142521012245.

79. Zakiyyan, N., Darr, C. M., Chen, B., Mathai, C., Gangopadhyay, K., McFarland, J., ... & Maschmann, M. R. (2021). Surface plasmon enhanced fluorescence temperature mapping of aluminum nanoparticle heated by laser. *Sensors, 21*(5), 1585. https://www.mdpi.com/1424-8220/21/5/1585.

80. Xia, T., Shao, Z., Yan, X., Liu, M., Yu, L., Wan, Y., ... & Zhao, D. (2021) Tailoring the triplet level of isomorphic Eu/Tb mixed MOFs for sensitive temperature sensing. *Chemical Communications, 57,* 3143–3146. https://pubs.rsc.org/en/content/articlehtml/2021/cc/d1cc00297j.

81. Zheng, B., Fan, J., Chen, B., Qin, X., Wang, J., Wang, F., ... & Liu, X. (2022). Rare-earth doping in nanostructured inorganic materials. *Chemical Reviews, 122,* 5519–5603. https://pubs.acs.org/doi/abs/10.1021/acs.chemrev.1c00644.

82. Sing, C. E., Kunzelman, J., & Weder, C. (2009). Time–temperature indicators for high temperature applications. *Journal of Materials Chemistry, 19,* 104–110. https://pubs.rsc.org/en/content/articlehtml/2009/jm/b813644k.

83. Rocha, J., Brites, C. D., & Carlos, L. D. (2016). Lanthanide organic framework luminescent thermometers. *Chemistry a European Journal, 22,* 14782–14795. https://doi.org/10.1002/chem.201600860.

84. Medintz, I. L., Uyeda, H. T., Goldman, E. R., & Mattoussi, H. (2005). Quantum dot bioconjugates for imaging, labelling and sensing. *Nature Materials, 4*(6), 435–446. https://doi.org/10.1038/nmat1390.

85. Beaulac, R., Archer, P. I., van Rijssel, J., Meijerink, A., & Gamelin, D. R. (2008). Exciton storage by Mn^{2+} in colloidal Mn^{2+}-doped CdSe quantum dots. *Nano Letters, 8*(9), 2949–2953. https://doi.org/10.1021/nl801847e.

86. Li, S., Zhang, K., Yang, J. M., Lin, L., & Yang, H. (2007). Single quantum dots as local temperature markers. *Nano Letters, 7*(10), 3102–3105. https://doi.org/10.1021/nl071606p.

87. Walker, G. W., Sundar, V. C., Rudzinski, C. M., Wun, A. W., Bawendi, M. G., & Nocera, D. G. (2003). Quantum-dot optical temperature probes. *Applied Physics Letters, 83*(17), 3555–3557. https://doi.org/10.1063/1.1620686.

88. Tang, L., Zhang, Y., Liao, C., Guo, Y., Lu, Y., Xia, Y., & Liu, Y. (2022). Temperature-dependent photoluminescence of CdS/ZnS core/shell quantum dots for temperature sensors. *Sensors, 22*(22), 8993. https://doi.org/10.3390/s22228993.

89. Napoli, A., Anzidei, M., Ciolina, F., Marotta, E., Marincola, B. C., Brachetti, G., ... & Catalano, C. (2013). MR-guided high-intensity focused ultrasound: current status of an emerging technology. *Cardiovascular and Interventional Radiology, 36*, 1190–1203. https://doi.org/10.1007/s00270-013-0592-4.

90. Rodrigues, D. B., Stauffer, P. R., Eisenbrey, J., Beckhoff, V., & Hurwitz, M. D. (2017). Oncologic applications of magnetic resonance guided focused ultrasound. *Advances in Radiation Oncology,* 69–108. https://dx.doi.org/10.1007/978-3-319-53235-6_4.

91. Kim, Y. S. (2015). Advances in MR image-guided high-intensity focused ultrasound therapy. *International Journal of Hyperthermia, 31*(3), 225–232. https://doi.org/10.3109/02656736.2014.976773.

92. Toccaceli, G., Delfini, R., Colonnese, C., Raco, A., & Peschillo, S. (2018). Emerging strategies and future perspective in neuro-oncology using transcranial focused ultrasonography technology. *World Neurosurgery, 117*, 84–91. https://doi.org/10.1016/j.wneu.2018.05.239.

93. Kuroda, K. (2005). Non-invasive MR thermography using the water proton chemical shift. *International Journal of Hyperthermia, 21*, 547–560. https://doi.org/10.1080/02656730500204495.

94. Ferrer, C. J., Bartels, L. W., van der Velden, T. A., Grüll, H., Heijman, E., Moonen, C. T., & Bos, C. (2020). Field drift correction of proton resonance frequency shift temperature mapping with multichannel fast alternating non-selective free induction decay readouts. *Magnetic Resonance in Medicine, 83*(3), 962–973. https://doi.org/10.1002/mrm.27985.

95. Lutz, N. W., & Bernard, M. (2020). Contactless thermometry by MRI and MRS: Advanced methods for thermotherapy and biomaterials. *Iscience, 23*(10). https://doi.org/10.1016%2Fj.isci.2020.101561.

96. Geim, A. K., & Grigorieva, I. V. (2013). Van der Waals heterostructures. *Nature, 499*, 419–425. https://www.nature.com/articles/nature12385.

97. Palacios-Berraquero, C., Barbone, M., Kara, D. M., Chen, X., Goykhman, I., Yoon, D., ... & Atatüre, M. (2016). Atomically thin quantum light-emitting diodes. *Nature Communications, 7*, 12978. https://www.nature.com/articles/ncomms12978.

98. Kianinia, M., Tawfik, S. A., Regan, B., Tran, T. T., Ford, M. J., Aharonovich, I., & Toth, M. (2017). Robust solid state quantum system operating at 800 K. *ACS Photonics, 4*(4), 768-773

99. Blakley, S., Liu, X., Fedotov, I., Cojocaru, I., Vincent, C., Alkahtani, M., ... & Zheltikov, A. (2019). Fiber-optic quantum thermometry with germanium-vacancy centers in diamond. *ACS Photonics, 6*(7), 1690–1693. https://dx.doi.org/10.1021/acsphotonics.9b00206.

100. Drijkoningen, S., Pobedinskas, P., Korneychuk, S., Momot, A., Balasubramaniam, Y., Van Bael, M. K., ... & Haenen, K. (2017). On the origin of diamond plates deposited at low temperature. *Crystal Growth & Design, 17*(8), 4306–4314. https://doi.org/10.1021/acs.cgd.7b00623.

101. Kikkeri, R., Laurino, P., Odedra, A., & Seeberger, P. H. (2010). Synthesis of carbohydrate-functionalized quantum dots in microreactors. *Angewandte Chemie International Edition, 49*(11), 2054–2057. https://doi.org/10.1002/anie.200905053.

102. Köster, S., Angile, F. E., Duan, H., Agresti, J. J., Wintner, A., Schmitz, C., ... & Weitz, D. A. (2008). Drop-based microfluidic devices for encapsulation of single cells. *Lab on a Chip, 8*(7), 1110–1115. https://pubs.rsc.org/en/content/articlehtml/2008/lc/b802941e.

103. Maltezos, G., Rajagopal, A., & Scherer, A. (2006). Evaporative cooling in microfluidic channels. *Applied Physics Letters, 89*(7). https://pubs.aip.org/aip/apl/article/89/7/074107/332574.

104. Kopp, M. U., Mello, A. J. D., & Manz, A. (1998). Chemical amplification: continuous-flow PCR on a chip. *Science, 280*(5366), 1046–1048. https://www.science.org/doi/abs/10.1126/science.280.5366.1046.

105. Weinert, F. M., Kraus, J. A., Franosch, T., & Braun, D. (2008). Microscale fluid flow induced by thermoviscous expansion along a traveling wave. *Physical Review Letters, 100*(16), 164501. https://journals.aps.org/prl/abstract/10.1103/PhysRevLett.100.164501.

106. Fan, W., Chen, X., Ge, Y., Jin, Y., Jin, Q., & Zhao, J. (2019). Single-cell impedance analysis of osteogenic differentiation by droplet-based microfluidics. *Biosensors and Bioelectronics, 145*, 111730. https://www.sciencedirect.com/science/article/pii/S0956566319308097.

107. Peterson, D. S., Rohr, T., Svec, F., & Fréchet, J. M. (2002). Enzymatic micro-reactor-on-a-chip: Protein mapping using trypsin immobilized on porous polymer monoliths molded in channels of microfluidic devices. *Analytical Chemistry, 74*(16), 4081–4088. https://pubs.acs.org/doi/abs/10.1021/ac020180q.

108. Stone, H. A., Stroock, A. D., & Ajdari, A. (2004). Engineering flows in small devices: Microfluidics toward a lab-on-a-chip. *Annual Review of Fluid Mechanics, 36*, 381–411. https://www.annualreviews.org/doi/abs/10.1146/annurev.fluid.36.050802.122124.

109. Suter, J. D., White, I. M., Zhu, H., & Fan, X. (2007). Thermal characterization of liquid core optical ring resonator sensors. *Applied Optics, 46*(3), 389–396. https://opg.optica.org/abstract.cfm?uri=ao-46-3-389.

110. Childs, P. R., Greenwood, J. R., & Long, C. A. (2000). Review of temperature measurement. *Review of Scientific Instruments, 71*(8), 2959–2978. https://pubs.aip.org/aip/rsi/article-abstract/71/8/2959/436066.

111. Ma, S., Pang, Y., Ji, Q., Zhao, X., Li, Y., Qin, Z., ... & Xu, Y. (2023). High-temperature sensing based on GAWBS in silica single-mode fiber. *Sensors, 23*, 1277. https://doi.org/10.3390/s23031277.

112. Runowski, M., Woźny, P., Stopikowska, N., Martín, I. R., Lavín, V., & Lis, S. (2020). Luminescent nanothermometer operating at very high temperature—Sensing up to 1000 K with upconverting nanoparticles (Yb^{3+}/Tm^{3+}). *ACS Applied Materials & Interfaces, 12*(39), 43933–43941. https://pubs.acs.org/doi/abs/10.1021/acsami.0c13011.

113. Debasu, M. L., Ananias, D., Pastoriza-Santos, I., Liz-Marzán, L. M., Rocha, J., & Carlos, L. D. (2013). All-in-one optical heater-thermometer nanoplatform operative from 300 to 2000 K based on Er^{3+} emission and blackbody radiation. *Advanced Materials, 25*(35), 4868–4874. https://onlinelibrary.wiley.com/doi/abs/10.1002/adma.201300892.

114. Gao, G., Busko, D., Kauffmann-Weiss, S., Turshatov, A., Howard, I. A., & Richards, B. S. (2018). Wide-range non-contact fluorescence intensity ratio thermometer based on Yb^{3+}/Nd^{3+} co-doped La_2O_3 microcrystals operating from 290 to 1230 K. *Journal of Materials Chemistry C, 6*(15), 4163–4170. https://pubs.rsc.org/en/content/articlehtml/2018/tc/c8tc00782a.

115. Fan, S., Wang, S., Yu, L., Sun, H., Gao, G., & Hu, L. (2017). Ion-redistribution induced efficient upconversion in β-NaYF$_4$: 20% Yb^{3+}, 2% Er^{3+} microcrystals with well controlled morphology and size. *Optics Express, 25*(1), 180–190. https://opg.optica.org/abstract.cfm?uri=oe-25-1-180.

116. Auzel, F. (2004). Upconversion and anti-stokes processes with f and d ions in solids. *Chemical Reviews, 104*(1), 139–174. https://pubs.acs.org/doi/full/10.1021/cr020357g.

117. Wang, F., & Liu, X. (2009). Recent advances in the chemistry of lanthanide-doped upconversion nanocrystals. *Chemical Society Reviews, 38*(4), 976–989. https://pubs.rsc.org/en/content/articlehtml/2009/cs/b809132n.

118. Zhou, J., Liu, Q., Feng, W., Sun, Y., & Li, F. (2015). Upconversion luminescent materials: advances and applications. *Chemical Reviews, 115*(1), 395–465. https://pubs.acs.org/doi/full/10.1021/cr400478f.

119. Xu, W., Song, Q., Zheng, L., Zhang, Z., & Cao, W. (2014). Optical temperature sensing based on the near-infrared emissions from Nd^{3+}/Yb^{3+} codoped $CaWO_4$. *Optics Letters, 39*(16), 4635–4638. https://opg.optica.org/abstract.cfm?uri=ol-39-16-4635.

120. Maia, L. J., Faria Filho, F. M., Jerez, V., Moura, A. L., & de Araújo, C. B. (2015). Structural and luminescence properties of Nd^{3+}/Yb^{3+} codoped $Al_4B_2O_9$ nanocrystalline powders. *Journal of Materials Chemistry C, 3*(44), 11689–11696. https://pubs.rsc.org/en/content/articlehtml/2015/tc/c5tc01696g.

121. Xu, W., Qi, H., Zheng, L., Zhang, Z., & Cao, W. (2015). Multifunctional nanoparticles based on the Nd^{3+}/Yb^{3+} codoped $NaYF_4$. *Optics Letters, 40*(23), 5678–5681. https://opg.optica.org/abstract.cfm?uri=ol-40-23-5678.

Chapter 10

Recent advancements in optical thermometry based on photoluminescent perovskite oxides

Sariga C. Lal and Subodh Ganesanpotti

10.1 INTRODUCTION

Temperature is recognized as one of the most measured physical variables, given its necessity in practically all experimental and applied fields of science, engineering, medicine, and various aspects of daily human life.[1–4] Monitoring temperature holds paramount importance for several reasons. Firstly, temperature significantly impacts medical care for both humans and animals as well as the realms of food, beverage, and agriculture. Secondly, numerous types of equipment and machinery necessitate specific temperature ranges for optimal functionality. Exposure to excessive heat can lead to overheating of components, resulting in malfunctions or even complete failure, especially in electronic devices like computers and servers. High temperatures may also cause damage to individual components within the equipment. Furthermore, monitoring temperature enables the optimization of energy usage and cost reduction. For instance, in buildings, temperature sensors are employed to regulate heating and cooling systems, thereby reducing energy consumption and associated expenses.[1–3,5] According to marketsandmarkets, the temperature sensor market is anticipated to grow from 7.4 billion USD in 2024 to 9.7 billion USD in 2029, reflecting a compound annual growth rate (CAGR) of 5.6% during the forecast period. The projected growth is attributed to the increasing integration of temperature control systems in food safety management, Industry 4.0 and the Internet of Things (IoT), along with the expanding use of temperature in portable and healthcare equipment over the next five years.[6]

Luminescence thermometry, a well-established concept dating back to the 1930s, has undergone progressive evolution as a real-time temperature sensing method owing to its fast response, high spatial, temporal, and temperature resolutions, remote detection, and minimized disturbance of the sample temperature during measurements.[2,7] Over the years, there has been a pronounced exponential surge in the number of publications and citations pertaining to this field. Figure 10.1 shows the number of articles published by subject area till 2023 and the total number of publications during the year 2000–2023.

DOI: 10.1201/9781032661537-10

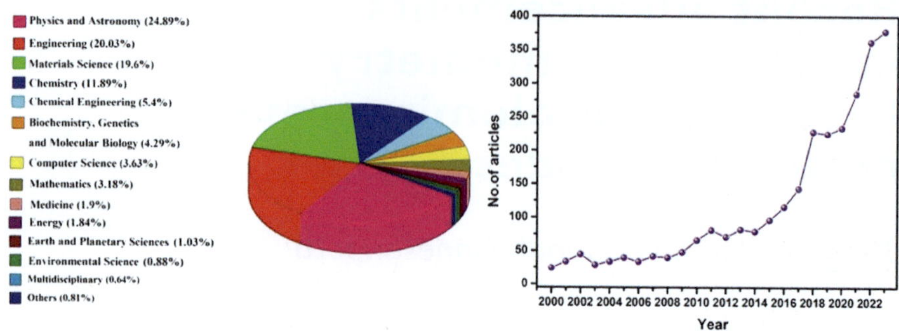

Figure 10.1 Piechart showing the number of articles on luminescence thermometry by subject area (left) published until 2023. The number of articles published during the year 2000 to 2023. (Data obtained from Scopus for keywords "optical thermometry".

The contactless operating mode of luminescence temperature sensors successfully mitigates inherent limitations associated with traditional contact thermometers such as liquid-filled and bimetallic thermometers, thermocouples, pyrometers, and thermistors. These sensors expand the scope of temperature measurements to nano- or sub-micrometer scales, high-speed moving objects, cellular interiors, and challenging environments, finding applications in various fields like microelectronics, micro-optics, microfluidics, and nanomedicine.[7]

Despite the dependence of optical thermometry (luminescence thermometry in this context) on monitoring various temperature-dependent parameters such as fluorescence intensity ratio (FIR), spectral line position, bandwidth, and rise or decay lifetime of excited levels, researchers commonly opt for FIR derived from thermally coupled energy levels (TCLs) or emissions originating from dual luminescent centers (comprising either lanthanide (Ln^{3+}) – lanthanide (Ln^{3+}) ions, lanthanide – transition metal (TM) ions, or lanthanide – ns^2 type ions like Tl^+, Sn^{2+}, Sb^{3+}, Bi^{3+}, Te^{4+} etc.).[2,8,9] FIR measurements exhibit the advantage of using a relatively uncomplicated device and operating without an external reference, displaying enhanced robustness against systematic errors arising from variations in excitation power. In contrast, spectral line position and bandwidth are the least explored temperature-dependent parameters in luminescence thermometry, which can be attributed to the requisite high spectral resolution of the spectrometer, the intricate treatment of experimental data, and generally lower thermal sensitivities compared to the FIR method.[8] Temperature sensing based on lifetime measurements demands a sophisticated device involving time-consuming data acquisition and treatment. Furthermore, this method becomes less reliable when the lifetimes are not single exponential. Despite the merits and demerits associated with different temperature-dependent parameters, most of the optical thermometers reported so far have focused on utilizing only one luminescent parameter for temperature sensing.[8] The evolving trend in optical thermometry emphasizes the importance of studying the temperature dependence of different luminescence parameters. In multiparametric thermometers, temperature exerts a simultaneous influence on various luminescence properties. The interplay of temperature dependencies across diverse luminescence properties paves the way for novel approaches to temperature sensing utilizing optical materials.[8,10] Recently, Maturi *et al.* reported a maximum temperature sensitivity of 50% K^{-1} achieved through a multiparametric temperature readout using the green fluorescent protein and Ag_2S nanoparticles.[10] The significance of multimode thermometry is underscored by the potential to broaden the dynamic range of temperature measurement and enhance thermometric parameters. The essence of multimode thermometry lies in its ability to switch between different modes, achieving maximum sensitivity at specific temperature ranges, making a current trend in the field.[8,10]

A luminescent material (host+activator or host+activator+sensitizer) is a critical part in optical thermometry. Host materials with different structures and compositions have been used to design optical thermometers so far. The ABX_3 perovskite structure, characterized by a three-dimensional arrangement of corner-sharing BX_6 octahedra (X=oxygen or halogens) with A-site cations positioned centrally in the cuboctahedral cavity can undergo modifications at either A- or B-sites. Double perovskites (represented by the formulae $AA'BB'O_6$, $A_2BB'O_6$, and $AA'B_2O_6$) and layered perovskites (Ruddleson-Popper, Aurivillius, and Dion-Jacobson phases) have garnered significant interest due to their diverse technological applications in ferroelectrics, superconductors, photonics, photocatalysts, and as host materials for luminescence centers.[4,11–16] Perovskite compounds serve as host lattices for lanthanide and transition metal ions, enabling efficient ultraviolet absorption through the excitation of the charge transfer state (CTS) and subsequent energy transfer to the activators. Photoluminescence observed in perovskites holds substantial potential for application in photovoltaic devices, photocatalysis, sensors, and lighting devices.[4,16,17] In recent years, several lanthanides and transition metal-activated perovskites have been probed for optical thermometry. Upon searching the keyword "Luminescence thermometry based on perovskites and double perovskites" on Google Scholar, roughly 8000 results were displayed, encompassing both academic publications and patents. Various perovskite materials, including both halide and oxide compositions, have been explored for multimode thermal sensing applications. Very recent reports have discussed certain halide perovskites such as $Cs_2NaInCl_6$: Sb^{3+}/Er^{3+}, $Cs_2NaErCl_6$: Yb^{3+}, $Cs_2AgInCl_6$: Bi^{3+}/Tb^{3+}, $Cs_3Bi_2Cl_9$: Er^{3+} metal halide perovskite, $Cs_2Ag_{0.6}Na_{0.4}In_{0.9}Bi_{0.1}Cl_6$: Tm^{3+} microcrystals, $CsPbCl_3$: Mn^{2+} quantum dots (QDs) and so on for multimode optical thermometry.[18–23] These reports explain multimode thermometer design incorporating both conventional and novel techniques. Beyond traditional oxide and halide perovskites, luminescence thermometry research extends to perovskite-like tin halides, heterometallic perovskite type metal-organic frameworks, and lanthanide-activated perovskite-type materials such as $LiTaO_3$, $LiNbO_3$, etc.[24–28] This chapter focuses specifically on the temperature sensing capabilities of perovskite oxides, providing a comprehensive review. The discussion encompasses the advantages and drawbacks of single and dual-emitting perovskites utilized for single-mode and multimode optical thermometry. Additionally, a concise overview of all developed techniques for optical thermometry based on perovskite oxides is presented. This chapter concludes by thoroughly examining the variation in thermometric figures of merit across different modes, their comparative analysis, and an outlook on potential enhancements to further refine thermometric characteristics.

10.2 SINGLE-MODE OPTICAL THERMOMETERS BASED ON PEROVSKITE OXIDES

Early luminescence thermometry systems utilized the shift in emission intensity and wavelength to measure temperature. In 1952, Bradley employed the intensity shift of ZnCdS:Ag, Cu phosphor to measure the temperature distribution on a flat wedge in a supersonic flow field.[29] Subsequently, Gross *et al.* utilized the intensity and wavelength shift of phosphors ZnS:Ag, Cl and CdS:Ag to develop phosphor-based detectors for thermal radiation.[30,31] Sholes and Small later used the decay lifetime of Cr^{3+} R lines ($^2E_g \rightarrow ^4A_{2g}$ spin-forbidden transitions) to monitor thermal dosimetry during hyperthermia treatments of malignant tumors.[32] The fundamental principle of luminescence thermometry is explained based on the FIR of emissions from TCLs of luminescent centers, especially rare earth ions. Various ratiometric methods for temperature sensing have been proposed, including dual-excited single-band ratiometric thermometry (SBR), time-resolved single-band ratiometric thermometry, and dual-activated ratiometric intensity measurements. For conventional ratiometric thermometry based on FIR of TCLs, a single luminescent center is sufficient.[2,33] However, to enhance the thermometric figures of merit (absolute, S_a and relative, S_r sensitivities), multiple activation centers are incorporated, with one acting as a reference signal and the other as a temperature probe. Several reports exist on optical thermometry based on perovskite oxides, working in single-mode and multi-mode configurations.

10.2.1 Perovskites activated with a single luminescent center

Perovskite oxides, activated with a single ion (typically Ln^{3+}, TM, or ns^2 type ions), are being explored for their suitability in FIR-based single-mode thermometry applications, owing to their multiple emission lines.

Examples include La_2MgTiO_6: Cr^{3+}, $(Ba, Sr)_2YTaO_6$: Eu^{3+}, Ba_2LaNbO_6: Mn^{4+}, $LiLaMgWO_6$: Er^{3+}, La_2MgTiO_6: Nd^{3+}, Gd_2ZnTiO_6: Pr^{3+}, $CaZrO_3$: Pb^{2+}, among others.[34–40] In most instances, either FIR based on activator emission or fluorescence decay time (FL) is employed to calculate thermometric figures of merit. Notably, in rare cases, FIR based on both the host and activator ions is utilized, as observed in Ba_2MgWO_6: Eu^{3+}, Ba_2MgWO_6: Er^{3+}, $(Ba, Sr)_2MgMoO_6$: Eu^{3+}, and $(Ba, Sr)_2ZnMoO_6$: Eu^{3+}.[41–43] The host emission in these cases stems from metal-to-ligand charge transfer transitions. Utilizing the FIR technique (Eu^{3+} to host emission), Ba_2ZnMoO_6: Eu^{3+} exhibits a maximum relative sensitivity (S_r) of 9.2% K^{-1} at 77 K within a temperature range of 73–173 K.[43] Similarly, the temperature dependence of the lifetime of the excited states of Mn^{4+} in $SrScLiTeO_6$ results in a maximum S_r of 7.86% K^{-1} at 473 K.[44] Recently, Arnab De *et al.* introduced an innovative approach using the Raman-PL intensity ratio (RPIR) to measure temperature in $BaTiO_3$: Eu^{3+}. The optical temperature sensor based on this technique operates effectively within a large dynamic range from 11 to 573 K, with maximum relative sensitivities (S_r) determined to be +6.4% K^{-1} at 65 K, –7.6% K^{-1} at 280 K below room temperature and +2.4% K^{-1} at 303 K above room temperature.[45] The sign reversal in S_r is attributed to the nonmonotonic temperature dependence of RPIR, given the mutually independent thermal dependence of Raman and PL signals.[45] Furthermore, reports indicate the utilization of emission intensity of a single band and SBR techniques. SBR is a successful technique involving the ratio of a single emission at two different excitation wavelengths. For example, Gd_2ZnTiO_6: Pr^{3+} achieves a maximum S_r of 3.64% K^{-1} at 298 K with a temperature uncertainty of 0.14 K via the SBR method.[39] Conversely, $BaLaTaO_6$: Mn^{4+} and $MgTiO_3$: Ni^{2+} exhibit a maximum S_r of 3.2% K^{-1} at 450 K (in the temperature range of 303–473 K) and 3.2% K^{-1} at 283 K (in the range of 100–500 K), respectively, based on the temperature dependence of emission intensity.[46,47] Optical temperature sensing has been explored with relaxor ferroelectric solid solutions featuring a perovskite phase, such as $Pb(Mg_{1/3}Nb_{2/3})O_3$-$PbTiO_3$:Pr^{3+}(PMN-PT: Pr^{3+}) and $0.94(Na_{1/2}Bi_{2/3}TiO_3)$-$0.06(BaTiO_3)$:$Eu^{3+}$(NBT-6BT:$Eu^{3+}$).[48,49] Transparent ferroelectric photoluminescent (PL) ceramics, endowed with commendable ferroelectric and piezoelectric properties, find applications involving the coupling between electrical and optical properties. By leveraging the intensity ratio of Pr^{3+} emission lines, PMN-PT: Pr^{3+} exhibits a remarkable maximum absolute temperature sensitivity (S_a) of 0.70% K^{-1} for FIR of transitions $^3P_1 \rightarrow {}^3H_5/{}^3P_2 \rightarrow {}^3H_5$ within the temperature range of 223–313 K.[48] Furthermore, a highly responsive cryogenic optical thermometer utilizing NBT-6BT: Eu^{3+} demonstrates temperature determination through the intensity ratio of emissions from host metastable states and Eu^{3+}. This approach achieves a maximum sensitivity (S_r) of 3.05% K^{-1} with an uncertainty of 0.16 K at 140 K, over a temperature range of 10–300 K.[49]

While the temperature dependence of a single emission center can yield excellent thermometric parameters, as illustrated in Table 10.1, this is not always the case, especially when dealing with rare earth-activated perovskites. In these instances, the emission intensity is not significantly affected by temperature, primarily because of the shielding effect exerted by the outer $5d6s^2$ orbitals on the $4f$ electrons, resulting in a weakened electron–lattice interaction. This diminished interaction leads to relatively low thermometric parameters, thereby restricting the performance of optical thermometers. A viable solution involves co-doping with other ions, such as transition metal (TM) ions, whose emission is notably influenced by temperature. This approach proves effective in achieving enhanced thermometric figures of merit.

10.2.2 Perovskites activated with multiple luminescent centers

In the quest for innovative optical sensor materials with high sensitivity, numerous researchers have directed their efforts toward dual-emitting-center phosphors. These materials predominantly rely on the different temperature quenching rates exhibited by the two activators, such as Ln^{3+}-Ln^{3+}, Ln^{3+}-TM, TM-TM, Ln^{3+}-ns^2 type ions or TM – ns^2 type ions. The robust and distinguishable emission peaks from the dual-emitting centers are valuable for generating precise FIR signals, effectively capturing temperature variations. Consequently, this strategic approach finds increasing application in developing optical thermometers.

As evident from Table 10.1, a significant number of co-doped perovskite oxides rely on the FIR method for thermometric calculations. In instances involving Ln^{3+}-Ln^{3+} combinations, the ratio of f-f emission transitions from one or both rare earth activators is a key consideration. For example, in the Y_2MgTiO_6: Tm^{3+}, Sm^{3+} double perovskite, the FIR of Sm^{3+} emission transitions, specifically $^4G_{5/2}{\rightarrow}^6H_{5/2}$ and $^4G_{5/2}{\rightarrow}^6H_{7/2}$, is utilized. In this scenario, a dipole–dipole interaction induces energy transfer from Tm^{3+} to Sm^{3+}.[50] Another example involves an upconverting phosphor, Ba_2LaNbO_6, co-doped with Er^{3+} and Yb^{3+}, where the FIR involving $^2H_{11/2}{\rightarrow}^4I_{15/2}$ and $^2F_{9/2}{\rightarrow}^4I_{15/2}$ transitions of Er^{3+} is employed for estimating thermometric properties, with Yb^{3+} ions serving as sensitizers for Er^{3+} luminescence.[51] In the case of Ca_2YNbO_6: Dy^{3+}, Eu^{3+}, both Dy^{3+} ($^4F_{9/2}{\rightarrow}^6H_{13/2}$) and Eu^{3+} ($^5D_0{\rightarrow}^7F_2$) emission lines are utilized to estimate thermometric figures of merit.[52] When rare earth and transition metal ions are co-doped, the rare earth emission serves as the reference signal, and the transition metal is used as the temperature probe, as seen in examples like La_2MgTiO_6: Eu^{3+}, Mn^{4+}, and $NaLaMgWO_6$: Eu^{3+}, Mn^{4+}.[13,53] For La_2MgTiO_6 activated with V^{5+} and Cr^{3+} (both transition metal ions), the intensity ratio of {host+V5+} emission to Cr^{3+} emission is characterized to evaluate temperature sensitivity, yielding a maximum S_r of 1.96% K^{-1} at 165 K within the temperature

Table 10.1 Perovskite oxides used for single-mode optical thermometry

			Perovskites activated with a single luminescent center		
Phosphor	Method	Temperature range	Maximum relative sensitivity and temperature	Temperature resolution (δT)	References
La_2MgTiO_6:Cr^{3+}	FIR	80–600 K	2.68% K^{-1} @220 K	0.015 K	71
Ba_2YTaO_6:Eu^{3+}	FL	303–563 K	0.18% K^{-1} @563 K	–	35
Sr_2YTaO_6:Eu^{3+}	FL	303–563 K	0.13% K^{-1} @563 K	–	35
Ba_2LaNbO_6:Mn^{4+}	FL	300–500 K	2.77% K^{-1} @397 K	–	36
Sr_2InSbO_6:Cr^{3+}	FL	135–460 K	0.73% K^{-1} @280 K	0.53K	72
$BaLaZnSbO_6$:Mn^{4+}	FL	13–473 K	1.87% K^{-1} @387K	–	73
Sr_2AlTaO_6:Cr^{3+}	FIR	293–473 K	2.18% K^{-1} @293 K	–	74
$SrScLiTeO_6$:Mn^{4+}	FL	23–498 K	7.86% K^{-1} @473 K	0.05 K	44
$CaTiO_3$:Pr^{3+}	FIR	20–200 K	0.82% K^{-1} @120 K	0.20 K	75
Sr_2GdSbO_6:Eu^{3+}	FIR	298–423 K	0.27% K^{-1} @298 K	–	76
$SrLaNaTeO_6$:Sm^{3+}	FIR	298–573 K	0.149% K^{-1} @498 K	–	77
Sr_2InTaO_6:Mn^{4+}	FL	298–418 K	3.27% K^{-1} @373 K	0.12 K	78
$Na_{0.5}Gd_{0.5}TiO_3$:Er^{3+}	FIR	298–473 K	1.2% K^{-1} @300 K	–	79
$ZnTiO_3$:Eu^{3+}	FIR	80–310 K	0.65% K^{-1} @310 K	–	80
Y_2MgTiO_6:Mn^{4+}	FIR	10–513 K	0.142% K^{-1} @153 K	–	81
Ba_2MgWO_6:Eu^{3+}, Li^+	FIR	77–200 K	1.5% K^{-1} @120 K	<1 K	41
$LaMg_{0.402}Nb_{0.598}O_3$:$Pr^{3+}$	FIR	298–523 K	0.7250% K^{-1} @473 K	–	82
Ba_2CaWO_6:Mn^{4+}	FIR	7–510 K	1.44% K^{-1} @150 K	–	83
$LiLaMgWO_6$:Er^{3+}	FIR	303–483 K	2.24% K^{-1} @483 K	–	37
$NaLaMgWO_6$:Er^{3+}	FIR	303–483 K	1.04% K^{-1} @303 K	–	84

(Continued)

Table 10.1 (Continued) Perovskite oxides used for single-mode optical thermometry

Phosphor	Method	Temperature range	Maximum relative sensitivity and temperature	Temperature resolution (δT)	References
		Perovskites activated with a single luminescent center			
$La_2MgHfO_6:Cr^{3+}$	FIR	84–264 K	1.12% K^{-1} @84 K	–	85
$NaGdMgTeO_6:Eu^{3+}$	FIR	300–500 K	0.23% K^{-1} @300 K	–	86
$Ba_2CaWO_6:Er^{3+}$	FIR	298–575 K	0.90% K^{-1} @386 K	–	11
$La_2MgTiO_6:Nd^{3+}$	FIR	77–698 K	0.83% K^{-1} @275 K	–	38
$Ba_2MgWO_6:Er^{3+}$	FIR	80–273 K	2.78% K^{-1} @198 K	0.019	42
$Ca_2InNbO_6:Mn^{4+}$	FL	303–423 K	3.30% K^{-1} @393 K	0.11 K	87
$Sr_2InNbO_6:Mn^{4+}$	FL	303–423 K	4.25% K^{-1} @363 K	0.09 K	87
$Ba_2MgMoO_6:Eu^{3+}$	FIR	73–273 K	6.2% K^{-1} @77 K	0.01–1 K	43
$Sr_2MgMoO_6:Eu^{3+}$	FIR	73–373 K	3.2% K^{-1} @348 K	0.01–1 K	43
$Ba_2ZnMoO_6:Eu^{3+}$	FIR	73–173 K	9.2% K^{-1} @77 K	<0.01 K	43
$Sr_2ZnMoO_6:Eu^{3+}$	FIR	73–373 K	1.8% K^{-1} @77 K	<0.01 K	43
$Sr_2CdTeO_6:Eu^{3+}$	FIR	300–500 K	0.28% K^{-1} @300 K	–	88
$LaGaO_3:Nd^{3+}$	FIR	400–700 K	1.4% K^{-1} @440 K	~1 K	89
$La_2MgTiO_6:Pr^{3+}$	FIR	80–500 K	1.28% K^{-1} @350 K	–	90
$Sr_2GaNbO_6:Eu^{3+}$	FIR	298–623 K	1.86% K^{-1} @298 K	0.016 K	91
$Sr_2GaNbO_6:Sm^{3+}$	FIR	298–623 K	1.70% K^{-1} @623 K	0.036 K	91
$Ca_2LaNbO_6:Pr^{3+}$	FIR	313–573 K	0.89% K^{-1} @523 K	–	92
$BaTiO_3:Eu^{3+}$	RPIR	11–300 K	+6.4% K^{-1} @65 K	~0.18 K	45
			−7.6% K^{-1} @280 K	~0.01 K	
		303–573 K	+2.4% K^{-1} @303 K	~0.074 K	

(Continued)

Table 10.1 (Continued) Perovskite oxides used for single-mode optical thermometry

Perovskites activated with a single luminescent center				
Phosphor	Method	Temperature range	Maximum relative sensitivity and temperature	Temperature resolution (δT)	References
$BaLaTaO_6:Mn^{4+}$	Emission intensity	303–473 K	3.2% K^{-1} @450 K	–	46
$Sr_2InTaO_6:Mn^{4+}$	FL	25–450 K	1.396% K^{-1} @375 K	0.005 K	93
$CaZrO_3:Pb^{2+}$	Emission intensity	303–443 K	1.32% K^{-1} @362 K	–	40
$Gd_2ZnTiO_6:Pr^{3+}$	SBR	298–573 K	3.64% K^{-1} @298 K	0.14 K	39
$La_2MgTiO_6:Er^{3+}$	FIR	303–483 K	1.107% K^{-1} @303 K	–	94
$CaZrO_3:Bi^{3+}$	Emission intensity	303–443 K	1.78% K^{-1} @440 K	–	95
$Ba_2LaTaO_6:Eu^{3+}$	FL	293–713 K	0.185% K^{-1} @513 K	2.70 K	96
$MgTiO_3:Ni^{2+}$	Emission intensity	100–500 K	3.2% K^{-1} @283 K	–	47
$CaTiO_3:Ni^{2+}$	Emission intensity	100–500 K	2.44% K^{-1} @303 K	–	47
$BaTiO_3:Ni^{2+}$	Emission intensity	100–500 K	1.18% K^{-1} @303 K	–	47
$SrTiO_3:Ni^{2+}$	Emission intensity	100–500 K	0.70% K^{-1} @303 K	–	47
$Sr_2GdTaO_6:Mn^{4+}$	FL	293–473 K	1.73% K^{-1} @453 K	0.289 K	97
$La_2MgTiO_6:Eu^{3+}$	FIR	77–450 K	3% K^{-1} @77 K	–	98
$BaTiO_3:Pr^{3+}$	FIR	313–413 K	2.77% K^{-1} @371 K	–	99
$LaGaO_3:Cr^{3+}$	FIR	300–600 K	2.5% K^{-1} @300 K	0.05 K	100
$Ca_2Y_{0.97}Bi_{0.03}SbO_6:Eu^{3+}$	FIR	303–483 K	0.968% K^{-1} @483 K	–	56
$Ca_2Y_{0.97}Bi_{0.03}SbO_6:Sm^{3+}$	FIR	303–483 K	0.864% K^{-1} @483 K	–	56
$CaTiO_3:Pr^{3+}$ (flux assisted)	FIR	298–523 K	~5.2% K^{-1} @298 K	–	101
$CaHfO_3:Cr^{3+}$	FIR	40–150 K	~2% K^{-1} @40 K	0.045 K	102

(Continued)

Table 10.1 (Continued) Perovskite oxides used for single-mode optical thermometry

		Perovskites activated with a single luminescent center			
Phosphor	Method	Temperature range	Maximum relative sensitivity and temperature	Temperature resolution (δT)	References
$BaLaMgSbO_6:Mn^{4+}$	FL	77–503 K	1.30% K^{-1} @491 K	0.01 K	103
$Sr_2InSbO_6:Eu^{3+}$	FL	300–600 K	0.289% K^{-1} @300 K	–	104
Sr_2EuSbO_6	FL	300–600 K	0.766% K^{-1} @300 K	–	104
$BaTiO_3:Pr^{3+}$	FIR	303–413 K	2.26% K^{-1} @413 K	–	105
$Ba_{0.9}Ca_{0.1}TiO_3:Pr^{3+}$	FIR	303–413 K	1.59% K^{-1} @413 K	–	105
$LaGdO_3:Er^{3+}$	FIR	298–900 K	1.5% K^{-1} @600 K	–	106
$LaScO_3:Cr^{3+}$	FIR	50–275°C	1.23% °C^{-1} @220°C	4.8°C	107
$0.94(Na_{1/2}Bi_{1/2}TiO_2)$-$0.06(BaTiO_3): Eu^{3+}$	FIR	10–300 K	3.05% K^{-1} @140 K	0.16 K	49

Perovskites Activated with Multiple Luminescent Centers

Phosphor	Method	Temperature range	Maximum relative sensitivity and temperature	Temperature resolution (δT)	References
$La_2MgTiO_6:Dy^{3+}/Mn^{4+}$	FIR	298–498 K	2.85% K^{-1} @448 K	–	108
$Ca_2GdNbO_6:Mn^{4+}/Bi^{3+}$	FIR	303–573 K	2.13% K^{-1} @498 K	–	109
$Ca_2GdNbO_6:Mn^{4+}/Sm^{3+}$	FIR	303–573 K	1.11% K^{-1} @448 K	–	109
$La_2MgSnO_6:Bi^{3+}/Sm^{3+}$	FIR	303–503 K	0.88% K^{-1} @503 K	–	110
$Ca_2YSbO_6:Bi^{3+}/Sm^{3+}$	FIR	303–503 K	0.973% K^{-1} @503 K	–	111
$Ba_2MgWO_6:Er^{3+},Yb^{3+}, K^{+}$	FIR	303–573 K	0.9% K^{-1} @303 K	–	112
$SrBaMgWO_6:Er^{3+},Yb^{3+}, K^{+}$	FIR	303–573 K	0.8% K^{-1} @303 K	–	112
$Sr_2MgWO_6:Er^{3+},Yb^{3+}, K^{+}$	FIR	303–573 K	0.6% K^{-1} @303 K	–	112
$Ba_2LaTaO_6:Bi^{3+}/Mn^{4+}$	FIR	80–473 K	3.81% K^{-1} @350 K	–	12

(Continued)

Table 10.1 (Continued) Perovskite oxides used for single-mode optical thermometry

Perovskites activated with multiple luminescent centers'

Phosphor	Method	Temperature range	Maximum relative sensitivity and temperature	Temperature resolution (δT)	References
$Ca_2InTaO_6:Eu^{3+}/Mn^{4+}$	FIR	303–483 K	3.49% K^{-1} @303 K	0.0705 K	113
$La_2MgTiO_6:Mn^{4+}/Eu^{3+}$	FIR	298–498 K	2.09% K^{-1} @498 K	0.24 K	13
$BaTiO_3:Er^{3+}/Yb^{3+}-SrTiO_3:Nb$	FIR	306–486 K	1.17% K^{-1} @306 K	0.7 K	114
$SrTiO_3:Eu^{3+}-ZnTiO_3:Mn^{4+}$	FIR	80–310 K	3.3% K^{-1} @215 K	0.02 K	115
$Sr_2GdTaO_6:Mn^{4+}/Sm^{3+}$	FIR	298–573 K	2.94% K^{-1} @298 K	–	116
$CaTiO_3:Er^{3+}/Yb^{3+}$	FIR	298–623 K	0.671% K^{-1} @383 K	–	117
$Ca_2YNbO_6:Dy^{3+}/Eu^{3+}$	FIR	300–475 K	3.140% K^{-1} @475 K	–	52
$Ba_2CaWO_6:Er^{3+}, Yb^{3+}, Na^+$	FIR	297–575 K	1.45% K^{-1} @297 K	–	118
$GdScO_3:Er^{3+}/Yb^{3+}$	FIR	299–473 K	1.1790% K^{-1} @299 K	0.008 K	119
$Y_2MgTiO_6:Tm^{3+}/Sm^{3+}$	FIR	300–700 K	1.8244% K^{-1} @300 K	–	50
$NaLaMgWO_6:Mn^{4+}/Tb^{3+}$	FIR	330–500 K	2.35% K^{-1} @500 K	–	120
$K_{0.5}Na_{0.5}NbO_3:Er^{3+}/Ta^{5+}$	FIR	273–543 K	1.5836% K^{-1} @273 K	–	121
$Ba_2LaNbO_6:Eu^{3+}/Mn^{4+}$	FIR	298–498 K	2.08% K^{-1} @398 K	–	122
$Ca_2LaNbO_6:Eu^{3+}/Mn^{4+}$	FIR	298–498 K	1.51% K^{-1} @455 K	–	122
$CaLaMgTaO_6:Bi^{3+}/Eu^{3+}$	FIR	303–573 K	1.33% K^{-1} @483 K	–	123
$La_2MgTiO_6:V^{5+}/Cr^{3+}$	FIR	80–343 K	1.96% K^{-1} @165 K	<0.05 K	54
$NaLaMgWO_6:Er^{3+}/Mn^{4+}$	FIR	303–523 K	1.31% K^{-1} @523 K and 503 K	0.3817 K	124
$NaLaMgWO_6:Eu^{3+}/Mn^{4+}$	FIR	303–523 K	0.86% K^{-1} @523 K	0.58 K	53
$La_3Li_3W_2O_{12}:Eu^{3+}/Mn^{4+}$	FIR	303–523 K	0.70% K^{-1} @443 K	0.72 K	125

(Continued)

Table 10.1 (Continued) Perovskite oxides used for single-mode optical thermometry

Perovskites activated with multiple luminescent centers'					
Phosphor	Method	Temperature range	Maximum relative sensitivity and temperature	Temperature resolution (δT)	References
Ba_2LaNbO_6:Er^{3+}/Yb^{3+}	FIR	293–473 K	1.46% K^{-1} @313.5 K	–	51
La_2LiSbO_6:Tb^{3+}/Mn^{4+}	FIR	303–523 K	0.946% K^{-1} @523 K	0.529 K	126
La_2LiSbO_6:Dy^{3+}/Mn^{4+}	FIR	303–523 K	0.796% K^{-1} @523 K	0.628 K	126
La_2MgGeO_6:Bi^{3+}/Mn^{4+}	FIR	293–473 K	3.027% K^{-1} @383 K	–	55
Gd_2ZnTiO_6:Er^{3+}/Yb^{3+}	FIR	298–573 K	0.9011% K^{-1} @539 K	–	127
Sr_2ScTaO_6:Eu^{3+}/Mn^{4+}	FIR	300–500 K	3.66% K^{-1} @300 K	–	128
Gd_2ZnTiO_6:Bi^{3+}/Mn^{4+}	FIR	313–473 K	2.4% K^{-1} @423 K	–	129
Gd_2ZnTiO_6:Er^{3+}/Yb^{3+}	FIR	300–480 K	2.188% K^{-1} @313 K	–	130
$BaTiO_3$:Ho^{3+}/Yb^{3+}	SHG/UCL	25–305°C	2.78% °C^{-1} @146°C	0.85°C	58
$LaAlO_3$:Er^{3+}/Yb^{3+}	FIR	100–673 K	2.44% K^{-1} @175 K	–	14
$ZnTiO_3$:Eu^{3+}/Mn^{4+}	FIR	80–310 K	2.7% K^{-1} @210 K	–	131
$ZnTiO_3$:Mn^{4+}, Eu^{3+}, Al^{3+}	FIR	80–310 K	1.85% K^{-1} @200 K	–	131
Ca_2YSbO_6:Bi^{3+}/Eu^{3+}	FIR	303–483 K	0.968% K^{-1} @483 K	–	132
Ca_2YSbO_6:Bi^{3+}/Sm^{3+}	FIR	303–483 K	0.864% K^{-1} @483 K	–	132
$Bi_4Ti_3O_{12}$:Ho^{3+}/Yb^{3+}	FIR	323–543 K	2.11% K^{-1} @323 K	0.01 K	24
La_2LiSbO_6:Bi^{3+}/Sm^{3+}	FIR	298–498 K	1.48% K^{-1} @498 K	–	9
Ca_2MgWO_6:Bi^{3+}/Eu^{3+}	FIR	298–523 K	8.52% K^{-1} @323 K	–	57
La_2LiNbO_6:Bi^{3+}/Eu^{3+}	FIR	298–498 K	0.90% K^{-1} @498 K	–	133
$LaGaO_3$:Sm^{3+}/Mn^{4+}	FIR	373–473 K	2.09% K^{-1} @423 K	–	134

(Continued)

Table 10.1 (Continued) Perovskite oxides used for single-mode optical thermometry

Perovskites activated with multiple luminescent centers'					
Phosphor	Method	Temperature range	Maximum relative sensitivity and temperature	Temperature resolution (δT)	References
La_2LiSbO_6:Eu^{3+}/Mn^{4+}	FIR	303–523 K	0.891% K^{-1} @523 K	0.56 K	135
$CaZrO_3$:Pb^{2+}/Sm^{2+}	FIR	303–443 K	1.13% K^{-1} @340 K	–	40
$CaZrO_3$:Bi^{3+}/Sm^{3+}	FIR	303–443 K	1.15% K^{-1} @443 K	–	95
$NaLaMgWO_6$:Mn^{4+}/Pr^{3+}/Bi^{3+}	FIR	298–573 K	3.39% K^{-1} @298 K	–	136
$LaGaO_3$:Cr^{3+}/Nd^{3+}	FIR	300–650 K	~2% K^{-1} @300 K	0.04 K	137
$LaGaO_3$:Cr^{3+}/Nd^{3+}	FIR	303–650 K	0.4896% K^{-1} @303 K	–	138
Ca_2LaSbO_6:Eu^{3+}/Mn^{4+}	FIR	273–473 K	2.60% K^{-1} @473 K	–	139
$BaTiO_3$:Er^{3+}/Yb^{3+}	SHG/UCL	330–575 K	~4% K^{-1} @400 K	0.07 K	59
$Li_2Zn_2Mo_3O_{12}$:Er^{3+}/Yb^{3+}	FIR	253–423 K	1.72% K^{-1} @253 K	–	140
$SrTiO_3$:Ni^{2+}/Er^{3+}	FIR	143–463 K	0.44% K^{-1} @303 K	–	47
$CaTiO_3$:Ni^{2+}/Er^{3+}	FIR	143–463 K	0.3% K^{-1} @303 K	–	47
$K_{0.3}La_{1.233}MgWO_6$:Eu^{3+}/Mn^{4+}	FIR	273–473 K	0.91% K^{-1} @323 K	–	141
$LaAlO_3$:Eu^{2+}/Eu^{3+}	FIR	293–473 K	1.193% K^{-1} @403 K	–	142
$La_{0.92}Gd_{0.8}AlO_3$:Eu^{2+}/Eu^{3+}	FIR	303–473 K	3.233% K^{-1} @303 K	–	143
$SrTiO_3$:Ni^{2+}/Er^{3+}	FIR	−150–210°C	0.80% K^{-1} @40°C	–	144
$Bi_4Ti_3O_{12}$:Er^{3+}/Yb^{3+}	FIR	313–543 K	0.95% K^{-1} @313 K	–	145
$La_2Ti_2O_7$:Er^{3+}/Yb^{3+}	FIR	333–553 K	0.63% K^{-1} @333 K	0.1 K	146

(Continued)

Table 10.1 (Continued) Perovskite oxides used for single-mode optical thermometry

Perovskites activated with multiple luminescent centers'					
Phosphor	Method	Temperature range	Maximum relative sensitivity and temperature	Temperature resolution (δT)	References
$BaTiO_3$:Ho^{3+}/Yb^{3+}	FIR	305–515 K	0.34% K^{-1}@305 K	–	147
$BaTiO_3$:Er^{3+}/Eu^{3+}	RPIR	10–300 K	–3% K^{-1}@300 K +6% K^{-1}@75 K	–	148
$NaLaMgWO_6$:Ho^{3+}/Yb^{3+}	FIR	293–553 K	0.46% K^{-1}@548 K	–	149
$NaLaMgWO_6$:Er^{3+}/Yb^{3+}	FIR	293–553 K	1.14% K^{-1}@298 K	–	149
$CaTiO_3$:Er^{3+}/Yb^{3+}/Li^+	FIR	303–523 K	1.217% K^{-1}@303 K	–	150
$PbTiO_3$:Er^{3+}/Yb^{3+}	FIR	298–548 K	1.225% K^{-1}@298 K	–	151
$LaScO_3$:Cr^{3+}/Yb^{3+}	FIR	50–275°C	1.7% °C^{-1}@220°C	0.08°C	107

range of 80–343 K.[54] Examples of ns^2 type ion – TM codoped perovskite phosphors include Ba_2LaTaO_6: Bi^{3+}, Mn^{4+} and La_2MgGeO_6: Bi^{3+}, Mn^{4+}.[12,55] To enhance temperature sensitivity, ions with distinct temperature-dependent characteristics are preferred, as in the case of ns^2 type ion and transition metal ion combination. Bi^{3+} ions exhibit relatively stable luminescence intensity with temperature, while Mn^{4+} ions, known for their strong electron-phonon coupling common to transition metal ions, display heightened susceptibility to temperature variation. In ns^2 type ions – Ln^{3+} combination in perovskites oxides for optical thermometry, examples such as Ca_2YSbO_6: (Eu^{3+}, Sm^{3+}), Bi^{3+} and La_2LiSbO_6: Sm^{3+}, Bi^{3+} utilize the intensity ratio of Ln^{3+} to Bi^{3+}.[9,56] Among codoped perovskite oxides, the maximum relative sensitivity of 8.52% K^{-1}@323 K is achieved for a Bi^{3+} and Eu^{3+} codoped Ca_2MgWO_6 phosphor over a temperature range of 298–523 K.[57]

A novel approach is demonstrated by Zheng et $al.$, where the simultaneous utilization of second harmonic generation (SHG) and upconversion luminescence (UCL) processes in $BaTiO_3$:Ho^{3+}, Yb^{3+} enables optical temperature sensing.[58] Under 976 nm laser excitation, the evolution of the SHG and UCL band intensity ratio is correlated with temperature, demonstrating a novel and effective strategy for nonlinear optical thermometry with high sensitivity of up to 2.78% °C^{-1} at 146°C. The SHG/UCL intensity ratio shows superior thermal sensitivity of ≈4% K^{-1} at 400 K and excellent temperature resolution ($\delta T = 0.07$ K) in $BaTiO_3$: Er^{3+}, Yb^{3+} over the range of 330–575 K.[59]

10.3 MULTIMODE OPTICAL THERMOMETERS BASED ON PEROVSKITE OXIDES

Implementing multimode thermal sensing can substantially expand the operational range and improve thermometric characteristics by seamlessly transitioning between multiple temperature-dependent parameters. This approach showcases optimal performance at specific temperature ranges. This section discusses the use of multimode thermometry in the context of single-doped and codoped perovskite oxides.

10.3.1 Perovskites activated with a single luminescent center

Multimode thermometry, utilizing perovskite oxides activated with a single luminescent center, primarily relies on the dependence of FIR and excited-state lifetime (FL) modes on temperature. Typically, FIR employs either the characteristic emission transition of rare earth or transition metal ions, with occasional consideration of host emission. The combined utilization of FIR, involving the intensity ratio of $^4F_{5/2} \rightarrow {}^4I_{9/2}$ transition of Nd^{3+} to host emission from the WO_6 group, and the ratio of emissions originating from TCLs of Nd^{3+} significantly expands the dynamic range of temperature measurement, as evidenced in Ba_2MgWO_6: Nd^{3+}.[60] FIR based on *f-f* Nd^{3+} transition to host emission, employed in the 77–248 K range, achieves a maximum relative sensitivity of 4.46% K^{-1} at 198 K. The utilization of emission from TCLs of Nd^{3+} in the range of 248–673 K yields a maximum sensitivity of 1.37% K^{-1} at 248 K.[60] In the case of $LaGaO_3$: Cr^{3+},[142] employing FIR and FL modes extends the dynamic range from 150 to 450 K, highlighting the effectiveness of multimode thermometry in enhancing temperature measurement capabilities, as detailed in Table 10.2. Rarely are SBR, RPIR, bandwidth, and emission intensity incorporated alongside FIR or FL for temperature measurement in single-ion activated perovskite oxides. Given the diverse measurement modes used in oxides, Vu *et al.* employ different synthesis techniques (mechanochemical and co-precipitation) to assess their impact on the sensitivity and operating range of Ba_2MgWO_6: Eu^{3+}. The maximum S_r obtained through co-precipitation is 1.17% K^{-1} at 248 K. In comparison, the mechanochemical technique yields 1.5% K^{-1} at 120 K in the temperature range of 80–300 K.[62] Jiawen *et al.* devised an excited-state thermometer based on the differing thermal change tendencies of excitation bands in double-perovskite La_2ZnTiO_6: Tb^{3+} phosphor.[63] The temperature-dependent excitation spectra at Tb^{3+} 543 nm emission reveal faster descents in the $4f^8 \rightarrow 4f^75d$ and Tb^{3+}–Ti^{4+} intervalence charge transfer (IVCT) excitation bands with rising temperature compared to Tb^{3+} 4f–4f excitation peaks. Two excitation intensity ratio (EIR) models – EIR-1 = *f-f/f-d* and EIR-2 = *f-f/* IVCT – are selected for temperature measurement. Consequently, maximum S_r of 0.734 and 0.727% K^{-1} at 573 K based on EIR 1 and EIR 2, respectively, are observed in the range 298–573 K.[63] Additionally, relaxor ferroelectric $Pb(In_{1/2}Nb_{1/2})O_3$-$Pb(Mg_{1/3}Nb_{2/3})O_3$–$PbTiO_3$ activated with Er^{3+} is explored for multimode thermometry based on TCLs and non-thermally coupled energy levels (NTCLs) of Er^{3+}. Maximum absolute sensitivities of 0.35% K^{-1} at 360 K and 0.25% K^{-1} at 327 K are obtained for TCLs and NTCLs, respectively.[64] Table 10.2 shows the perovskite oxides activated with single luminescent centers for multimode optical thermometry.

10.3.2 Perovskites activated with multiple luminescent centers

Table 10.2 presents an overview of perovskite oxides activated with multiple luminescent centers for optical thermometry. The reported combinations of ions codoped in oxide perovskites for multimode thermometry encompass RE/RE, RE/TM, TM/TM, and RE/ns^2 type ions. Subodh *et al.* introduced a six-mode thermometry using $SrLaLiTeO_6$:Mn^{4+}, Eu^{3+} double perovskites.[65] The diverse modes include FIR of Mn^{4+} to Eu^{3+} emissions, TCLs of Eu^{3+}-based FIR, SBR, the ratio of charge transfer absorption edge

Table 10.2 Perovskite oxides used for multimode optical thermometry

		Perovskites activated with a single luminescent center			
Phosphor	Method	Temperature range	Maximum relative sensitivity and temperature	Temperature resolution (δT)	References
$Ba_2MgWO_6{:}Nd^{3+}$	$FIR = \dfrac{^4F_{5/2} \rightarrow \,^4I_{9/2}}{\text{host emission}}$	77–248 K	4.46% K^{-1} @198 K	0.005K	60
	$FIR_{TCLs} = \dfrac{^4F_{5/2} \rightarrow \,^4I_{9/2}}{^4F_{3/2} \rightarrow \,^4I_{9/2}}$	248–673 K	1.37% K^{-1} @248 K	0.027 K	
$Ba_2Y_{2/3}TeO_6{:}Eu^{3+}$	RPIR	298–873 K	0.52% K^{-1} @573 K	0.17 K	152
	FIR	300–500 K	0.18% K^{-1} @300 K	–	153
	FL	300–500 K	0.136% K^{-1} @500 K	–	
$Sr_2LuSbO_6{:}Mn^{4+}$	FIR	38–608 K	2.893% K^{-1} @63 K	–	154
	FL	38–578 K	0.285% K^{-1} @578 K	–	
$Sr_2LuNbO_6{:}Mn^{4+}$	FIR	12–523 K	0.55% K^{-1} @136 K	–	154
	FL	12–523 K	1.55% K^{-1} @519 K	0.35 K	
$BaLaLiTeO_6{:}Er^{3+}$	FIR_{TCLs} (downconversion)	298–573 K	1.07% K^{-1} @298 K	0.47 K	155
	FIR_{TCLs} (upconversion)	298–478 K	1.24% K^{-1} @298 K	0.41 K	
$SrLaLiTeO_6{:}Er^{3+}$	FIR_{TCLs} (downconversion)	298–573 K	1.20% K^{-1} @298 K	0.44 K	155
	FIR_{TCLs} (upconversion)	298–478 K	1.13% K^{-1} @298 K	0.44 K	
$CaLaLiTeO_6{:}Er^{3+}$	FIR_{TCLs} (downconversion)	298–573 K	1.12% K^{-1} @298 K	0.46 K	155
	FIR_{TCLs} (upconversion)	298–478 K	1.11% K^{-1} @298 K	0.46 K	

(Continued)

Table 10.2 (Continued)　Perovskite oxides used for multimode optical thermometry

	Perovskites activated with a single luminescent center				
Phosphor	Method	Temperature range	Maximum relative sensitivity and temperature	Temperature resolution (δT)	References
$MgLaLiTeO_6:Er^{3+}$	FIR_{TCLs} (downconversion)	298–573 K	1.15% K^{-1} @298 K	0.45 K	155
	FIR_{TCLs} (upconversion)	298–478 K	1.03% K^{-1} @298 K	0.49 K	
$Ca_2MgWO_6:Dy^{3+}$	$FIR_1 = \dfrac{^4F_{9/2} \to {}^6H_{13/2}}{\text{host emission}}$	80–313 K	4.38% K^{-1} @213 K	0.005 K	156
	$FIR_4 = \dfrac{^4F_{9/2} \to {}^6H_{15/2}}{\text{host emission}}$	80–300 K	3% K^{-1} @253 K	–	
$Sr_2MgWO_6:Dy^{3+}$	$FIR_2 = \dfrac{^4F_{9/2} \to {}^6H_{13/2}}{\text{host emission}}$	80–313 K	3.04% K^{-1} @233 K	0.04	156
	$FIR_5 = \dfrac{^4F_{9/2} \to {}^6H_{15/2}}{\text{host emission}}$	80–300 K	2.63% K^{-1} @213 K	–	
$Ba_2MgWO_6:Dy^{3+}$	$FIR_3 = \dfrac{^4F_{9/2} \to {}^6H_{13/2}}{\text{host emission}}$	80–273 K	2.26% K^{-1} @193 K	0.03	156
	$FIR_6 = \dfrac{^4F_{9/2} \to {}^6H_{15/2}}{\text{host emission}}$	80–313 K	2.03% K^{-1} @193 K	–	
$Sr_2ScSbO_6:Mn^{4+}$	FIR	23–648 K	1.21% K^{-1} @83 K	–	157
	FL	23–648 K	0.948% K^{-1} @648 K	–	
$GdAlO_3:Eu^{3+}$	FIR	293–793 K	2.96% K^{-1} @293 K	–	158
	FL	620–793 K	2.28% K^{-1} @793 K	–	

(Continued)

Table 10.2 (Continued) Perovskite oxides used for multimode optical thermometry

			Perovskites activated with a single luminescent center		
Phosphor	Method	Temperature range	Maximum relative sensitivity and temperature	Temperature resolution (δT)	References
$LiNbO_3{:}Tb^{3+}$	SBR	80–460 K	1.43% K^{-1} @200 K	–	159
	FL	80–410 K	2.36% K^{-1} @260 K	–	
$LiNb_{0.8}Ta_{0.2}O_3{:}Tb^{3+}$	SBR	80–460 K	0.97% K^{-1} @230 K	–	159
	FL	80–410 K	2.13% K^{-1} @290 K	–	
$SrTiO_3{:}Mn^{4+}$	FIR	13–200 K	1.75% K^{-1} @13 K	0.05 K	16
	FL	13–300 K	3.04% K^{-1} @280 K	0.10 K	
$Sr_2TiO_4{:}Mn^{4+}$	FIR	13–200 K	8.99% K^{-1} @40 K	0.05 K	16
	FL	13–300 K	3.70% K^{-1} @290 K	0.10 K	
$Sr_3Ti_2O_7{:}Mn^{4+}$	FIR	13–200 K	10.37% K^{-1} @40 K	0.05 K	16
	FL	13–300 K	4.68% K^{-1} @280 K	0.013 K	
$K_{0.5}Na_{0.5}NbO_3{:}Pr^{3+}$	$FIR = \dfrac{I_{SHG}}{^3P_0 \rightarrow {}^3H_6}$	300–560 K	5.93% K^{-1} @560 K	0.03 K	160
	$FIR = \dfrac{^3P_0 \rightarrow {}^3H_6}{^3P_0 \rightarrow {}^3F_2}$	300–420 K	6.41% K^{-1} @420 K	0.105 K	
$Gd_2ZnTiO_6{:}Mn^{4+}$	Bandwidth	303–483 K	0.34% K^{-1} @423 K	–	161
	FL	298–473 K	2.43% K^{-1} @423 K	–	161
$Gd_2ZnTiO_6{:}Pr^{3+}$	FIR	293–433 K	1.67% K^{-1} @330 K	–	162
	FL	433–593 K	1.48% K^{-1} @593 K	–	

(Continued)

Table 10.2 (Continued) Perovskite oxides used for multimode optical thermometry

Perovskites activated with a single luminescent center					
Phosphor	Method	Temperature range	Maximum relative sensitivity and temperature	Temperature resolution (δT)	References
Ba_2MgWO_6:Eu^{3+}	FIR (Mechanochemical route)	80–300 K	1.5% K^{-1} @120 K	–	62
	FIR (Co-precipitation route)	80–300 K	1.17% K^{-1} @248 K	–	
$LiLaMgWO_6$:Pr^{3+}	$FIR_1 = \dfrac{{}^1D_2 \rightarrow {}^3H_4}{{}^3P_0 \rightarrow {}^3H_4}$	298–573 K	~2.0% K^{-1} @298 K	–	163
	$FIR_2 = \dfrac{{}^1D_2 \rightarrow {}^3H_4}{{}^3P_0 \rightarrow {}^3F_2}$	298–573 K	3.25% K^{-1} @298 K	0.15	
La_2MgTiO_6:Er^{3+}	$FIR_{TCLs} = \dfrac{{}^2H_{11/2} \rightarrow {}^4I_{15/2}}{{}^4S_{3/2} \rightarrow {}^4I_{13/2}}$	77–398 K	3.08% K^{-1} @148 K	<0.1 K	164
	$FIR_{NTCLs} = \dfrac{{}^2H_{11/2} \rightarrow {}^4I_{15/2}}{{}^2H_{9/2} \rightarrow {}^4I_{15/2}}$	77–398 K	1.9% K^{-1} @150 K	<0.1 K	
Ca_2LaNbO_6:Sm^{3+}	FIR	323–573 K	0.23% K^{-1} @353 K	2.19 K	165
	SBR	323–573 K	0.28% K^{-1} @473 K	1.79 K	
$NaLaCaWO_6$:Eu^{3+}	FIR	298–523 K	2.23% K^{-1} @298 K	–	166
	SBR	298–448 K	1.43% K^{-1} @298 K	–	
$LaGaO_3$:Cr^{3+}	FIR	150–300 K	2.07% K^{-1} @150 K	0.24 K	61
	FL	300–450 K	1.38% K^{-1} @450 K	0.36 K	
$Pb(In_{1/2}Nb_{1/2})O_3$-$Pb(Mg_{1/3}Nb_{2/3})O_3$-$PbTiO_3$:Er^{3+}	FIR_{TCLs}	160–360 K	Not calculated	–	64
	FIR_{NTCLs}	160–360 K	Not calculated	–	

(Continued)

Table 10.2 (Continued) Perovskite oxides used for multimode optical thermometry

	Perovskites activated with a single luminescent center				
Phosphor	Method	Temperature range	Maximum relative sensitivity and temperature	Temperature resolution (δT)	References
$La_2ZnTiO_6{:}Tb^{3+}$	$EIR = \dfrac{^7F_6 \rightarrow {}^5D_3}{IVCT}$	298–573 K	0.727% K^{-1} @573 K	–	63
	$EIR = \dfrac{^7F_6 \rightarrow {}^5D_3}{4f^8 \rightarrow 4f^7 5d}$	298–573 K	0.734% K^{-1} @573 K	–	
$Na_2La_2Ti_3O_{10}{:}Pr^{3+}$	FIR_{TCLs}	125–300 K	2.43% K^{-1} @125 K	–	167
	FIR_{NTCLs}	300–550 K	1.96% K^{-1} @443 K	–	
$Sr_2InSbO_6{:}Sm^{3+}$	Emission intensity	298–523 K	10.9% K^{-1} @523 K	–	168
	FL	298–523 K	2.15% K^{-1} @523 K	~0.019 K	168
$Ba(Zr_{0.16}Mg_{0.28}Ta_{0.56})O_3{:}Pr^{3+}$	FIR_{TCLs}	293–873 K	0.25% K^{-1} @373 K	–	169
	FIR_{NTCLs}	293–873 K	1.05% K^{-1} @483 K	–	
	FL	303–873 K	0.17% K^{-1} @873 K	–	
$SrZn_{0.33}Nb_{0.67}O_3{:}Pr^{3+}$	$FIR_1 = \dfrac{^1D_2 \rightarrow {}^3H_4}{^3P_0 \rightarrow {}^3H_4}$	300–500 K	0.37% K^{-1} @400 K	–	170
	$FIR_2 = \dfrac{^1D_2 \rightarrow {}^3H_4}{^3P_0 \rightarrow {}^3F_2}$	300–500 K	0.35% K^{-1} @400 K	–	

(Continued)

Table 10.2 (Continued) Perovskite oxides used for multimode optical thermometry

		Perovskites activated with a single luminescent center			
Phosphor	Method	Temperature range	Maximum relative sensitivity and temperature	Temperature resolution (δT)	References
$SrZn_{0.33}Nb_{0.67}O_3{:}Pr^{3+}, Ga^{3+}$	$FIR_1 = \dfrac{^1D_2 \rightarrow {}^3H_4}{^3P_0 \rightarrow {}^3H_4}$	300–500 K	0.83% K^{-1} @500 K	–	170
	$FIR_2 = \dfrac{^1D_2 \rightarrow {}^3H_4}{^3P_0 \rightarrow {}^3F_2}$	300–500 K	0.70% K^{-1} @500 K	–	
$SrTiO_3{:}Tb^{3+}$	FIR	77–330 K	2.20% K^{-1} @210 K	–	171
	FL	77–330 K	8.83% K^{-1} @180 K	–	
$Pb(In_{1/2}Nb_{1/2})$ O_3-$Pb(Mg_{1/3}Nb_{2/3})$ O_3-$PbTiO_3{:}Pr^{3+}$	FIR	300–480 K	1.73% K^{-1} @460 K	–	172
	FL	303–463 K	1.25% K^{-1} @463 K	–	

Perovskites Activated with Multiple Luminescent Centers

Phosphor	Method	Temperature range	Maximum relative sensitivity and temperature	Temperature resolution (δT)	References
$SrLaLiTeO_6{:}Eu^{3+}/Mn^{4+}$	$FIR = \dfrac{^2E_g \rightarrow {}^4A_{2g}}{^5D_0 \rightarrow {}^7F_2}$	200–500 K	0.81% K^{-1} @500 K	–	65
	$FIR_{TCLs} = \dfrac{^5D_1 \rightarrow {}^7F_1}{^5D_0 \rightarrow {}^7F_2}$	240–500 K	0.76% K^{-1} @500 K	–	
	SBR	240–500 K	0.93% K^{-1} @420 K	–	
	$FIR = \dfrac{I_{charge\ transfer\ band\ edge}}{I_{Eu^{3+}}}$	200–360 K	1.97% K^{-1} @200 K	–	
	FL	80–500 K	0.33% K^{-1} @340 K	–	
	Bandwidth	300–500 K	0.20% K^{-1} @500 K	–	

(Continued)

Table 10.2 (Continued) Perovskite oxides used for multimode optical thermometry

| Perovskites activated with multiple luminescent centers | | | | | |
Phosphor	Method	Temperature range	Maximum relative sensitivity and temperature	Temperature resolution (δT)	References
$Ca_2LaTaO_6{:}Bi^{3+}/Eu^{3+}$	FIR	293–510 K	0.76% K^{-1} @360 K	–	173
	FL	293–510 K	1.04% K^{-1} @293 K	–	
$Sr_2LuTaO_6{:}Tb^{3+}/Mn^{4+}$	$FIR = \dfrac{^5D_4 \to {}^7F_4}{^2E_g \to {}^7A_{2g}}$	313–573 K	1.98% K^{-1} @543 K	–	174
	$FIR = \dfrac{^5D_4 \to {}^7F_5}{^2E \to {}^4A_{2g}}$	313–573 K	5.95% K^{-1} @573 K	–	
$La_2MgTiO_6{:}Mn^{4+}/Cr^{3+}$	$FIR\ I = \dfrac{\text{host emission}}{^2E_g \to {}^4A_{2g}(Mn^{4+})}$	80–600 K	1.74% K^{-1} @220 K	0.23 K	71
	$FIR\ II = \dfrac{\text{host emission}}{^2E_g \to {}^4A_{2g}(Cr^{3+})}$	80–600 K	1.70% K^{-1} @220 K	<0.1 K	
	$FIR\ III = \dfrac{^2E_g \to {}^4A_{2g}(Cr^{3+})}{^2E_g \to {}^4A_{2g}(Mn^{4+})}$	80–600 K	1.12% K^{-1} @550 K	<0.1 K	
$Sr_2LaNbO_6{:}Er^{3+}/Ho^{3+}$	FIR_{TCLs}	303–693 K	0.838% K^{-1} @303 K	–	175
	FIR_{NTCLs}	303–693 K	1.591% K^{-1} @641 K	–	
$Li_4AlSbO_6{:}Eu^{3+}/Mn^{4+}$	FIR	80–460 K	2.47% K^{-1} @456 K	–	176
	FL	80–460 K	3.07% K^{-1} @416 K	–	

(Continued)

Table 10.2 (Continued) Perovskite oxides used for multimode optical thermometry

			Perovskites activated with multiple luminescent centers		
Phosphor	Method	Temperature range	Maximum relative sensitivity and temperature	Temperature resolution (δT)	References
$CaZrO_3:Er^{3+},Yb^{3+},Mo^{6+}$	FIR_{TCLs} (downconversion)	303–523 K	0.697% K^{-1} @303 K	–	67
	FIR_{TCLs} (upconversion)	303–523 K	0.445% K^{-1} @303 K	–	
$Ba_2GdNbO_6:Eu^{3+}/Mn^{4+}$	Emission intensity	303–483 K	1.78% K^{-1} @423 K	–	177
	FIR	303–483 K	2.72% K^{-1} @483 K	–	
	FL	303–483 K	1.73% K^{-1} @430 K	–	
$Sr_2YNbO_6:Bi^{3+}/Eu^{3+}$	FIR	303–573 K	0.898% K^{-1} @453 K	–	69
	Chromaticity coordinates	303–573 K	0.208% K^{-1} @453 K	–	
$SrGdLiTeO_6:Mn^{4+}/Tb^{3+}$	FIR	298–573 K	1.49% K^{-1} @560 K	0.14 K	178
	FL	298–573 K	1.88% K^{-1} @573 K	0.19 K	
$Ca_2InTaO_6:Sm^{3+}/Mn^{4+}$	$\text{FIR I}=\dfrac{^4G_{5/2} \rightarrow {}^6H_{5/2}}{^2E_g \rightarrow {}^4A_{2g}}$	303–483 K	3.30% K^{-1} @303 K	0.0745 K	113
	$\text{FIR II}=\dfrac{^4G_{5/2} \rightarrow {}^6H_{7/2}}{^2E_g \rightarrow {}^4A_{2g}}$	303–483 K	3.32% K^{-1} @303 K	0.0741 K	
	$\text{FIR III}=\dfrac{^4G_{5/2} \rightarrow {}^6H_{9/2}}{^2E_g \rightarrow {}^4A_{2g}}$	303–483 K	3.80% K^{-1} @303 K	0.0648 K	113

(Continued)

Table 10.2 (Continued) Perovskite oxides used for multimode optical thermometry

Phosphor	Method	Temperature range	Maximum relative sensitivity and temperature	Temperature resolution (δT)	References
	Perovskites activated with multiple luminescent centers				
Ca_2InTaO_6:Pr^{3+}/Mn^{4+}	$FIR\ I = \dfrac{^3P_0 \rightarrow {}^3H_4}{^2E_g \rightarrow {}^4A_{2g}}$	303–483 K	2.03% K^{-1} @303 K	0.1214 K	113
	$FIR\ II = \dfrac{^3P_1 \rightarrow {}^3H_5}{^2E_g \rightarrow {}^4A_{2g}}$	303–483 K	2.71% K^{-1} @303 K	0.0908 K	
	$FIR\ III = \dfrac{^3D_2 \rightarrow {}^3H_4}{^2E_g \rightarrow {}^4A_{2g}}$	303–483 K	2.46% K^{-1} @303 K	0.1001 K	
	$FIR\ IV = \dfrac{^3P_0 \rightarrow {}^3F_2}{^2E_g \rightarrow {}^4A_{2g}}$	303–483 K	1.49% K^{-1} @303 K	0.1648 K	
$CaTiO_3$:Er^{3+}/Yb^{3+}	FIR_{TCLs}	303–618 K	1.14% K^{-1} @303 K	–	179
	FIR_{NTCLs}	303–618 K	1.66% K^{-1} @303 K	–	
$ZnTiO_3$:Eu^{3+}/Mn^{4+}	$FIR\ I = \dfrac{^5D_0 \rightarrow {}^7F_1}{^2E_g \rightarrow {}^4A_{2g}}$	80–310 K	3.2% K^{-1} @210 K	0.010 K	180
	$FIR\ II = \dfrac{^5D_0 \rightarrow {}^7F_0}{^2E_g \rightarrow {}^4A_{2g}}$	80–310 K	2.5% K^{-1} @210 K	0.013 K	
$SrLaMgNbO_6$:Tb^{3+}/Mn^{4+}	FIR	303–543 K	1.93% K^{-1} @543 K	–	181
	FL	303–543 K	3.27% K^{-1} @513 K	–	

(Continued)

Table 10.2 (Continued) Perovskite oxides used for multimode optical thermometry

Perovskites activated with multiple luminescent centers					
Phosphor	Method	Temperature range	Maximum relative sensitivity and temperature	Temperature resolution (δT)	References
$LaMgGeO_6:Er^{3+}/Yb^{3+}$	FIR_{TCLs}	293–473 K	1.130% K^{-1} @303 K	–	182
	FIR_{NTCLs}	293–473 K	0.623% K^{-1} @303 K	–	
	FL	293–473 K	2.914% K^{-1} @293 K	–	
$Ba_2LuNbO_6:Er^{3+}/Yb^{3+}$	FIR_{TCLs}	303–573 K	1.53% K^{-1} @303 K	0.19 K	183
	FIR_{NTCLs}	303–573 K	1.81% K^{-1} @303 K	0.16 K	
$SrLaLiWO_6:Er^{3+}/Yb^{3+}$	FIR_{TCLs} (downconversion)	303–573 K	0.91% K^{-1} @450 K	0.14 K	68
	FIR_{TCLs} (upconversion)	303–573 K	0.98% K^{-1} @303 K	0.13 K	
	FL	303–573 K	0.39% K^{-1} @543 K	–	
$ZnTiO_3:Eu^{3+}/Mn^{4+}$	FIR	80–305 K	3.15% K^{-1} @305 K	0.011 K	184
	FL	80–305 K	1.4% K^{-1} @290 K	–	184
$SrTiO_3:Er^{3+}/Yb^{3+}$	FIR_{TCLs}	303–618 K	1.28% K^{-1} @303 K	–	185
	FIR_{NTCLs}	303–618 K	0.98% K^{-1} @303 K	–	
$La_2CaSnO_6:Eu^{3+}/Mn^{4+}$	$FIR\ (\lambda_{exc}=318\,nm)=\dfrac{^2E_g \to {}^4A_{2g}}{^5D_0 \to {}^7F_2}$	80–440 K	2.954% K^{-1} @440 K	–	186
	$FIR\ (\lambda_{exc}=394\,nm)=\dfrac{^2E_g \to {}^4A_{2g}}{^5D_0 \to {}^7F_2}$	80–440 K	0.74% K^{-1} @440 K	–	

(Continued)

Table 10.2 (Continued) Perovskite oxides used for multimode optical thermometry

Perovskites activated with multiple luminescent centers					
Phosphor	Method	Temperature range	Maximum relative sensitivity and temperature	Temperature resolution (δT)	References
Ca_2InSbO_6:Dy^{3+}/Mn^{4+}	FIR	298–573 K	2.16% K^{-1} @498 K	0.3K	187
	FL	298–573 K	1.890% K^{-1} @423 K	–	
$SrLaLiTeO_6$:Dy^{3+}/Mn^{4+}	FIR	298–673 K	1.60% K^{-1} @673 K	0.3125 K	188
	FL	298–673 K	2.18% K^{-1} @673 K	–	
La_2MgTiO_6:Pr^{3+}/Dy^{3+}	FIR	298–548 K	2.357% K^{-1} @498 K	–	189
	FL	298–548 K	1.814% K^{-1} @473 K	–	
$La_2Mg_{1.33}Ta_{0.67}O_6$:$Eu^{3+}$/$Mn^{4+}$	FIR	298–448 K	2.72% K^{-1} @448 K	–	190
	FL	298–448 K	3.42% K^{-1} @390 K	–	
$SrGdLiTeO_6$:Mn^{4+}/Eu^{3+}	FIR	300–550 K	4.9% K^{-1} @550 K	0.1 K	191
	FL	298–573 K	0.229% K^{-1} @573 K	–	
$CaGdMgSbO_6$:Sm^{3+}/Mn^{4+}	FIR	298–573 K	1.38% K^{-1} @548 K	–	192
	FL	298–573 K	1.23% K^{-1} @523 K	–	
La_2ZnTiO_6:Bi^{3+}/Eu^{3+}	FIR	293–473 K	1.50% K^{-1} @293 K	~0.3 K	193
	FL	293–573 K	1.23% K^{-1} @573 K	~0.08 K	
$SrGdLiTeO_6$:Sm^{3+}/Mn^{4+}	FIR	298–573 K	8.69% K^{-1} @573 K	0.32 K	70
	FL	298–598 K	1.3% K^{-1} @520 K	–	
$BaLaMgNbO_6$:Dy^{3+}/Mn^{4+}	FIR	230–470 K	1.82% K^{-1} @457 K	–	194
	FL	230–470 K	2.43% K^{-1} @437 K	–	

(Continued)

Table 10.2 (Continued) Perovskite oxides used for multimode optical thermometry

	Perovskites activated with multiple luminescent centers				
Phosphor	Method	Temperature range	Maximum relative sensitivity and temperature	Temperature resolution (δT)	References
$Bi_4Ti_3O_{12}$:Pr^{3+}/Er^{3+}	$FIR\ I = \dfrac{^2H_{11/2} \rightarrow\ ^4I_{15/2}}{^1D_2 \rightarrow\ ^3H_4}$	298–568 K	1.03% K^{-1} @478 K	–	195
	$FIR\ II = \dfrac{^2H_{11/2} \rightarrow\ ^4I_{15/2}}{^4S_{3/2} \rightarrow\ ^4I_{15/2}}$	298–568 K	1.1% K^{-1} @298 K	–	195
$GdZnTiO_6$:Tm^{3+}/Yb^{3+}	FIR_{TCLs}	290–473 K	Not calculated	–	196
	FIR_{NTCLs}	290–473 K	Not calculated	–	
La_2MgGeO_6:Bi^{3+}/Er^{3+}	FIR_{TCLs}	293–473 K	1.23% K^{-1} @293 K	–	197
	FIR_{NTCLs}	293–473 K	0.99% K^{-1} @293 K	–	
La_2MgGeO_6:Bi^{3+}/Eu^{3+}	FIR	293–473 K	2.389% K^{-1} @363 K	–	198
	FL	293–433 K	4.09% K^{-1} @383 K	–	
Gd_2ZnTiO_6:Er^{3+}/Tm^{3+}/Yb^{3+}	FIR_{TCLs}	300–473 K	0.41% K^{-1} @300 K	–	199
	FIR_{NTCLs}	300–473 K	0.69% K^{-1} @473 K	–	
Sr_2YTaO_6:Er^{3+}/Yb^{3+}	FIR_{TCLs}	293–473 K	1.32% K^{-1} @293 K	0.07 K	200
	FIR_{NTCLs}	293–473 K	0.58% K^{-1} @293 K	0.17 K	
Ca_2MgWO_6:Er^{3+}/Yb^{3+}	FIR	303–573 K	0.92% K^{-1} @303 K	0.31 K	201
	FL	303–573 K	0.11% K^{-1} @573 K	–	
$LiNbO_3$:Tm^{3+}/Yb^{3+}	FIR_{TCLs}	180–260 K	0.69% K^{-1} @180 K	–	66
	FIR_{NTCLs}	80–140 K	1.25% K^{-1} @80 K	0.50 K	

(Continued)

Table 10.2 (Continued) Perovskite oxides used for multimode optical thermometry

Phosphor	Method	Temperature range	Maximum relative sensitivity and temperature	Temperature resolution (δT)	References
		Perovskites activated with multiple luminescent centers			
$GdScO_3$:Nd^{3+}, Er^{3+}, Cr^{3+}	FIR	100–500 K	0.96% K^{-1} @300 K	–	202
	FL	100–500 K	0.66% K^{-1} @266 K	–	
$BiLaWO_6$:Er^{3+},Tm^{3+},Yb^{3+}	FIR_{TCLs}	30–300 K	0.427% K^{-1} @30K	–	203
	FIR_{NTCLs}	30–300 K	0.7278% K^{-1} @110 K	–	
$CaTiO_3$:Er^{3+}, Nd^{3+},Yb^{3+}	$FIR\ I = \dfrac{^2H_{11/2(I)} \rightarrow\ ^4I_{15/2}}{^4S_{3/2(2)} \rightarrow\ ^4I_{15/2}}$	303–573 K	1.09% K^{-1} @303 K	0.4 K	204
	$FIR\ II = \dfrac{^2H_{11/2(I)} \rightarrow\ ^4I_{15/2}}{^4S_{3/2(I)} \rightarrow\ ^4I_{15/2}}$	303–573 K	0.95% K^{-1} @303 K	–	
$LiTaO_3$:Ti^{4+}/Eu^{3+}	FIR	303–443 K	5.425% K^{-1} @303 K	0.14 K	205
	FL	303–443 K	3.637% K^{-1} @303 K	0.027 K	

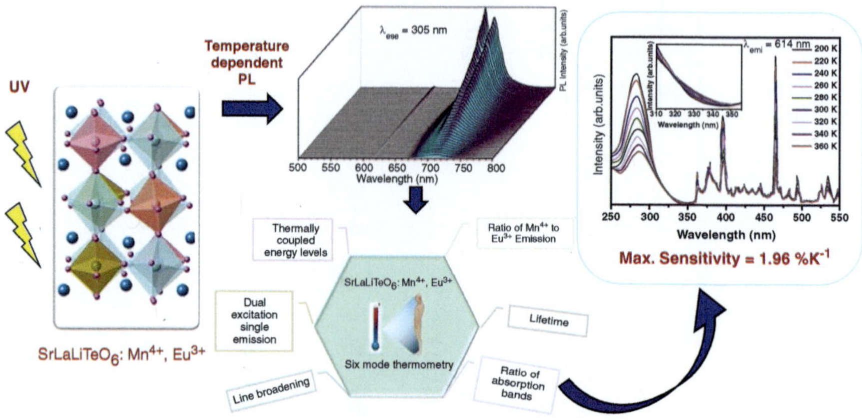

Figure 10.2 A six-mode optical thermometry is realized using SrLaLiTeO$_6$: Mn^{4+}, Eu^{3+}. An illustration showing different methods adopted to detect temperature using SrLaLiTeO$_6$: Mn^{4+}, Eu^{3+} phosphor. Maximum relative sensitivity is achieved when the ratio of absorption bands in the excitation spectrum is considered, as shown here. (Reprinted from Ref. [65] with permission from Elsevier.)

of the host to 4f-4f absorption of Eu^{3+} ions, excited-state lifetime, and line broadening methods (FWHM) as illustrated in Figure 10.2. This phosphor thermometer exhibits a broad dynamic temperature range from 80 to 500 K, achieving a maximum S_r of 1.97% K^{-1} at 200 K.[65] In LiNbO$_3$: Tm^{3+}, Yb^{3+}, a dynamic range of 80–260 K is achieved by leveraging TCLs and NTCLs of Tm^{3+} ions, with a maximum S_r of 1.25% K^{-1} at 80 K.[66] Another multimode approach involves utilizing the FIR from TCLs of Er^{3+} during both upconversion and downconversion processes, as demonstrated in CaZrO$_3$: Er^{3+}, Yb^{3+}, Mo^{6+}, and SrLaLiWO$_6$: Er^{3+}/Yb^{3+}.[67,68] Attempts to enhance relative sensitivity are made by utilizing FIR from the same energy levels but under different processes, resulting in slight changes in sensitivities, as detailed in Table 10.2.

A novel approach is introduced in Sr$_2$YNbO$_6$: Bi^{3+}/Eu^{3+}, where chromaticity coordinates are employed for temperature detection. Utilizing the temperature-dependent variation of the x-coordinate, maximum absolute and relative sensitivities are found to be 0.075% K^{-1} (at 423 K) and 0.208% K^{-1} (at 453 K), respectively, within a temperature range of 303–573 K. The intensity ratio of emissions between Bi^{3+} ($^3P_1\rightarrow{}^1S_0$) and Eu^{3+} ($^5D_0\rightarrow{}^7F_2$) within the same system results in a sensitivity (S_r) of 0.898% K^{-1} at 453 K.[69] The multimode approach using the FIR and FL methods in SrGdLiTeO$_6$: Sm^{3+}/Mn^{4+} achieves a remarkable maximum S_r in a wide temperature range of 298–573 K, reaching 8.69% K^{-1} at 573 K, with a minimum temperature resolution of 0.32 K.[70] In another double perovskite Ba$_2$LaLiWO$_6$: Er^{3+}, Yb^{3+}, luminescence intensities are effectively enhanced by Sr^{2+} substitution

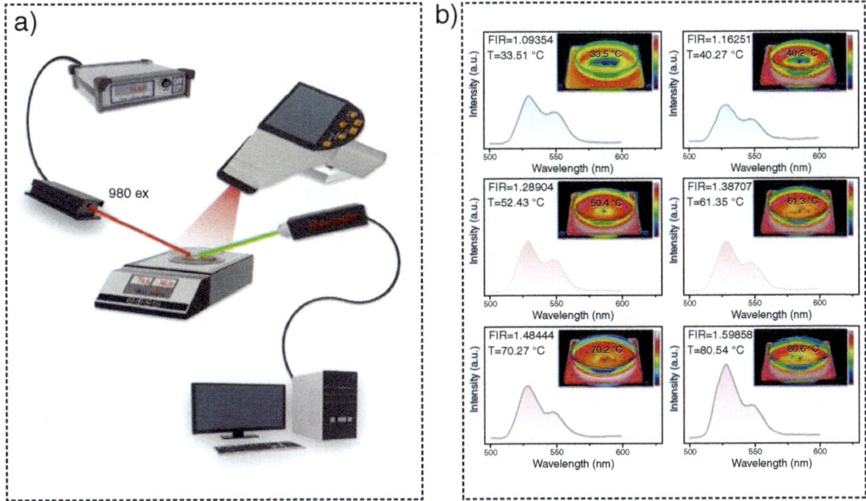

Figure 10.3 (a) Experimental setup for water temperature test using SrLaLiWO$_6$:Er^{3+}, Yb^{3+} phosphor attached to pork tissue (b) Some experimental test results showing the thermal images at different temperatures. (Reprinted from Ref. [68] with permission from Elsevier.)

in the Ba^{2+} site. The phosphor shows excellent temperature sensing performances in three modes: downconversion FIR, upconversion FIR, and fluorescence lifetime (FL) within the temperature range of 303–573 K.[68] As proof of concept, the researchers carried out an experiment to examine the temperature sensing performance of Sr$_2$LaLiWO$_6$: Er^{3+}, Yb^{3+} by measuring the temperature of a pork tissue.[68] The results of the experiment are shown in Figure 10.3. In contrast to single-mode thermometry, multimode thermometry broadens the operational range and unequivocally enhances the thermometric characteristics of perovskite oxide phosphors. Carefully selected combinations of host and activator can yield intriguing outcomes, opening avenues for the practical utilization of these materials in temperature sensors.

10.4 CURRENT TRENDS IN PEROVSKITE OXIDES-BASED OPTICAL THERMOMETRY

Luminescence thermometry is a semi-contact method advantageous for environments inaccessible to other methods. It works across a broad temperature range, from cryogenic to extremely high temperatures, and is suitable for harsh conditions. Different applications of temperature sensors necessitate diverse probes and approaches. However, the complexity and interdisciplinary nature of modern applications require extensive scientific knowledge, which can be challenging. Consequently, researchers are focusing on improving the thermometric figure of merit, mainly the absolute and relative sensitivity of luminescence thermometry by employing innovative techniques.

New methods have been developed utilizing perovskite oxides for high-performance optical thermometry. A novel approach using the intensity ratio of second harmonic generation (SHG) and upconversion luminescence (UCL) processes in $BaTiO_3$:Ho^{3+}, Yb^{3+} for optical temperature measurement is introduced by Zheng *et al.* Notably, this band intensity ratio follows a sigmoidal pattern with temperature changes, enabling it to detect phase shifts between non-centrosymmetric and centrosymmetric systems. Crucially, this research introduces a pioneering and highly effective approach to nonlinear optical thermometry, characterized by a remarkable sensitivity reaching 2.78% °C^{-1}, particularly relevant in optical temperature sensing.[58] De *et al.* proposed another strategy involving the use of the Raman signal of the host crystal in conjunction with the PL signal of the rare earth ions in $BaTiO_3$: Eu^{3+}, Er^{3+} perovskites.[45,148]

Multiple innovative strategies have also been proposed to improve the thermometric performance of perovskite oxides. A fundamental discovery is the enhanced thermal stability and luminescence of Eu^{3+} ions in La_2CaSnO_6:Eu^{3+}, Mn^{4+} phosphor when excited at a wavelength corresponding to the $^5H_3$ (318 nm) energy level than excited under wavelengths 299 nm (charge transfer band) and 394 nm ($^7F_0 \rightarrow {}^5L_6$). This finding enabled the development of a La_2CaSnO_6:Eu^{3+}, Mn^{4+} temperature probe that exhibits reverse thermal quenching and superior sensitivity. The probe's maximum absolute sensitivity is 0.08523 K^{-1} (0.0082 K^{-1} when $\lambda_{exc} = 394$ nm), and relative sensitivity is 2.954% K^{-1} (0.74% K^{-1} when $\lambda_{exc} = 394$ nm), demonstrating its potential for high-precision, non-contact temperature sensing.[186] Hence, the selective use of a high-energy excited state as the excitation light source significantly enhances the thermal stability of Eu^{3+} ions, making La_2CaSnO_6:Eu^{3+}, Mn^{4+} a promising material for temperature sensing applications. A chemometrics model is applied to $Na_{0.5}Gd_{0.5}TiO_3$: Er^{3+} phosphors, aiming to overcome limitations of traditional FIR techniques, particularly for materials with low FIR sensitivity. The model enables quick and accurate temperature determination with high correlation coefficients in a single step by analyzing photoluminescence spectra changes with temperature and effectively circumvents the significant temperature errors that arise from the monotonous variation of FIR at high temperatures.[79] Incorporating Li^+ as a charge compensator in Eu^{3+} activated zinc titanate composites significantly improves their photoluminescence, thermal stability, and temperature sensing capabilities. This enhancement is attributed to the ability of charge compensators to stabilize the charge in the crystal lattice. This stabilization facilitates more effective energy transfer, leading to increased luminescent emissions, essential for precise temperature measurement in optical thermometry.[80] The introduction of Al^{3+} ions as a charge compensator in the $ZnTiO_3$ lattice, doped with Eu^{3+}/Mn^{4+}, significantly enhances the emission intensity of Eu^{3+}, thereby improving the thermometer's sensitivity. The presence of Al^{3+} helps in balancing the charge when Eu^{3+} replaces Zn^{2+} in the lattice, reducing defects and promoting better luminescence.[131] Co-doping Mg^{2+} in $CaBa_2WO_6$:Er^{3+} yields similar results as Mg^{2+} ions play a key role in stabilizing the crystal structure and enhancing the luminescent properties essential for temperature sensing.[11] Zhang *et al.* demonstrated the significance of phase transition in enhancing the luminescence thermometry of Ta^{5+}-triggered $K_{0.5}Na_{0.5}NbO_3$:Er^{3+} transparent ceramics, as illustrated in Figure 10.4. The phase transition from orthorhombic to tetragonal and then to cubic, caused by varying concentrations of Ta^{5+}, plays a crucial role. This transition is directly linked to improvements in luminescent temperature sensing properties. Specifically, the cubic-phase region demonstrates the best relative sensitivity of 1.58% K^{-1} at 273 K within the temperature range of 273–543 K.[121] This connection between

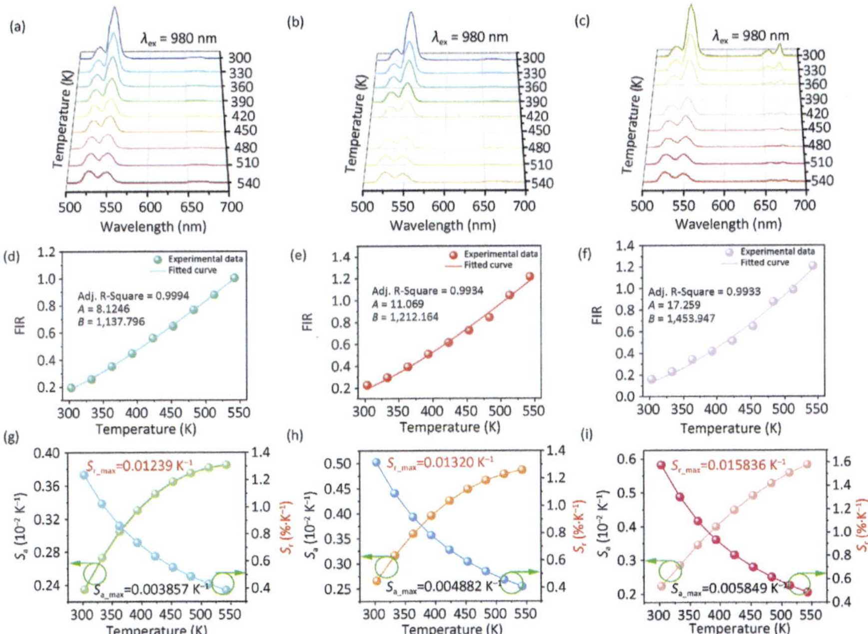

Figure 10.4 Temperature-dependent upconversion of $K_{0.5}Na_{0.5}NbO_3$: $0.003Er^{3+}$, xTa^{5+} ($x=0$, $x=0.36$, and $=0.65$) ceramics in the range of 273–543 K. (a) Orthorhomic phase ($x=0$), (b) Tetragonal phase ($x=0.36$) and (c) Cubic phase ($x=0.65$). Temperature-dependent FIR values of (d) orthorhombic, (e) tetragonal, and (f) cubic phases. Relative sensitivity (S_r) and absolute sensitivity (S_a) as a function of the temperature for (g) orthorhombic, (h) tetragonal, and (i) cubic phases. (Reprinted from Ref. [121] with permission from Elsevier.)

phase transition and luminescence thermometry is essential for developing highly sensitive temperature monitoring materials in chemical reactions and other industrial contexts. Back *et al.* gave an in-depth analysis of ratiometric luminescent thermometers, explicitly focusing on the $LaGaO_3$:Nd^{3+} system, and explored using phase transition as a control marker for enhanced reliability in optical thermometry. This system exhibits a notable relative sensitivity and maintains a low-temperature uncertainty, approximately 1 K, across an extensive temperature range, reaching up to 850 K. This novel approach highlights the possibility of designing customizable thermometers for various applications, underscoring its significance in the field of optical thermal sensors.[89] The influence of mechanochemical and co-precipitation synthesis routes on the sensitivity and operating range of a $Ba_2Mg_{1-x}Eu_xWO_6$ optical thermometer is investigated by Vu *et al.* The higher sensitivity in optical thermometry is via the co-precipitation route due to smaller, more homogeneous crystallite sizes and uniform distribution of Eu^{3+} dopants. This results in a larger surface area, efficient energy transfer, and stronger luminescence. The lower sintering temperature also preserves the structural integrity of the dopants and the lattice, further enhancing sensitivity.[62]

Understanding the thermal quenching mechanism is essential for effectively employing materials in temperature detection applications. With increasing temperature, the $^3P_0$ and $^1D_2$ luminescence in Pr^{3+} doped La_2MgTiO_6 have various de-excitation channels. The significant reduction in the intensity of $^3P_0$ luminescence primarily results from electron transfer occurring through the crossover channel connecting the $^3P_0$ multiplet and the $[Ti^{4+}–e^-]$ state. In contrast, the $^1D_2$ luminescence maintains exceptional thermal stability over a wide temperature range. The luminescence intensity of the Pr^{3+} $4f$ excited multiplet is influenced by both the electron population probability originating from higher-energy states and the electron de-excitation rate to lower-lying multiplets. The electron population dynamics and de-excitation rates which play a direct role in determining the luminescence efficiency of Pr^{3+}, in turn affect optical temperature sensing properties of the material based on the FIR of emission from $^1D_2$ and $^3P_0$ states.[90] Wu *et al.* recently reported a universal strategy for improving absolute sensitivity for temperature detection in the Er^{3+}/Yb^{3+} doped Gd_2ZnTiO_6. This strategy involved manipulating the FIR effect. By effectively controlling this ratio (e.g., reversing the ratio), the researchers were able to significantly enhance the absolute sensitivity of Er^{3+}-activated Gd_2ZnTiO_6 by factors of 4.6.[130] A study on the temperature-dependent emission spectra of $LaGaO_3$:Cr^{3+} ranging from 300 to 600 K noted two key features: a broad band linked to fluorescence from the $^4T_2$ excited state, which increases with temperature, and sharp peaks associated with phosphorescence from the ^2E state, decreasing with temperature. The luminescence intensity ratio (LIR) method, enhanced by a novel two-step deconvolution process (double-deconvolution), was used to separate these overlapping emissions. The analysis showed a solid fit to the Boltzmann distribution, with a relative sensitivity at room temperature of about 2.5% K^{-1}, suggesting the compound's high potential for *in vivo* biomedical thermometry applications.[100] The sensitivity of temperature measurement in the Ca_2MgWO_6:Er^{3+}/Yb^{3+} material is significantly enhanced by adding Eu^{3+}. The absolute sensitivity improved from 81.83×10^{-4} K^{-1} at 423 K to 88.12×10^{-4} K^{-1} at 398 K, primarily due to the energy transfer between Er^{3+} and Eu^{3+}, which boosts the FIR.[206] A reliable Boltzmann cryogenic thermometer is proposed using $CaHfO_3$:Cr^{3+}. This material shows its effectiveness in temperature measurement by adhering to the Boltzmann law within the 40–150 K range. It exhibits a high relative sensitivity of approximately 2% K^{-1} at 40 K and remarkable thermal resolution, ranging between 0.045 and 0.77 K across the same temperature span. Furthermore, its performance surpasses a ruby thermometer, which relies on the intensity ratio of two R-lines.[102] The analysis of temperature sensing using the FIR method in $CaTiO_3$:Er^{3+}-Nd^{3+}-Yb^{3+} phosphors focused on different combinations of thermally coupled Stark levels under co-excitation at wavelengths 808 and 980 nm, spanning temperatures from 303 to 573 K. As a result, the study reported a maximum sensitivity of approximately 10.9×10^{-3} K^{-1} at room temperature (303 K) and an outstanding thermal resolution of about 0.4 K. This high sensitivity and resolution were attributed to the Stark splitting of thermally coupled levels ($^2H_{11/2}(1)$ & $^4S_{3/2}(2)$) as opposed to the conventional $^2H_{11/2}$ and $^4S_{3/2}$ levels.[204] Using heat and incident infrared (IR) excitation as physical inputs, Kumar *et al.* have demonstrated that the $SrTiO_3$:Er^{3+}, Yb^{3+} phosphor possesses the capability to serve as a platform for designing diverse elementary logic gates, including AND, INHIBIT, and DEMULTIPLEX photonic molecular logic gates. The logical output of these gates is determined by the thermal variation of observables, such as the FIR of thermally coupled levels of Er^{3+} (FIR_{TC}), FIR of non-thermally coupled levels (FIR_{NTC}), S_a, and S_r. The excitation source, operating at a wavelength of 980 nm, is assumed to be ON for implementing these logic gates. The temperature change, $\delta T = 105$ K, is established as one of the input parameters for the realization of all molecular logic gates. In each case, the offset values of FIR_{TC}, FIR_{NTC}, S_r, and S_a are carefully selected based on their respective changes in value over a temperature rise of 105 K.[185]

A thermometer is proposed by Eldridge and Chambers recently using a perovskite oxide Cr:$GdAlO_3$ as sensing element at the tip of a Yttrium Aluminium Garnet (YAG) optical fiber thermometer for luminescence decay-based temperature measurements at high temperatures, up to 1200°C. The strong crystal field at the Cr^{3+} dopant site in the $GdAlO_3$ structure, along with thermal population of the broadband emitting $^4T_2$ level by the long-lived underlying ^2E level, minimizes thermal quenching even at high temperatures. This allows for the retention of high luminescence intensity, outperforming rare earth-doped thermographic phosphors. The spin-allowed broadband emission from Cr:$GdAlO_3$ also reduces interference from background thermal radiation, improving signal-to-noise ratios for

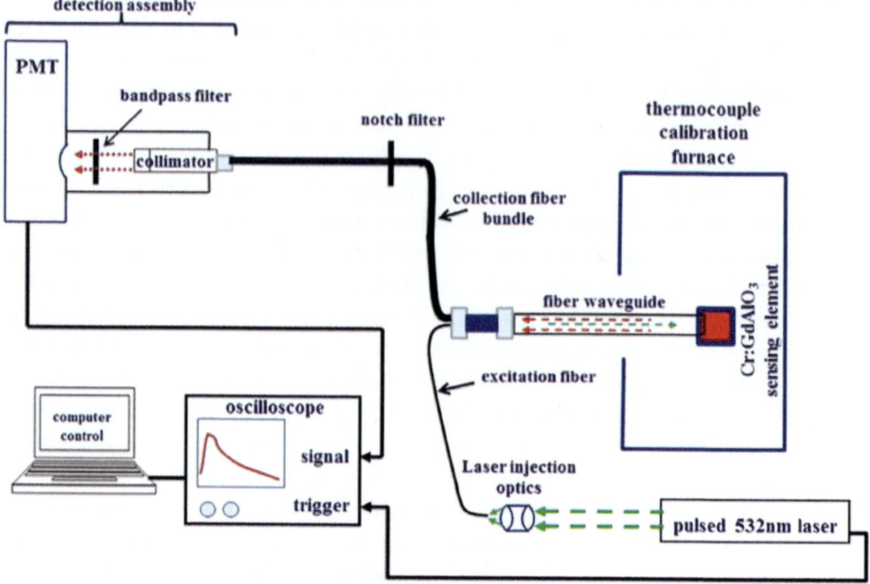

Figure 10.5 Schematic diagram showing the setup for decay time versus temperature calibration of optical fiber thermometer using Cr^{3+} doped $GdAlO_3$. (Reprinted from Ref. [207] with permission from IOP Publishing Ltd.)

decay curve analysis. The study achieved a temperature accuracy of ±5°C in the 800–1200°C range, even with a simplified temperature-independent correction for background fiber emission.[207] A schematic diagram showing the setup for decay time versus temperature calibration of the optical fiber thermometer using Cr^{3+} doped $GdAlO_3$ is shown in Figure 10.5.

The reports mentioned above indicate that perovskite oxide materials hold promise for experimentation and the creation of optical thermometers by discovering innovative temperature measurement techniques and strategies to improve thermal sensitivity. These discoveries pave the way for the advancement of temperature sensing technology and the exploration of new possibilities in this field.

10.5 MAJOR CHALLENGES AND OUTLOOK TO OVERCOME THE CHALLENGES

Despite being widely used for optical thermometry applications, perovskite oxides have not been able to achieve a relative sensitivity higher than 10% K^{-1}.[168] This limitation persists even after extensive efforts to account for temperature variations across multiple luminescent parameters and the implementation of various strategies to enhance thermal sensitivity and temperature resolution. Utilizing emissions from multiple luminescent centers in the temperature probe addresses the limitations of TCL-based thermometry, although it has not led to a significant enhancement in relative sensitivity. Additionally, there is a tendency to avoid discussing the errors that can arise when determining temperature using these probes. Moreover, maintaining consistent temperature resolution throughout the entire temperature measurement range has proven challenging. In brief, simultaneously achieving high relative sensitivity and a wide dynamic range has not been accomplished so far. The fundamental idea behind optical thermometry is straightforward and compact. Nevertheless, putting it into practical use is highly challenging and frequently unfeasible, especially given that optical thermometers are designed for use in demanding and harsh environments. Consequently, developing tailor-made test designs for optical thermometers specific to different applications becomes a costly endeavor compared to traditional thermometer solutions.

Another issue is the size of these oxides. In various applications where temperature measurements are conducted at the nanoscale, luminescent temperature probes must also be in the nano-size range. For example, there is a significant demand for precise temperature measurements at intra- and inter-cellular levels. Furthermore, while luminescence thermometry has potential applications in theranostics, luminescence bioimaging, and even magnetic resonance imaging and X-ray computed tomography, most of the perovskite oxides explored for optical thermometry are not in nanosize, with only a few exceptions.

Numerous studies have highlighted the impact of crystal structure and morphology on the temperature-sensing capabilities of perovskite oxides. As structural changes typically entail symmetry breaking, one can anticipate an effective variation in thermal response in the luminescent intensity ratio. Carefully modulating the structure and choosing appropriate luminescent centers may have a positive impact on the thermometric properties of the perovskite oxides. Composites containing perovskite phases as in Eu^{3+} activated zinc titanate and $Pb(In_{1/2}Nb_{1/2})O_3$-$Pb(Mg_{1/3}Nb_{2/3})O_3$ – $PbTiO_3$: Pr^{3+} ferroelectric relaxors, provide valuable insights into the necessity of selecting an appropriate temperature probe for multifunctional applications.[80,172] Ferroelectric relaxors hold promise in applications where there is a need to combine electrical and optical properties. Recent developments, as discussed in Section 10.4, in temperature-read-out techniques and the construction of high-sensitivity probes demonstrate that there is ample room for improvement, even in fundamental methodological aspects. Additionally, many unexplored applications could benefit from luminescence thermometry, but adapting this method for novel uses often necessitates expanding the background scientific knowledge base. Addressing the challenges above and ensuring the reliability, repeatability, and reproducibility of temperature sensing, perovskite oxides are poised to become the unequivocal preference for the scientific and engineering community in the field of optical thermometry.

REFERENCES

1. Wang, X.; Liu, Q.; Bu, Y.; Liu, C. S.; Liu, T.; Yan, X. Optical Temperature Sensing of Rare-Earth Ion Doped Phosphors. *RSC Adv.* **2015**, *5* (105), 86219–86236. DOI: 10.1039/c5ra16986k.

2. Dramićanin, M. D. Trends in Luminescence Thermometry. *J. Appl. Phys.* **2020**, *128* (4), 040902. DOI: 10.1063/5.0014825.

3. Brites, C. D. S.; Balabhadra, S.; Carlos, L. D. Lanthanide-Based Thermometers: At the Cutting-Edge of Luminescence Thermometry. *Adv. Opt. Mater.* **2019**, *7* (5), 1801239. DOI: 10.1002/adom.201801239.

4. Lal, S. C.; Jawahar, I. N.; Ganesanpotti, S. Optical Thermometry Based on Eu^{3+} Activated Double Perovskites: State of the Art and Challenges. In *Horizons in World Physics*, 1st ed.; Nova Science Publishers, 2022; pp. 1–55. DOI: 10.52305/DKIM5863.

5. *Greystone Energy Systems Inc.* https://us.greystoneenergy.com/the-importance-of-temperature-monitoring/ (accessed 2024-01-16).

6. *MARKETSANDMARKETS.* https://www.marketsandmarkets.com/Market Reports/temperature-sensor-market-522.html (accessed 2024-01-16).

7. Brites, C. D. S.; Millan, A.; Carlos, L. D. Lanthanides in Luminescent Thermometry. In *Handbook on the Physics and Chemistry of Rare Earths*, 1st ed.; Elsevier, 2016; pp. 339–427.

8. Kolesnikov, I. E.; Mamonova, D. V.; Kurochkin, M. A.; Kolesnikov, E. Y.; Lähderanta, E. Multimode Luminescence Thermometry Based on Emission and Excitation Spectra. *J. Lumin.* **2021**, *231*, 117828. DOI: 10.1016/j.jlumin.2020.117828.

9. Song, M.; Zhao, W.; Xue, J.; Wang, L.; Wang, J. Color-Tunable Luminescence and Temperature Sensing Properties of a Single-Phase Dual-Emitting La_2LiSbO_6:Bi^{3+}, Sm^{3+} Phosphor. *J. Lumin.* **2021**, *235*, 118014. DOI: 10.1016/j.jlumin.2021.118014.

10. Maturi, F. E.; Brites, C. D. S.; Ximendes, E. C.; Mills, C.; Olsen, B.; Jaque, D.; Ribeiro, S. J. L.; Carlos, L. D. Going Above and Beyond: A Tenfold Gain in the Performance of Luminescence Thermometers Joining Multiparametric Sensing and Multiple Regression. *Laser Photon Rev.* **2021**, *15* (11), 2100301. DOI: 10.1002/LPOR.202100301.

11. Xu, J.; Bu, Y.; Wang, J.; Meng, L.; Wang, X.; Yan, X. Site-Dependent Photoluminescence and Optical Thermometric Behaviors of Double-Perovskite $CaBa_2WO_6$:Er^{3+}. *Chem.Phys.Lett.* **2020**, *749*, 137410. DOI: 10.1016/j.cplett.2020.137410.

12. Zhu, X.; Wang, L.; Shi, Q.; Guo, H.; Qiao, J.; Cui, C.; Ivanovskikh, K. V.; Huang, P. High Sensitivity Dual-Mode Ratiometric Optical Thermometry Based on Bi^{3+}/Mn^{4+} Co-Doped Ba_2LaTaO_6. *J. Lumin.* **2023**, *262*, 119949. DOI: 10.1016/j.jlumin.2023.119949.

13. Long, J.; Xu, Y.; Cheng, K.; Liu, X.; Huang, W.; Deng, C. A Novel Multifunctional Double Perovskite Structure Phosphor La_2MgTiO_6:Mn^{4+}, Eu^{3+}. *Opt. Mater.* **2023**, *141*, 113967. DOI: 10.1016/j.optmat.2023.113967.

14. Voiculescu, A. M.; Hau, S.; Stanciu, G.; Avram, D.; Gheorghe, C. Optical Thermometry through Infrared Excited Green Upconversion Emissions of Er^{3+}-Yb^{3+} Co-Doped $LaAlO_3$ Phosphors. *J. Lumin.* **2022**, *242*, 118602. DOI: 10.1016/j.jlumin.2021.118602.

15. Huang, Y.; Bai, G.; Zhao, Y.; Xie, H.; Yang, X.; Xu, S. Yb/Ho Codoped Layered Perovskite Bismuth Titanate Microcrystals with Upconversion Luminescence: Fabrication, Characterization, and Application in Optical Fiber Ratiometric Thermometry. *Inorg. Chem.* **2020**, *59* (19), 14229–14235. DOI: 10.1021/acs.inorgchem.0c02015.

16. Piotrowski, W. M.; Bolek, P.; Brik, M. G.; Zych, E.; Marciniak, L. Frontiers of Deep-Red Emission of Mn^{4+} Ions with Ruddlesden–Popper Perovskites. *Inorg. Chem.* **2023**, *62* (51), 21164–21172. DOI: 10.1021/acs.inorgchem.3c03113.

17. Colella, S.; Mazzeo, M.; Rizzo, A.; Gigli, G.; Listorti, A. The Bright Side of Perovskites. *J. Phys. Chem. Lett.* **2016**, *7* (21), 4322–4334. DOI: 10.1021/acs.jpclett.6b01799.

18. Song, R.; Xu, S.; Li, Y.; Gao, Y.; Yu, H.; Cao, Y.; Zhang, X.; Chen, B. Designing Multi-Mode Optical Thermometers via Sb^{3+}/Er^{3+} Co-Doped $Cs_2NaInCl_6$ Lead-Free Double Perovskite Microcrystals. *J. Alloys Compd.* **2023**, *961*, 171126. DOI: 10.1016/j.jallcom.2023.171126.

19. Chen, W.; Jin, M.; Tang, J.; Li, Y.; Chen, C.; Xiang, J.; Li, Z.; Guo, C. Self-Calibrated Multi-Mode Optical Thermometry in up-Converting Phosphor $Cs_2NaErCl_6$: Yb^{3+}. *J. Lumin.* **2023**, *263*, 120132. DOI: 10.1016/j.jlumin.2023.120132.

20. Song, R.; Xu, S.; Li, Y.; Zhang, Q.; Gao, Y.; Yu, H.; Cao, Y.; Li, X.; Zhang, S.; Chen, B. Bi^{3+} and Tb^{3+} Co-Doped $Cs_2AgInCl_6$ Lead-Free Double Perovskite Nanocrystals for Detection of Temperature and Copper Ions. *Spectrochim. Acta A Mol. Biomol. Spectrosc.* **2023**, *288*, 122181. DOI: 10.1016/j.saa.2022.122181.

21. Zhao, C.; Gao, Y.; Zhou, D.; Zhu, F.; Chen, J.; Qiu, J. High-Efficiency Dual-Mode Luminescence of Metal Halide Perovskite $Cs_3Bi_2Cl_9$:Er^{3+} and Its Use in Optical Temperature Measurement with High Sensitivity. *J. Alloys Compd.* **2023**, *944*, 169134. DOI: 10.1016/j.jallcom.2023.169134.

22. Yang, S.; Wei, Y.; Wang, Q.; Sun, Z.; Li, A. H. Tm^{3+}-Doped $Cs_2Ag_{0.6}Na_{0.4}In_{0.9}Bi_{0.1}Cl_6$ Microcrystals for Thermometry. *J. Alloys Compd.* **2023**, *968*, 172212. DOI: 10.1016/j.jallcom.2023.172212.

23. Chang, Q.; Zhou, X.; Jiang, S.; Xiang, G.; Li, L.; Li, Y.; Jing, C.; Ling, F.; Wang, Y.; Xiao, P. Dual-Mode Luminescence Temperature Sensing Performance of Manganese (II) Doped $CsPbCl_3$ Perovskite Quantum Dots. *Ceram. Int.* **2022**, *48* (22), 33645–33652. DOI: 10.1016/j.ceramint.2022.07.310.

24. Huang, Y.; Bai, G.; Zhao, Y.; Xie, H.; Yang, X.; Xu, S. Yb/Ho Codoped Layered Perovskite Bismuth Titanate Microcrystals with Upconversion Luminescence: Fabrication, Characterization, and Application in Optical Fiber Ratiometric Thermometry. *Inorg. Chem.* **2020**, *59* (19), 14229–14235. DOI: 10.1021/acs.inorgchem.0c02015.

25. Yakunin, S.; Benin, B. M.; Shynkarenko, Y.; Nazarenko, O.; Bodnarchuk, M. I.; Dirin, D. N.; Hofer, C.; Cattaneo, S.; Kovalenko, M. V. High-Resolution Remote Thermometry and Thermography Using Luminescent Low-Dimensional Tin-Halide Perovskites. *Nat. Mater.* **2019**, *18* (8), 846–852. DOI: 10.1038/s41563-019-0416-2.

26. Ptak, M.; Dziuk, B.; Stefańska, D.; Hermanowicz, K. The Structural, Phonon and Optical Properties of $[CH_3NH_3]M_{0.5}Cr_XAl_{0.5-x}(HCOO)_3$ (M=Na, K; $X=0$, 0.025, 0.5) Metal-Organic Framework Perovskites for Luminescence Thermometry. *Phys. Chem. Chem. Phys.* **2019**, *21* (15), 7965–7972. DOI: 10.1039/C9CP01043B.

27. Yang, X.; Lin, S.; Ma, D.; Long, S.; Zhu, Y.; Li, H.; Wang, B. Up-Conversion Luminescence of $LiTaO_3:Er^{3+}$ Phosphors for Optical Thermometry. *Ceram. Int.* **2020**, *46* (1), 1178–1182. DOI: 10.1016/j.ceramint.2019.09.088.

28. Xing, L.; Yang, W.; Ma, D.; Wang, R. Effect of Crystallinity on the Optical Thermometry Sensitivity of Tm^{3+}/Yb^{3+} Codoped $LiNbO_3$ Crystal. *Sens. Actuators B Chem.* **2015**, *221*, 458–462. DOI: 10.1016/j.snb.2015.06.132.

29. Bradley III, C. L. A Temperature-Sensitive Phosphor Used to Measure Surface Temperatures in Aerodynamics. *Rev. Sci. Instrum.* **1953**, *24* (3), 219–220. DOI: 10.1063/1.1770668.

30. Gross, G. E. *Midwest Research Institute Final Report* (M.R.I. Project No.2300-P); June 1960.

31. Gravitt, J. C.; Gross, G. E.; Beachy, J. E.; McAllum, W. E. *Midwest Research Institute Final Report* (M.R.I. Project No.2423-P); June 1963.

32. Sholes, R. R.; Small, J. G. Fluorescent Decay Thermometer with Biological Applications. *Rev. Sci. Instrum.* **1980**, *51* (7), 882–884. DOI: 10.1063/1.1136350.

33. Allison, S. W.; Gillies, G. T. Remote Thermometry with Thermographic Phosphors: Instrumentation and Applications. *Rev. Sci. Instrum.* **1997**, *68* (7), 2615–2650. DOI: 10.1063/1.1148174.

34. Stefańska, D.; Vu, T. H. Q.; Dereń, P. J. Multiple Ways for Temperature Detection Based on La_2MgTiO_6 Double Perovskite Co-Doped with Mn^{4+} and Cr^{3+} Ions. *J. Alloys Compd.* **2023**, *938*, 168653. DOI: 10.1016/j.jallcom.2022.168653.

35. Hua, Y.; Wang, T.; Li, H.; Yu, J. S.; Li, L. Charge Transfer Band Excited $(Sr,Ba)_2YTaO_6:Eu^{3+}$ Reddish-Orange-Emitting Phosphors for Luminescence Lifetime Thermometry and Flexible Anti-Counterfeiting Labels. *J. Alloys Compd.* **2023**, *930*, 167454. DOI: 10.1016/j.jallcom.2022.167454.

36. Wang, W.; Li, Q.; Chen, L.; Wang, Y.; Zhong, R.; Dong, H.; Qiu, Y.; Hu, Y.; Zhang, X. Red Emitting $Ba_2LaNbO_6: Mn^{4+}$ Phosphor for the Lifetime-Based Optical Thermometry. *J. Lumin.* **2023**, *257*, 119683. DOI: 10.1016/j.jlumin.2023.119683.

37. Ran, W.; Noh, H. M.; Park, S. H.; Lee, B. R.; Kim, J. H.; Jeong, J. H.; Shi, J.; Liu, G. Simultaneous Bifunctional Application of Solid-State Lighting and Ratiometric Optical Thermometer Based on Double Perovskite $LiLaMgWO_6:Er^{3+}$ Thermochromic Phosphors. *RSC Adv.* **2019**, *9* (13), 7189–7195. DOI: 10.1039/c8ra10242b.

38. Vu, T. H. Q.; Bondzior, B.; Stefańska, D.; Dereń, P. J. Influence of Temperature on Near-Infrared Luminescence, Energy Transfer Mechanism and the Temperature Sensing Ability of $La_2MgTiO_6:Nd^{3+}$ Double Perovskites. *Sens. Actuators A Phys.* **2021**, *317*, 112453. DOI: 10.1016/j.sna.2020.112453.

39. Wang, J.; Li, J.; Lei, R.; Zhao, S.; Xu, S. Single Band Ratiometric Luminescence Thermometry Based on Pr^{3+} Doped Oxides Containing Charge Transfer States. *J. Mater. Chem. C* **2022**, *10* (34), 12413–12421. DOI: 10.1039/d2tc02336a.

40. Tian, X.; Wang, C.; Dou, H.; Wu, L. Photoluminescence Origin and Non-Contact Thermometric Properties in Pb^{2+}-Activated $CaZrO_3$ Perovskite Phosphor. *J. Alloys Compd.* **2022**, *892*, 162250. DOI: 10.1016/j.jallcom.2021.162250.

41. Stefańska, D.; Bondzior, B.; Vu, T. H. Q.; Miniajluk-Gaweł, N.; Dereń, P. J. The Influence of Morphology and Eu^{3+} Concentration on Luminescence and Temperature Sensing Behavior of Ba_2MgWO_6 Double Perovskite as a Potential Optical Thermometer. *J. Alloys Compd.* **2020**, *842*, 155742. DOI: 10.1016/j.jallcom.2020.155742.

42. Vu, T. H. Q.; Bondzior, B.; Stefańska, D.; Dereń, P. J. An Er^{3+} Doped Ba_2MgWO_6 Double Perovskite: A Phosphor for Low-Temperature Thermometry. *Dalton Trans.* **2022**, *51* (20), 8056–8065. DOI: 10.1039/d2dt00554a.

43. Bondzior, B.; Stefańska, D.; Vu, T. H. Q.; Dereń, P. J. Optimization of Eu^{3+}-to-Host Emission Ratio in Double-Perovskite Molybdenites for Highly Sensitive Temperature Sensors. *J. Phys. Chem. C* **2022**, *126* (31), 13247–13255. DOI: 10.1021/acs.jpcc.2c02924.

44. Li, H.; Li, L.; Mei, L.; Zhou, X.; Cao, Z.; Ling, F.; Wang, Y.; Jiang, S.; Xiang, G.; Su Yu, J.; Zhang, X. Customizing Dual-Functional Platforms for Solid-State Lighting and Luminescence Lifetime Thermometry with High Sensing Sensitivity. *J. Lumin.* **2023**, *263*, 120026. DOI: 10.1016/j.jlumin.2023.120026.

45. De, A.; Dwij, V.; Sathe, V.; Hernández-Rodríguez, M. A.; Carlos, L. D.; Ranjan, R. Synergistic Use of Raman and Photoluminescence Signals for Optical Thermometry with Large Temperature Sensitivity. *Physica B Condens. Matter.* **2022**, *626*, 413455. DOI: 10.1016/j.physb.2021.413455.

46. Hong, Y.; Li, H.; Luo, D.; Yang, H.; Zhou, B.; Zhang, W.; Ding, J.; Zhou, J.; Wu, Q. Mn^{4+}-Activated Double Perovskite Red Phosphor for Multifunctional Applications. *J. Lumin.* **2022**, *244*, 118752. DOI: 10.1016/j.jlumin.2022.118752.

47. Matuszewska, C.; Marciniak, L. The Influence of Host Material on NIR II and NIR III Emitting Ni^{2+}-Based Luminescent Thermometers in $ATiO_3$: Ni^{2+} (A = Sr, Ca, Mg, Ba) Nanocrystals. *J. Lumin.* **2020**, *223*, 117221. DOI: 10.1016/j.jlumin.2020.117221.

48. Qin, Y.; Han, F.; Yan, P.; Wang, Y.; Zhang, Y.; Zhang, S. Fluorescence Intensity Ratio (FIR) Analysis of the Temperature Sensing Properties in Transparent Ferroelectric PMN-PT:Pr^{3+} Ceramic. *Ceram. Int.* **2021**, *47* (17), 24092–24097. DOI: 10.1016/j.ceramint.2021.05.119.

49. De, A.; Mishra, A.; Khatua, D. K.; Dwij, V.; Sathe, V.; Jena, S.; Ranjan, R. Optical Temperature Sensing by Tuning Photoluminescence in a Wide (Visible to Near Infrared) Wavelength Range in a Eu^{3+}-Doped Bi-Based Relaxor Ferroelectric. *Opt. Lett.* **2022**, *47* (3), 489–492. DOI: 10.1364/ol.441377.

50. Liu, H.; Guo, J.; Li, X.; Zhang, Z.; Zeng, C.; Xu, J.; Xiong, Z. Luminescence and Temperature Sensing Properties of $Y_{2-x-y}Tm_xSm_yMgTiO_6$ Phosphors. *J. Lumin.* **2024**, *267*, 120392. DOI: 10.1016/j.jlumin.2023.120392.

51. Peng, S.; Lai, F.; Xiao, Z.; Cheng, H.; Jiang, Z.; You, W. Upconversion Luminescence and Temperature Sensing Properties of Er^{3+}/Yb^{3+} Doped Double-Perovskite Ba_2LaNbO_6 Phosphor. *J. Lumin.* **2022**, *242*, 118569. DOI: 10.1016/j.jlumin.2021.118569.

52. Shi, Y.; Cui, R.; Liu, X.; Yang, L.; Li, Y.; Zhang, J.; Deng, C. Ca_2YNbO_6: Dy^{3+}, Sm^{3+}/Eu^{3+} Warm White Phosphors with Tunable Light Color and FIR Optical Temperature Measurement Characteristics. *Opt. Mater.* **2023**, *136*, 113495. DOI: 10.1016/j.optmat.2023.113495.

53. Zhou, H.; Guo, N.; Zhu, M.; Li, J.; Miao, Y.; Shao, B. Photoluminescence and Ratiometric Optical Thermometry in Mn^{4+}/Eu^{3+} Dual-Doped Phosphor via Site-Favorable Occupation. *J. Lumin.* **2020**, *224*, 117311. DOI: 10.1016/j.jlumin.2020.117311.

54. Stefańska, D.; Bondzior, B.; Vu, T. H. Q.; Grodzicki, M.; Dereń, P. J. Temperature Sensitivity Modulation through Changing the Vanadium Concentration in a La_2MgTiO_6:V^{5+},Cr^{3+} double Perovskite Optical Thermometer. *Dalton Trans.* **2021**, *50* (28), 9851–9857. DOI: 10.1039/d1dt00911g.

55. Chen, Y.; Zhou, L.; Shen, Y.; Lei, L.; Ye, R.; Chen, L.; Deng, D.; Xu, S. A Ratiometric Optical Thermometer Based on Bi^{3+} and Mn^{4+} Co-Doped La_2MgGeO_6 Phosphor with High Sensitivity and Signal Discriminability. *J. Alloys Compd.* **2021**, *887*, 161283. DOI: 10.1016/j.jallcom.2021.161283.

56. Hua, Y.; Qiu, X.; Sonne, C.; Brown, R. J. C.; Kim, K.-H. *Construction of Novel Luminescent Thermometers Using Single-Doped Phosphors with Color-Tuning Properties.* **2022**. Available at SSRN. DOI: 10.2139/ssrn.3989701 (accessed on 2024-01-03).

57. Cui, M.; Wang, J.; Li, J.; Huang, S.; Shang, M. An Abnormal Yellow Emission and Temperature-Sensitive Properties for Perovskite-Type Ca_2MgWO_6 Phosphor via Cation Substitution and Energy Transfer. *J. Lumin.* **2019**, *214*, 116588. DOI: 10.1016/j.jlumin.2019.116588.

58. Zheng, T.; Runowski, M.; Martín, I. R.; Lis, S.; Vega, M.; Llanos, J. Nonlinear Optical Thermometry—A Novel Temperature Sensing Strategy via Second Harmonic Generation (SHG) and Upconversion Luminescence in $BaTiO_3$:Ho^{3+},Yb^{3+} Perovskite. *Adv. Opt. Mater.* **2021**, *9* (12), 2100386. DOI: 10.1002/adom.202100386.

59. Zheng, T.; Runowski, M.; Woźny, P.; Barszcz, B.; Lis, S.; Vega, M.; Llanos, J.; Soler-Carracedo, K.; Martín, I. R. Boltzmann vs. Non-Boltzmann (Non-Linear) Thermometry-Yb^{3+}-Er^{3+} Activated Dual-Mode Thermometer and Phase Transition Sensor via Second Harmonic Generation. *J. Alloys Compd.* **2022**, *906*, 164329. DOI: 10.1016/j.jallcom.2022.164329.

60. Vu, T. H. Q.; Bondzior, B.; Stefańska, D.; Dereń, P. J. Low-Temperature Optical Thermometer Based on the Luminescence of the Double Perovskite Ba_2MgWO_6: Nd^{3+}. *J. Lumin.* **2023**, *257*, 119750. DOI: 10.1016/j.jlumin.2023.119750.

61. Mondal, A.; Manam, J. Structural, Optical and Temperature Dependent Photoluminescence Properties of Cr^{3+}-Activated $LaGaO_3$ Persistent Phosphor for Optical Thermometry. *Ceram. Int.* **2020**, *46* (15), 23972–23984. DOI: 10.1016/j.ceramint.2020.06.174.

62. Vu, Q. T. H.; Bondzior, B.; Stefańska, D.; Miniajluk-Gaweł, N.; Winiarski, M. J.; Dereń, P. J. On How the Mechanochemical and Co-Precipitation Synthesis Method Changes the Sensitivity and Operating Range of the $Ba_2Mg_{1-x}Eu_xWO_6$ Optical Thermometer. *Sci. Rep.* **2021**, *11* (1), 22847. DOI: 10.1038/s41598-021-02309-9.

63. Jiawen, L.; Jiawen, W.; Ruoshan, L.; Shiqing, X. Ratiometric Luminescence Thermometry Based on Diverse Thermal Response from Excited States in La_2ZnTiO_6: Tb^{3+}. *J. Lumin.* **2022**, *249*, 119047. DOI: 10.1016/j.jlumin.2022.119047.

64. Liu, Z.; Chen, D. Upconversion Photoluminescence and Dual-Mode Temperature Sensing Properties of PIN-PMN-PT:Er^{3+} Ceramic. *J. Alloys Compd.* **2020**, *815*, 152092. DOI: 10.1016/j.jallcom.2019.152092.

65. Lal, S.C.; Jawahar, I.N.; Ganesanpotti, S. A Six-Mode Optical Thermometry Rooted from the Distinct Thermal Behavior of SrLaLiTeO$_6$: Mn^{4+}, Eu^{3+} Double Perovskites and Their Potential Application in Wavelength Detection. *J. Sci.: Adv. Mater. Devices.* **2023**, *8* (2), 100544. DOI: 10.1016/j.jsamd.2023.100544.

66. Liu, Z.; Long, S.; Zhu, Y.; Wang, W.; Wang, B. Optical Thermometry Based on Thermolabile Intrinsic Polarons in Tm^{3+} and Yb^{3+} Co-Doped Congruent Lithium Niobate Single Crystal. *J. Alloys Compd.* **2021**, *867*, 158986. DOI: 10.1016/j.jallcom.2021.158986.

67. Nam, H. T.; Tam, P. D.; Van Hai, N.; Van, H. N. Multifunctional Optical Thermometry Using Dual-Mode Green Emission of CaZrO$_3$:Er/Yb/Mo Perovskite Phosphors. *RSC Adv.* **2023**, *13* (21), 14660–14674. DOI: 10.1039/d3ra02759g.

68. Dong, M.; Yin, T.; Guo, G.; Liu, Z.; Wang, F.; Wang, C.; Guan, L.; Li, X. Three-Mode Optical Thermometry Based on Multiresponsive Ba$_{1-x}$Sr$_x$LaLiWO$_6$:Er^{3+}, Yb^{3+} Phosphors. *Ceram. Int.* **2024**, *50* (1), 1050–1058. DOI: 10.1016/j.ceramint.2023.10.198.

69. Xu, S. J.; Lei, J. H.; Li, L. J.; Chen, J. Y.; Chen, L. P.; Guo, H. Dual-Mode Optical Thermometry of Sr$_2$YNbO$_6$:Bi^{3+},Eu^{3+} Phosphors Designed by Response Surface Methodology. *J. Lumin.* **2023**, *255*, 119615. DOI: 10.1016/j.jlumin.2022.119615.

70. Wu, Z.; Li, L.; Wang, Y.; Ling, F.; Cao, Z.; Jiang, S.; Xiang, G.; Zhou, X.; Hua, Y.; Yu, J. S. High-Sensitivity Luminescent Thermometer Based on Mn^{4+}/Sm^{3+} Dual-Emission Centers in Double-Perovskite Tellurate. *Ceram. Int.* **2022**, *48* (19), 27664–27671. DOI: 10.1016/j.ceramint.2022.06.064.

71. Stefańska, D.; Vu, T. H. Q.; Dereń, P. J. Multiple Ways for Temperature Detection Based on La$_2$MgTiO$_6$ Double Perovskite Co-Doped with Mn^{4+} and Cr^{3+} Ions. *J. Alloys Compd.* **2023**, *938*, 168653. DOI: 10.1016/j.jallcom.2022.168653.

72. Zhao, W.; Li, L.; Wu, Z.; Wang, Y.; Cao, Z.; Ling, F.; Jiang, S.; Xiang, G.; Zhou, X.; Hua, Y. Bifunctional Near-Infrared-Emitting Phosphors of Chromium (III)-Activated Antimonates for NIR Light-Emitting Diodes and Low-Temperature Thermometry. *J. Alloys Compd.* **2023**, *965*, 171370. DOI: 10.1016/j.jallcom.2023.171370.

73. Wang, J.; Zhao, X.; Song, M.; Zhao, J.; Xue, J. Efficient Lifetime-Based Optical Thermometry Using BaLaZnSbO$_6$:Mn^{4+} Red-Emitting Phosphors. *J. Lumin.* **2023**, *263*, 120019. DOI: 10.1016/j.jlumin.2023.120019.

74. Han, M.; Chen, S.; Li, J.; Gao, Z.; Zhang, Y.; Shen, Y.; Tian, Y.; Deng, D. Sr$_2$(Ga/Al)TaO$_6$:Cr^{3+} Phosphor with Tunable near-Infrared Emitting for Light-Emitting Diodes and Optical Thermometer. *J. Alloys Compd.* **2024**, *973*, 172927. DOI: 10.1016/j.jallcom.2023.172927.

75. Sharma, S. K. Exploring the Possibility of Using Stark Lines of Pr^{3+} in CaTiO$_3$ Perovskite for Low Temperature Luminescence Thermometry. *J. Phys. Chem. C* **2023**, *127* (47), 22934–22942. DOI: 10.1021/acs.jpcc.3c06338.

76. Tai, Y.; Cui, R.; Zhang, J.; Wang, C.; Zhao, T.; Zhang, B.; Deng, C. Luminescence Properties and Optical Sensing Behaviors of Sr_2GdSbO_6:Eu^{3+} Phosphors. *J. Rare Earths.* **2023.** Article ASAP. DOI: 10.1016/j.jre.2023.06.009 (accessed 2024-01-04).

77. Ruan, F.; Fan, G.; Li, N.; Zhou, J.; Li, Y.; Fan, D.; Chen, Q. Anomalous Thermal Quenching Behavior of Novel Orange-Red $SrLa_{1-x}NaTeO_6$:xSm^{3+} Phosphors for Optical Temperature Sensing. *J. Lumin.* **2024,** *265,* 120223. DOI: 10.1016/j.jlumin.2023.120223.

78. Hua, Y.; Yu, J. S. Deep-Red-Emitting Phosphors of Mn^{4+}-Activated Tantalite for High-Sensitivity Lifetime Thermometry and Security Films. *Dalton Trans.* **2023,** *52* (31), 10751–10759. DOI: 10.1039/d3dt01384g.

79. Yuan, L.; Wang, X.; Cui, R.; Deng, C. Quick and Accurate Optical Thermometry Based on Chemometrics Model Strategy in $Na_{0.5}Gd_{0.5}TiO_3$:Er^{3+}. *Ceram. Int.* **2023,** *49* (16), 26786–26793. DOI: 10.1016/j.ceramint.2023.05.215.

80. Hu, F.; Wu, Y.; Sun, C.; Xu, C.; Zhu, B.; Wang, Q.; Shi, Q.; Li, S.; Zhang, D. Influence of Charge Compensation on the Photoluminescence and Temperature Sensing of Zinc Titanate Composites. *Ceram. Int.* **2023,** *49* (8), 11779–11787. DOI: 10.1016/j.ceramint.2022.12.022.

81. Cai, P.; Qin, L.; Chen, C.; Wang, J.; Bi, S.; Kim, S. Il; Huang, Y.; Seo, H. J. Optical Thermometry Based on Vibration Sidebands in Y_2MgTiO_6:Mn^{4+} Double Perovskite. *Inorg. Chem.* **2018,** *57* (6), 3073–3081. DOI: 10.1021/acs.inorgchem.7b02938.

82. Zhang, H.; Gao, Z.; Li, G.; Zhu, Y.; Liu, S.; Li, K.; Liang, Y. A Ratiometric Optical Thermometer with Multi-Color Emission and High Sensitivity Based on Double Perovskite $LaMg_{0.402}Nb_{0.598}O_3$: Pr^{3+} Thermochromic Phosphors. *Chem. Eng. J.* **2020,** *380,* 122491. DOI: 10.1016/j.cej.2019.122491.

83. Wang, Z.; Chen, X.; Luo, M.; Qin, L. The Optical Properties of Mn^{4+} Ions in Ba_2CaWO_6 with Double Perovskite Structure for Optical Thermometer Application. *Mater. Res. Bull.* **2022,** *156,* 112004. DOI: 10.1016/j.materresbull.2022.112004.

84. Ran, W.; Noh, H. M.; Park, S. H.; Lee, B. R.; Kim, J. H.; Jeong, J. H.; Shi, J. Er^{3+}-Activated $NaLaMgWO_6$ Double Perovskite Phosphors and Their Bifunctional Application in Solid-State Lighting and Non-Contact Optical Thermometry. *Dalton Trans.* **2019,** *48* (13), 4405–4412. DOI: 10.1039/c9dt00315k.

85. Wang, Y.; Suo, H.; Li, L.; Wang, G.; Deng, W.; Ding, W.; Wang, Z.; Li, P.; Zhang, Z. Cr^{3+}-Doped Double-Perovskites for near-Infrared Luminescent Ratiometric Thermometry. *Physica B Condens. Matter.* **2022,** *625,* 413496. DOI: 10.1016/j.physb.2021.413496.

86. Sreelekshmi, A. K.; Lal, S. C.; Ganesanpotti, S. Probing the Multifunctionality of Double Layered Perovskite $NaGdMgTeO_6$:Eu^{3+} in Ratiometric Phosphor Thermometry and Solid-State Lighting. *J. Alloys Compd.* **2022,** *905,* 164138. DOI: 10.1016/j.jallcom.2022.164138.

87. Hua, Y.; Li, H.; Wang, T.; Yu, J. S.; Li, L. Customization of Novel Double-Perovskite $(Ca,Sr)_2InNbO_6$:Mn^{4+} Red-Emitting Phosphors for Luminescence Lifetime Thermometers with Good Relative Sensing Sensitivity. *J. Alloys Compd.* **2022,** *925,* 166498. DOI: 10.1016/j.jallcom.2022.166498.

88. Suraja, N. J.; Mahesh, A.; Sibi, K. S.; Ganesanpotti, S. Insights into the Crystal Structure and Multifunctional Optical Properties of A_2CdTeO_6 (A=Ba, Sr, Ca) Double Perovskites. *J. Alloys Compd.* **2021,** *865,* 158902. DOI: 10.1016/j.jallcom.2021.158902.

89. Back, M.; Ueda, J.; Xu, J.; Murata, D.; Brik, M. G.; Tanabe, S. Ratiometric Luminescent Thermometers with a Customized Phase-Transition-Driven Fingerprint in Perovskite Oxides. *ACS Appl. Mater. Interfaces.* **2019**, *11* (42), 38937–38945. DOI: 10.1021/acsami.9b13010.

90. Shi, R.; Lin, L.; Dorenbos, P.; Liang, H. Development of a Potential Optical Thermometric Material through Photoluminescence of Pr^{3+} in La_2MgTiO_6. *J. Mater. Chem. C.* **2017**, *5* (41), 10737–10745. DOI: 10.1039/c7tc02661g.

91. Chen, Q.; Yang, X.; Zhang, G.; Ma, Q.; Han, S.; Ma, B. Color-Tunable Eu^{3+}- or Sm^{3+}-Doped Perovskite Phosphors as Optical Temperature-Sensing Materials. *Opt. Mater.* **2021**, *111*, 110585. DOI: 10.1016/j.optmat.2020.110585.

92. Zhang, A.; Sun, Z.; Wang, Z.; Jia, M.; Choi, B. C.; Fu, Z.; Jeong, J. H.; Park, S. H. Self-Calibrated Ratiometric Thermometers and Multi-Mode Anti-Counterfeiting Based on Ca_2LaNbO_6:Pr^{3+} Optical Material. *Scr. Mater.* **2022**, *211*, 114515. DOI: 10.1016/j.scriptamat.2022.114515.

93. Liu, W.; Zhao, D.; Zhang, R. J.; Yao, Q. X.; Zhu, S. Y. Fluorescence Lifetime-Based Luminescent Thermometry Material with Lifetime Varying over a Factor of 50. *Inorg. Chem.* **2022**, *61* (41), 16468–16476. DOI: 10.1021/acs.inorgchem.2c02707.

94. Hua, Y.; Yu, J. S. Strong Green Emission of Erbium(III)-Activated La_2MgTiO_6 Phosphors for Solid-State Lighting and Optical Temperature Sensors. *ACS Sustain. Chem. Eng.* **2021**, *9* (14), 5105–5115. DOI: 10.1021/acssuschemeng.0c09375.

95. Tian, X.; Dou, H.; Wu, L. Bi^{3+}-Based Luminescent Thermometry in Perovskite-Type $CaZrO_3$ Phosphor. *J. Mater. Sci. Mater. Electron.* **2020**, *31* (5), 3944–3950. DOI: 10.1007/s10854-020-02942-6.

96. Hua, Y.; Wang, T.; Yu, J. S.; Du, P. Tailoring of Strong Orange-Red-Emitting Materials for Luminescence Lifetime Thermometry, Anti-Counterfeiting, and Solid-State Lighting Applications. *Mater. Today Chem.* **2022**, *25*, 100945. DOI: 10.1016/j.mtchem.2022.100945.

97. Wu, Z.; Li, L.; Li, H.; Mei, L.; Xia, W.; Yi, Y.; Hua, Y. Designing Bifunctional Platforms for LED Devices and Luminescence Lifetime Thermometers: A Case of Non-Rare-Earth Mn^{4+} Doped Tantalate Phosphors. *Dalton Trans.* **2022**, *51* (23), 9062–9071. DOI: 10.1039/d2dt01120d.

98. Bondzior, B.; Stefańska, D.; Vũ, T. H. Q.; Miniajluk-Gaweł, N.; Dereń, P. J. Red Luminescence with Controlled Rise Time in La_2MgTiO_6: Eu^{3+}. *J. Alloys Compd.* **2021**, *852*, 157074. DOI: 10.1016/j.jallcom.2020.157074.

99. Tang, W.; Zuo, C.; Li, Y.; Ma, C.; Yuan, X.; Wen, Z.; Cao, Y. Exploiting Intervalence Charge-Transfer Engineering to Finely Control $(Ba,Sr)TiO_3$:Pr^{3+} Luminescence Thermometers. *J. Lumin.* **2021**, *236*, 118103. DOI: 10.1016/j.jlumin.2021.118103.

100. Mullins, A. L.; Ćirić, A.; Ristić, Z.; Williams, J. A. G.; Evans, I. R.; Dramićanin, M. D. Double-Deconvolution Method for the Separation of Thermalised Emissions from Chromium-Doped Lanthanum Gallate and its Potential in Luminescence-Based Thermometry. *J. Lumin.* **2022**, *246*, 118847. DOI: 10.1016/j.jlumin.2022.118847.

101. Tian, X.; Lian, S.; Ji, C.; Huang, Z.; Wen, J.; Chen, Z.; Peng, H.; Wang, S.; Li, J.; Hu, J.; Peng, Y. Enhanced Photoluminescence and Ultrahigh Temperature Sensitivity from NaF Flux Assisted $CaTiO_3$: Pr^{3+} Red Emitting Phosphor. *J. Alloys Compd.* **2019**, *784*, 628–640. DOI: 10.1016/j.jallcom.2019.01.087.

102. Back, M.; Ueda, J.; Brik, M. G.; Tanabe, S. Pushing the Limit of Boltzmann Distribution in Cr^{3+}-Doped $CaHfO_3$ for Cryogenic Thermometry. *ACS Appl. Mater. Interfaces* **2020**, *12* (34), 38325–38332. DOI: 10.1021/acsami.0c08965.

103. Li, G.; Li, G.; Mao, Q.; Pei, L.; Yu, H.; Liu, M.; Chu, L.; Zhong, J. Efficient Luminescence Lifetime Thermometry with Enhanced Mn^{4+}-Activated $BaLaCa_{1-x}Mg_xSbO_6$ Red Phosphors. *Chem. Eng. J.* **2022**, *430*, 132923. DOI: 10.1016/j.cej.2021.132923.

104. Li, L.; Li, X.; Xia, W.; Wang, Y.; Ling, F.; Jiang, S.; Xiang, G.; Zhou, X.; Li, D.; Hua, Y. Investigation on the Optical Sensing Behaviors in Single Eu^{3+}-Activated Sr_2InSbO_6 Phosphors under Green Light Excitation. *J. Alloys Compd.* **2022**, *906*, 164322. DOI: 10.1016/j.jallcom.2022.164322.

105. Tang, W.; Zuo, C.; Li, Y.; Ma, C.; Yuan, X.; Cao, Y.; Wen, Z. Inner Mechanism of Pr^{3+} Luminescence Thermometers Based on the Intervalence Charge Transfer State. *Opt. Mater. Express.* **2022**, *12* (2), 727–737. DOI: 10.1364/ome.445603.

106. Gutiérrez-Cano, V.; Rodríguez, F.; González, J. A.; Valiente, R. Upconversion and Optical Nanothermometry in $LaGdO_3$: Er^{3+} Nanocrystals in the RT to 900 K Range. *J. Phys. Chem. C.* **2019**, *123* (49), 29818–29828. DOI: 10.1021/acs.jpcc.9b06959.

107. Elzbieciak-Piecka, K.; Suta, M.; Marciniak, L. Structurally Induced Tuning of the Relative Sensitivity of $LaScO_3$:Cr^{3+} Luminescent Thermometers by Co-Doping Lanthanide Ions. *Chem. Eng. J.* **2021**, *421*, 129757. DOI: 10.1016/j.cej.2021.129757.

108. Long, J.; Xu, Y.; Huang, W.; Deng, C. Dual-Mode Optical Thermometry Based on La_2MgTiO_6: Mn^{4+}, Dy^{3+} Double Perovskite Phosphors. *J. Mater. Sci. Mater. Electron.* **2023**, *34* (22), 1613. DOI: 10.1007/s10854-023-11024-2.

109. Sun, J.; Sun, Z.; Li, Y.; Jin, Z.; Ma, L.; Lu, R.; Zhang, X. Realization of Plant Growth Lighting and Temperature Detecting Based on Novel Bi^{3+}, Sm^{3+} and Mn^{4+} Doped Ca_2GdNbO_6 Double Perovskite Phosphors. *Opt. Mater.* **2023**, *145*, 114394. DOI: 10.1016/j.optmat.2023.114394.

110. Xu, Q.; Qian, W.; Muhammad, R.; Chen, X.; Yu, X.; Song, K. Photoluminescence and Temperature Sensing Properties of Bi^{3+}/Sm^{3+} Co-Doped La_2MgSnO_6 Phosphor for Optical Thermometer. *Crystals.* **2023**, *13* (7), 991. DOI: 10.3390/cryst13070991.

111. Han, M.; Chen, Y.; Fu, J.; Lin, J.; Wen, Y.; Chen, S.; Li, J.; Deng, D. Photoluminescence and Temperature Dependent Properties of Bi^{3+}, Sm^{3+} Co-Doped Ca_2YSbO_6 Phosphor for Dual-Emitting Optical Thermometer. *Opt. Mater.* **2023**, *138*, 113720. DOI: 10.1016/j.optmat.2023.113720.

112. Dong, G.; Zhang, K.; Dong, M.; Li, X.; Liu, Z.; Zhang, L.; Fu, N.; Guan, L.; Li, X.; Wang, F. Effect of Sr^{2+} Ions on the Structure, up-Conversion Emission and Thermal Sensing of Er^{3+}, Yb^{3+} Co-Doped Double Perovskite $Ba_{(2-x)}Sr_xMgWO_6$ Phosphors. *Phys. Chem. Chem. Phys.* **2023**, *25* (8), 6214–6224. DOI: 10.1039/d2cp05190g.

113. Hua, Y.; Yu, J. S.; Li, L. Rare-Earth and Transition Metal Ion Single-/Co-Doped Double-Perovskite Tantalate Phosphors: Validation of Suitability for Versatile Applications. *J. Adv. Ceram.* **2023**, *12* (5), 954–971. DOI: 10.26599/JAC.2023.9220731.

114. Chen, H.; Shen, J.; Du, X.; Chen, W.; Guo, J.; Bian, T.; Liang, Y.; An, Y.; Wu, Z.; Liu, W.; Zhang, Y. In Situ Probing the Heating Effect and Phase Transition in Perovskite Heterostructures. *Mater. Des.* **2023**, *232*, 112093. DOI: 10.1016/j.matdes.2023.112093.

115. Yuan, J.; Zhao, G.; Ren, S.; Wu, Y.; Hu, F.; Wang, Q.; Shi, Q.; Yang, B.; Li, S.; Zhang, D. Multimode Fluorescence Intensity Ratio Thermometer Based on Synergistic Luminescence from Eu^{3+} to Mn^{4+} of $SrTiO_3$:Eu^{3+}-$ZnTiO_3$:Mn^{4+} Nanocomposites. *Ceram. Int.* **2023**, *49* (11), 17699–17708. DOI: 10.1016/j.ceramint.2023.02.136.

116. Han, B.; Zhu, J.; Chu, C.; Yang, X.; Wang, Y.; Li, K.; Hou, Y.; Li, K.; Copner, N.; Teng, P. Sm^{3+}-Mn^{4+} Activated Sr_2GdTaO_6 Red Phosphor for Plant Growth Lighting and Optical Temperature Sensing. *Sens. Actuators A Phys.* **2023**, *349*, 114089. DOI: 10.1016/j.sna.2022.114089.

117. Singh, P.; Jain, N.; Shukla, S.; Tiwari, A. K.; Kumar, K.; Singh, J.; Pandey, A. C. Luminescence Nanothermometry Using a Trivalent Lanthanide Co-Doped Perovskite. *RSC Adv.* **2023**, *13* (5), 2939–2948. DOI: 10.1039/d2ra05935e.

118. Li, L.; Zhou, Z.; Huang, F.; Peng, S.; Huang, Y.; Wang, G.; Li, X.; Chen, F. F.; Yang, C.; Li, X. X.; Yu, Y. Improving Luminescence and Thermometric Performance of Ba_2CaWO_6:Er^{3+} by Tri-Doping with Yb^{3+} and Na^+. *J. Rare Earths.* **2023**, *41* (1), 42–50. DOI: 10.1016/j.jre.2022.02.005.

119. Kachhap, S.; Giri, N. K.; Shruti; Prakash, R.; Singh, S. K. Photon Upconversion-Based Non-Invasive Temperature Sensing Using $Gd_{1-x-y}Yb_xEr_yScO_3$ Perovskite Nanocrystals. *J. Alloys Compd.* **2023**, *936*, 168192. DOI: 10.1016/j.jallcom.2022.168192.

120. Liu, H.; Yang, X.; Xiao, S. Color Tunability of $NaLaMgWO_6$:Mn^{4+}, Tb^{3+} Phosphor and Its Potential Application in Temperature Sensor and Wavelength Detector. *Appl. Phys. B* **2023**, *129* (12), 186. DOI: 10.1007/s00340-023-08133-5.

121. Zhang, Q.; Yang, Z.; Wang, M.; Liu, Y.; Lei, S.; Luo, L.; Poelman, D. A Strategy for Enhancing Luminescence Thermometry via Ta^{5+}-Triggered $K_{0.5}Na_{0.5}NbO_3$: Er^{3+} Transparent Ceramics. *J. Mater.* **2023**, Article ASAP. DOI: 10.1016/j.jmat.2023.10.011 (accessed 2024-01-04).

122. Wang, P.; Mao, J.; Zhao, L.; Jiang, B.; Xie, C.; Lin, Y.; Chi, F.; Yin, M.; Chen, Y. Double Perovskite A_2LaNbO_6:Mn^{4+}, Eu^{3+} (A = Ba, Ca) Phosphors: Potential Applications in Optical Temperature Sensing. *Dalton Trans.* **2019**, *48* (27), 10062–10069. DOI: 10.1039/c9dt01524h.

123. Xu, C.; Li, C.; Deng, D.; Yu, H.; Wang, L.; Shen, C.; Jing, X.; Xu, S. Double Perovskite Structure $CaLaMgTaO_6$: Bi^{3+}, Eu^{3+} Co-Doped Phosphors for Optical Temperature Measurement. *J. Lumin.* **2021**, *236*, 118096. DOI: 10.1016/j.jlumin.2021.118096.

124. Zhang, Y.; Guo, N.; Shao, B.; Li, J.; Ouyang, R.; Miao, Y. Photoluminescence and Optical Temperature Measurement of Mn^{4+}/Er^{3+} Co-Activated Double Perovskite Phosphor through Site-Advantageous Occupation. *Spectrochim. Acta A Mol. Biomol. Spectrosc.* **2021**, *259*, 119797. DOI: 10.1016/j.saa.2021.119797.

125. Tang, Q.; Guo, N.; Xin, Y.; Li, W.; Shao, B.; Ouyang, R. Luminous Tuning in Eu^{3+}/Mn^{4+} Co-Doped Double Perovskite Structure by Designing the Site-Occupancy Strategy for Solid-State Lighting and Optical Temperature Sensing. *Mater. Res. Bull.* **2022**, *149*, 111704. DOI: 10.1016/j.materresbull.2021.111704.

126. Song, Y.; Guo, N.; Li, J.; Xin, Y.; Lü, W.; Miao, Y. Dual-Emissive Ln^{3+}/Mn^{4+} Co-Doped Double Perovskite Phosphor: Via Site-Beneficial Occupation. *Mater. Adv.* **2021**, *2* (4), 1402–1412. DOI: 10.1039/d0ma00841a.

127. Liao, J.; Kong, L.; Wang, M.; Huang, J. Sol-Gel Synthesis and Optical Temperature Sensing Behavior of Double-Perovskite Gd_2ZnTiO_6 :Yb^{3+}/Er^{3+} Phosphors. *ECS J. Solid State Sci. Technol.* **2019**, *8* (11), R149–R156. DOI: 10.1149/2.0181911jss.

128. Yun, X.; Zhou, J.; Zhu, Y.; Li, X.; Liu, S.; Xu, D. A Potentially Multifunctional Double-Perovskite Sr_2ScTaO_6:Mn^{4+}, Eu^{3+} Phosphor for Optical Temperature Sensing and Indoor Plant Growth Lighting. *J. Lumin.* **2022**, *244*, 118724. DOI: 10.1016/j.jlumin.2021.118724.

129. Zhang, M.; Jia, M.; Sheng, T.; Fu, Z. Multifunctional Optical Thermometry Based on the Transition Metal Ions Doped Down-Conversion Gd_2ZnTiO_6: Bi^{3+}, Mn^{4+} Phosphors. *J Lumin* **2021**, *229*. DOI: 10.1016/j.jlumin.2020.117653.

130. Wu, Y.; Xu, S.; Xiao, Z.; Lai, F.; Huang, J.; Fu, J.; Ye, X.; You, W. A Universal Strategy to Enhance the Absolute Sensitivity for Temperature Detection in Bright Er^{3+}/Yb^{3+} Doped Double Perovskite Gd_2ZnTiO_6 Phosphors. *Mater. Chem. Front.* **2020**, *4* (4), 1182–1191. DOI: 10.1039/c9qm00726a.

131. Zhu, B.; Yang, Q.; Zhang, W.; Cui, S.; Yang, B.; Wang, Q.; Li, S.; Zhang, D. A High Sensitivity Dual-Mode Optical Thermometry Based on Charge Compensation in $ZnTiO_3$:M (M$=$$Eu^{3+}$, Mn^{4+}) Hexagonal Prisms. *Spectrochim. Acta A Mol. Biomol. Spectrosc.* **2022**, *274*, 121101. DOI: 10.1016/j.saa.2022.121101.

132. Hua, Y.; Qiu, X.; Sonne, C.; Brown, R. J. C.; Kim, K. H. Construction of Novel Luminescent Thermometers Based on Dual-Emission Centers of Rare-Earth and Bismuth Ions. *Chemosphere.* **2022**, *303*, 135150. DOI: 10.1016/j.chemosphere.2022.135150.

133. Song, M.; Ran, W.; Ren, Y.; Wang, L.; Zhao, W. Characterizations and Photoluminescence Properties of a Dual-Functional La_2LiNbO_6:Bi^{3+}, Eu^{3+} Phosphor for WLEDs and Ratiometric Temperature Sensing. *J. Alloys Compd.* **2021**, *865*, 158825. DOI: 10.1016/j.jallcom.2021.158825.

134. Li, Z.; Yu, X.; Wang, T.; Wang, S.; Guo, L.; Cui, Z.; Yan, G.; Feng, W.; Zhao, F.; Chen, J.; Xu, X.; Qiu, J. Highly Sensitive Optical Thermometer of Sm^{3+}, Mn^{4+} Activated $LaGaO_3$ Phosphor for the Regulated Thermal Behavior. *J. Am. Ceram. Soc.* **2022**, *105* (4), 2804–2812. DOI: 10.1111/jace.18286.

135. Song, Y.; Guo, N.; Li, J.; Ouyang, R.; Miao, Y.; Shao, B. Photoluminescence and Temperature Sensing of Lanthanide Eu^{3+} and Transition Metal Mn^{4+} Dual-Doped Antimoniate Phosphor through Site-Beneficial Occupation. *Ceram. Int.* **2020**, *46* (14), 22164–22170. DOI: 10.1016/j.ceramint.2020.05.293.

136. Wang, M.; Han, Z.; Huang, J.; Liao, J.; Sun, Y.; Huang, H.; Wen, H.-R. $NaLaMgWO_6$:$Mn^{4+}/Pr^{3+}/Bi^{3+}$ Bifunctional Phosphors for Optical Thermometer and Plant Growth Illumination Matching Phytochrome P_R and P_{FR}. *Spectrochim. Acta A Mol. Biomol. Spectrosc.* **2021**, *259*, 119915. DOI: 10.1016/j.saa.2021.119915.

137. Mullins, A. L.; Ćirić, A.; Zeković, I.; Williams, J. A. G.; Dramićanin, M. D.; Evans, I. R. Dual-Emission Luminescence Thermometry Using $LaGaO_3$:Cr^{3+}, Nd^{3+} Phosphors. *J. Mater. Chem. C.* **2022**, *10* (28), 10396–10403. DOI: 10.1039/d2tc02011d.

138. Xu, H.; Bai, G.; He, K.; Tao, S.; Lu, Z.; Zhang, Y.; Xu, S. Multifunctional Optical Sensing Applications of Luminescent Ions Doped Perovskite Structured $LaGaO_3$ Phosphors in Near-Infrared Spectroscopy. *Mater. Today Phys.* **2022**, *28*, 100872. DOI: 10.1016/j.mtphys.2022.100872.

139. Fan, H.; Lu, Z.; Meng, Y.; Chen, P.; Zhou, L.; Zhao, J.; He, X. Optical Temperature Sensor with Superior Sensitivity Based on Ca_2LaSbO_6: Mn^{4+}, Eu^{3+} Phosphor. *Opt. Laser Technol.* **2022**, *148*, 107804. DOI: 10.1016/j.optlastec.2021.107804.

140. Shi, H.; Han, F.; Wang, X.; Ren, X.; Lei, R.; Huang, L.; Zhao, S.; Xu, S. Highly Precise FIR Thermometer Based on the Thermally Enhanced Upconversion Luminescence for Temperature Feedback Photothermal Therapy. *RSC Adv.* **2022**, *12* (14), 8274–8282. DOI: 10.1039/d1ra09451c.

141. Long, Z.; Xiao, S.; Yang, X. Mn^{4+}, Eu^{3+} Co-Doped $K_{0.3}La_{1.233}MgWO_6$: A Potentially Multifunctional Luminescent Material. *ACS Appl. Electron Mater.* **2020**, *2* (12), 3889–3897. DOI: 10.1021/acsaelm.0c00730.

142. Chen, B.; Li, C.; Deng, D.; Ruan, F.; Wu, M.; Wang, L.; Zhu, Y.; Xu, S. Temperature Sensitive Properties of Eu^{2+}/Eu^{3+} Dual-Emitting $LaAlO_3$ Phosphors. *J. Alloys Compd.* **2019**, *792*, 702–712. DOI: 10.1016/j.jallcom.2019.04.013.

143. Li, C.; Chen, B.; Deng, D.; Yu, H.; Li, H.; Shen, C.; Wang, L.; Xu, S. A Ratiometric Optical Thermometer with Tunable Sensitivity and Superior Signal Discriminability Based on Eu^{2+}/Eu^{3+} Co-Doped $La_{1-y}Gd_yAlO_3$ Phosphors. *J. Lumin.* **2020**, *221*, 117036. DOI: 10.1016/j.jlumin.2020.117036.

144. Matuszewska, C.; Elzbieciak-Piecka, K.; Marciniak, L. Transition Metal Ion-Based Nanocrystalline Luminescent Thermometry in $SrTiO_3$:Ni^{2+},Er^{3+} Nanocrystals Operating in the Second Optical Window of Biological Tissues. *J. Phys. Chem. C* **2019**, *123* (30), 18646–18653. DOI: 10.1021/acs.jpcc.9b04002.

145. Zhao, Y.; Bai, G.; Hua, Y.; Yang, Q.; Chen, L.; Xu, S. Optical Thermometry Based on Upconversion Emission of Yb^{3+}/Er^{3+} Codoped Bismuth Titanate Microcrystals. *J. Lumin.* **2020**, *221*, 117037. DOI: 10.1016/j.jlumin.2020.117037.

146. Liu, Y.; Bai, G.; Pan, E.; Hua, Y.; Chen, L.; Xu, S. Upconversion Fluorescence Property of Er^{3+}/Yb^{3+} Codoped Lanthanum Titanate Microcrystals for Optical Thermometry. *J. Alloys Compd.* **2020**, *822*, 153449. DOI: 10.1016/j.jallcom.2019.153449.

147. Singh, P.; Yadav, R. S.; Singh, P.; Rai, S. B. Upconversion and Downshifting Emissions of Ho^{3+}-Yb^{3+} Co-Doped $ATiO_3$ Perovskite Phosphors with Temperature Sensing Properties in Ho^{3+}-Yb^{3+} Co-Doped $BaTiO_3$ Phosphor. *J. Alloys Compd.* **2021**, *855*, 157452. DOI: 10.1016/j.jallcom.2020.157452.

148. De, A.; Ranjan, R. Large Temperature Tuning of the Emission Color of a Phosphor by Dual Use of Raman and Photoluminescence Signals. *Mater. Horiz.* **2020**, *7* (4), 1101–1105. DOI: 10.1039/c9mh01796h.

149. Zhang, J.; Jin, C. Electronic Structure, Upconversion Luminescence and Optical Temperature Sensing Behavior of Yb^{3+}-Er^{3+}/Ho^{3+} Doped $NaLaMgWO_6$. *J. Alloys Compd.* **2019**, *783*, 84–94. DOI: 10.1016/j.jallcom.2018.12.281.

150. Pattnaik, S.; Rai, V. K. Impact of Charge Compensation on Optical and Thermometric Behaviour of Titanate Phosphors. *Mater. Res. Bull.* **2020**, *125*, 110761. DOI: 10.1016/j.materresbull.2019.110761.

151. Liao, J.; Han, Z.; Huang, J.; Fu, B.; Sun, Y.; Yuan, H.; Wen, H. Sol-Gel Synthesis and Optical Temperature Sensing Properties of $PbTiO_3$:Yb^{3+}/Er^{3+} Phosphors. *J. Phys. Chem.Solids.* **2022**, *162*, 110515. DOI: 10.1016/j.jpcs.2021.110515.

152. Lal, S. C.; Nima, A. M.; Jawahar, I. N.; Ganesanpotti, S. Raman and Photoluminescence Intensity Ratio: A Tool for High Temperature Optical Thermometry in $Ba_2Y_{2/3}TeO_6$: Eu^{3+} Double Perovskites. *Ceram. Int.* **2023**, *49* (11), 19528–19532. DOI: 10.1016/j.ceramint.2023.02.191.

153. Lal, S. C.; Isuhak Naseemabeevi, J.; Ganesanpotti, S. Distortion Induced Structural Characteristics of $Ba_2R_{2/3}TeO_6$ (R = Y, Gd, Tb, Dy, Ho, Er, Tm, Yb and Lu) Double Perovskites and Their Multifunctional Optical Properties for Lighting and Ratiometric Temperature Sensing. *Mater. Adv.* **2021**, *2* (4), 1328–1342. DOI: 10.1039/d0ma00471e.

154. Li, H.; Li, L.; Mei, L.; Zhao, W.; Zhou, X.; Wang, Y.; Hua, Y.; Du, P. Thermally Stable Rare-Earth-Free Double Perovskite Phosphors toward Dual-Mode Optical Thermometry and Dual-Functional Lighting Sources. *J. Mater. Chem. C* **2023**, *11* (37), 12637–12648. DOI: 10.1039/d3tc02471g.

155. Zhu, J.; Yang, T.; Li, H.; Xiang, Y.; Song, R.; Zhang, H.; Wang, B. Improving the up/down-Conversion Luminescence via Cationic Substitution and Dual-Functional Temperature Sensing Properties of Er^{3+} Doped Double Perovskites. *Chem. Eng. J.* **2023**, *471*, 144550. DOI: 10.1016/j.cej.2023.144550.

156. Vu, T. H. Q.; Stefańska, D.; Dereń, P. J. Effect of A-Cation Radius on the Structure, Luminescence, and Temperature Sensing of Double Perovskites A_2MgWO_6 Doped with Dy^{3+} (A = Ca, Sr, Ba). *Inorg. Chem.* **2023**, *62* (49), 20020–20029. DOI: 10.1021/acs.inorgchem.3c02798.

157. Li, H.; Li, L.; Mei, L.; Hua, Y.; Cao, Z.; Ling, F.; Wang, Y.; Jiang, S.; Xiang, G.; Zhou, X. A Dual-Functional Platform for Plant Cultivation and Wide-Range Optical Thermometry Based on Vibration Sidebands. *Ceram. Int.* **2023**, *49* (11), 18084–18094. DOI: 10.1016/j.ceramint.2023.02.177.

158. Lojpur, V.; Ćulubrk, S.; Medić, M.; Dramicanin, M. Luminescence Thermometry with Eu^{3+} Doped $GdAlO_3$. *J. Lumin.* **2016**, *170*, 467–471. DOI: 10.1016/j.jlumin.2015.06.032.

159. Long, S.; Liu, Z.; Yang, X.; Wang, B. Highly Sensitive Dual-Mode Thermometry over a Wide Temperature Range Based on Bandgap Engineering in $LiNb_{1-x}Ta_xO_3$:Tb^{3+}. *Ceram. Int.* **2023**, *49* (18), 30696–30704. DOI: 10.1016/j.ceramint.2023.07.026.

160. Soler-Carracedo, K.; Zheng, T.; Runowski, M.; Luo, L.; Martín, I. R. Pr^{3+}-Doped Perovskite Niobate Ceramics towards Improving Performance of Optical Temperature Sensor by Second Harmonic Generation (SHG) Combined with Lanthanide Luminescence. *Ceram. Int.* **2023**, *49* (9), 14177–14182. DOI: 10.1016/j.ceramint.2023.01.004.

161. Zheng, T.; Luo, L.; Du, P.; Lis, S.; Rodríguez-Mendoza, U. R.; Lavín, V.; Runowski, M. Highly-Efficient Double Perovskite Mn^{4+}-Activated Gd_2ZnTiO_6 Phosphors: A Bifunctional Optical Sensing Platform for Luminescence Thermometry and Manometry. *Chem. Eng. J.* **2022**, *446*, 136839. DOI: 10.1016/j.cej.2022.136839.

162. Gao, Y.; Cheng, Y.; Hu, T.; Ji, Z.; Lin, H.; Xu, J.; Wang, Y. Broadening the Valid Temperature Range of Optical Thermometry through Dual-Mode Design. *J. Mater. Chem. C.* **2018**, *6* (41), 11178–11183. DOI: 10.1039/C8TC03851A.

163. Wang, J.; Chen, N.; Li, J.; Feng, Q.; Lei, R.; Wang, H.; Xu, S. A Novel High-Sensitive Optical Thermometer Based on the Multi-Color Emission in Pr^{3+} Doped $LiLaMgWO_6$ Phosphors. *J. Lumin.* 2021, *238*, 118240. DOI: 10.1016/j.jlumin.2021.118240.

164. Vu, T. H. Q.; Bondzior, B.; Stefańska, D.; Dereń, P. J. Exploration of the Temperature Sensing Ability of La_2MgTiO_6: Er^{3+} Double Perovskites Using Thermally Coupled and Uncoupled Energy Levels. *Materials* 2021, *14* (19), 5557. DOI: 10.3390/ma14195557.

165. Zhang, A.; Sun, Z.; Jia, M.; Fu, Z.; Choi, B. C.; Jeong, J. H.; Park, S. H. Sm^{3+}-Doped Niobate Orange-Red Phosphors with a Double-Perovskite Structure for Plant Cultivation and Temperature Sensing. *J. Alloys Compd.* 2022, *889*, 161671. DOI: 10.1016/j.jallcom.2021.161671.

166. Zhou, X.; Zhao, S.; Li, S.; Wang, Y.; Li, L.; Jiang, S.; Xiang, G.; Jing, C.; Li, J.; Yao, L. Luminescent Properties of Eu^{3+}-Doped $NaLaCaWO_6$ Red Phosphors and Temperature Sensing Derived from the Excited State of Charge Transfer Band. *J. Lumin.* 2022, *248*, 118964. DOI: 10.1016/j.jlumin.2022.118964.

167. Gao, Y.; Huang, F.; Lin, H.; Xu, J.; Wang, Y. Intervalence Charge Transfer State Interfered Pr^{3+} Luminescence: A Novel Strategy for High Sensitive Optical Thermometry. *Sens. Actuators B Chem.* 2017, *243*, 137–143. DOI: 10.1016/j.snb.2016.11.143.

168. Xia, W.; Li, L.; Hua, Y.; Ling, F.; Wang, Y.; Cao, Z.; Jiang, S.; Xiang, G.; Zhou, X.; Yu, J. S. Realizing Dual-Mode Luminescent Thermometry with Excellent Sensing Sensitivity in Single-Phase Samarium (III)-Doped Antimonite Phosphors. *J. Alloys Compd.* 2022, *917*, 165435. DOI: 10.1016/j.jallcom.2022.165435.

169. Tang, W.; Zuo, C.; Li, Y.; Ma, C.; Yuan, X.; Wen, Z.; Cao, Y. A Wide Temperature Range Dual-Mode Luminescence Thermometer Based on Pr^{3+}-Doped $Ba(Zr_{0.16}Mg_{0.28}Ta_{0.56})O_3$ transparent Ceramic. *J. Mater. Chem. C.* 2021, *9* (42), 15112–15120. DOI: 10.1039/d1tc03330a.

170. Cui, Y.; Gao, Y.; Meng, Z.; Hu, T.; Chen, Y.; Chen, Y.; Zeng, Q. Improving the Temperature-Sensing Performance of the $SrZn_{0.33}Nb_{0.67}O_3$:Pr^{3+} Phosphor via Ga^{3+} Doping. *Mater. Adv.* 2022, *3* (7), 3267–3277. DOI: 10.1039/d1ma01247a.

171. Piotrowski, W.; Kuchowicz, M.; Dramićanin, M.; Marciniak, L. Lanthanide Dopant Stabilized Ti^{3+} State and Supersensitive Ti^{3+}-Based Multiparametric Luminescent Thermometer in $SrTiO_3$:Ln^{3+} (Ln^{3+}=Lu^{3+}, La^{3+}, Tb^{3+}) Nanocrystals. *Chem. Eng. J.* 2021, *428*, 131165. DOI: 10.1016/j.cej.2021.131165.

172. Liu, Z.; Wang, R.; Chen, D. Dual-Mode Optical Temperature Sensing Properties of PIN-PMN-PT:Pr^{3+} Ceramic Based on Fluorescence Intensity Ratios and Lifetimes. *J. Mater. Sci. Mater. Electron.* 2022, *33* (7), 3748–3756. DOI: 10.1007/s10854-021-07566-y.

173. Fan, Z.; Fan, X.; Xue, J.; Wang, Y. Designing Dual Mode of Non-Contact Optical Thermometers in Double Perovskite Ca_2LaTaO_6:Bi^{3+}, Eu^{3+} Phosphors. *Mater. Today Chem.* 2023, *30*, 101528. DOI: 10.1016/j.mtchem.2023.101528.

174. Zhang, Y.; Sun, B.; Liu, J.; Zhang, Z.; Liu, H. Luminescence and Energy Transfer Performances of Tb^{3+}/Mn^{4+} Co-Doped Sr_2LuTaO_6 Double-Perovskite Phosphors for a Highly Sensitive Optical Thermometer. *Dalton Trans.* 2023, *52* (37), 13304–13315. DOI: 10.1039/d3dt02270f.

175. Fu, Y.; Wang, X.; Liu, S.; Wang, H.; Tian, Y.; Xing, M.; Pang, Q.; Xin, F.; Luo, X. Multicolor Upconversion Emission and Highly Optical Temperature Sensing Based on Lanthanide-Doped Double Perovskite Sr_2LaNbO_6 Phosphors. *Ceram. Int.* **2023**, *49* (6), 9574–9583. DOI: 10.1016/j.ceramint.2022.11.127.

176. Wu, H.; Chen, Y.; Cai, P.; Pu, X.; Wang, X.; Ai, Q.; Si, J.; Liu, Z. High-Sensitivity Eu^{3+}/Mn^{4+} Co-Doped Li_4AlSbO_6 Perovskite Type Thermometry with Efficient Dual Emission. *J. Lumin.* **2023**, *257*, 119769. DOI: 10.1016/j.jlumin.2023.119769.

177. Luo, Y.; Chen, Y.; Li, L.; Chen, J.; Pang, T.; Chen, L.; Guo, H. Three-Mode Fluorescence Thermometers Based on Double Perovskite Ba_2GdNbO_6:Eu^{3+}, Mn^{4+} Phosphors. *Ceram. Int.* **2023**, *49* (23), 38007–38014. DOI: 10.1016/j.ceramint.2023.09.131.

178. Li, L.; Li, X.; Wu, Z.; Hua, Y.; Zhou, X.; Wang, Y.; Cao, Z.; Jiang, S.; Xiang, G.; Yu, J. S. Designing Dual-Emission Phosphors for Temperature Warning Indication and Dual-Mode Luminescence Thermometry. *Dalton Trans.* **2023**, *52* (43), 15798–15806. DOI: 10.1039/d3dt01525d.

179. Kumar, I.; Kumar, A.; Kumar, S.; Nair, B.; Swart, H. C.; Gathania, A. K. *NIR Light-Triggered Green Emitting Perovskite-Based Phosphor for Optical Thermometry and Molecular Logic Gate Applications*, **2023**. Available at SSRN. DOI: 10.2139/ssrn.4662074 (accessed on 2024-01-05).

180. Hu, F.; Ren, S.; Wu, Y.; Sun, C.; Zhu, B.; Wang, Q.; Li, S.; Zhang, D. Dual-Mode Optical Thermometer Based on Fluorescence Intensity Ratio of Eu^{3+}/Mn^{4+} Co-Doping Zinc Titanate Phosphors. *Spectrochim. Acta A Mol. Biomol. Spectrosc.* **2023**, *288*, 122127. DOI: 10.1016/j.saa.2022.122127.

181. Zhu, B.; Lei, W.; Shi, Q.; Guo, H.; Qiao, J.; Cui, C.; Huang, P. Tb^{3+}/Mn^{4+} Co-Doped $SrLaMgNbO_6$ Green/Red Dual-Emitting Phosphors for Optical Thermometry. *Ceram. Int.* **2023**, *49* (22), 35285–35292. DOI: 10.1016/j.ceramint.2023.08.201.

182. Guo, J.; Shen, Y.; Chen, L.; Deng, D.; Xu, S. A Multi-Mode Optical Thermometry Based on the up-Conversion La_2MgGeO_6: Er^{3+}, Yb^{3+} Phosphor. *J. Lumin.* **2024**, *266*, 120331. DOI: 10.1016/j.jlumin.2023.120331.

183. Chen, Y.; Luo, Y.; Chen, J.; Wang, Q.; Guo, H. *Dual Mode Temperature Sensing of Ba_2LuNbO_6:Er^{3+}, Yb^{3+} Materials*, **2023**. Available at SSRN. DOI: 10.2139/ssrn.4671076 (accessed on 2024-01-05).

184. Hu, F.; Zhang, D.; Wu, Y.; Sun, C.; Xu, C.; Wang, Q.; Xie, Y.; Shi, Q.; Li, S.; Wang, K. Optical Behaviors of Mn^{4+}-Modified Cubic Type $ZnTiO_3$:Eu^{3+} Nanocrystals: Application in Optical Thermometers Based on Fluorescence Intensity Ratio and Lifetime. *Spectrochim. Acta A Mol. Biomol. Spectrosc.* **2024**, *304*, 123339. DOI: 10.1016/j.saa.2023.123339.

185. Kumar, I.; Kumar, A.; Kumar, S.; Nair, G. B.; Swart, H. C.; Gathania, A. K. Simultaneous Realization of FIR-Based Multimode Optical Thermometry and Photonic Molecular Logic Gates in Er^{3+} and Yb^{3+} Co-Doped $SrTiO_3$ Phosphor. *Phys. Scr.* **2023**, *98* (10), 105532. DOI: 10.1088/1402-4896/acfa2b.

186. Duan, F.; Wang, L.; Shi, Q.; Guo, H.; Qiao, J.; Cui, C.; Huang, P. Suitable Selection of High-Energy State Excitation to Enhance the Thermal Stability of Eu^{3+} and the Sensitivity of La_2CaSnO_6:Eu^{3+}, Mn^{4+} Temperature Measuring Materials. *J. Mater. Chem. C.* **2023**, *11* (42), 14705–14713. DOI: 10.1039/d3tc02906a.

187. Liu, W.; Zhao, D.; Zhang, R. J.; Jia, L.; Zhu, S. Y. Design and Fabrication of a Bimodal Emitting Phosphor by Co-Doping Two Different Activators for High Sensitivity Temperature Sensing. *Ceram. Int.* **2023**, *49* (15), 25823–25830. DOI: 10.1016/j.ceramint.2023.05.129.

188. Wu, Z.; Li, L.; Tian, G.; Wang, Y.; Ling, F.; Cao, Z.; Jiang, S.; Xiang, G.; Li, Y.; Zhou, X. High-Sensitivity and Wide-Temperature-Range Dual-Mode Optical Thermometry under Dual-Wavelength Excitation in a Novel Double Perovskite Tellurate Oxide. *Dalton Trans.* **2021**, *50* (33), 11412–11421. DOI: 10.1039/d1dt01147b.

189. Zhang, H.; Liang, Y.; Yang, H.; Liu, S.; Li, H.; Gong, Y.; Chen, Y.; Li, G. Highly Sensitive Dual-Mode Optical Thermometry in Double-Perovskite Oxides via Pr^{3+}/Dy^{3+} Energy Transfer. *Inorg. Chem.* **2020**, *59* (19), 14337–14346. DOI: 10.1021/acs.inorgchem.0c02118.

190. Song, M.; Wang, L.; Wang, J.; Du, P. Constructing Double Perovskite Eu^{3+}/Mn^{4+}-Codoped $La_2Mg_{1.33}Ta_{0.67}O_6$ Phosphors for High Sensitive Dual-Mode Optical Thermometers. *J. Lumin.* **2022**, *252*, 119347. DOI: 10.1016/j.jlumin.2022.119347.

191. Li, L.; Tian, G.; Deng, Y.; Wang, Y.; Cao, Z.; Ling, F.; Li, Y.; Jiang, S.; Xiang, G.; Zhou, X. Constructing Ultra-Sensitive Dual-Mode Optical Thermometers: Utilizing FIR of Mn^{4+}/Eu^{3+} and Lifetime of Mn^{4+} Based on Double Perovskite Tellurite Phosphor. *Opt. Express.* **2020**, *28* (22), 33747–33757. DOI: 10.1364/oe.409242.

192. Liao, J.; Wang, M.; Kong, L.; Chen, J.; Wang, X.; Yan, H.; Huang, J.; Tu, C. Dual-Mode Optical Temperature Sensing Behavior of Double-Perovskite $CaGdMgSbO_6$:Mn^{4+}/Sm^{3+} Phosphors. *J. Lumin.* **2020**, *226*, 117492. DOI: 10.1016/j.jlumin.2020.117492.

193. Wang, J.; Lei, R.; Zhao, S.; Huang, F.; Deng, D.; Xu, S.; Wang, H. Color Tunable Bi^{3+}/Eu^{3+} Co-Doped La_2ZnTiO_6 Double Perovskite Phosphor for Dual-Mode Ratiometric Optical Thermometry. *J. Alloys Compd.* **2021**, *881*, 160601. DOI: 10.1016/j.jallcom.2021.160601.

194. Lin, Y.; Zhao, L.; Jiang, B.; Mao, J.; Chi, F.; Wang, P.; Xie, C.; Wei, X.; Chen, Y.; Yin, M. Temperature-Dependent Luminescence of $BaLaMgNbO_6$:Mn^{4+}, Dy^{3+} Phosphor for Dual-Mode Optical Thermometry. *Opt. Mater.* **2019**, *95*, 109199. DOI: 10.1016/j.optmat.2019.109199.

195. Wang, C.; Huang, Y.; Heydari, E.; Yang, X.; Xu, S.; Bai, G. Dual-Mode Optical Ratiometric Thermometer Based on Rare Earth Ions Doped Perovskite Oxides with Tunable Luminescence. *Ceram. Int.* **2022**, *48* (9), 12578–12584. DOI: 10.1016/j.ceramint.2022.01.125.

196. Wu, Y.; Xu, S.; Lai, F.; Liu, B.; Huang, J.; Ye, X.; You, W. Intense Near-Infrared Emission, Upconversion Processes and Temperature Sensing Properties of Tm^{3+} and Yb^{3+} Co-Doped Double Perovskite Gd_2ZnTiO_6 Phosphors. *J. Alloys Compd.* **2019**, *804*, 486–493. DOI: 10.1016/j.jallcom.2019.07.036.

197. Chen, Y.; Lin, J.; Fu, J.; Ye, R.; Lei, L.; Shen, Y.; Deng, D.; Xu, S. Dual-Functions of Non-Contact Optical Thermometry and Anti-Counterfeiting Based on La_2MgGeO_6: Bi^{3+}, Er^{3+} Phosphors. *J. Lumin.* **2022**, *252*, 119404. DOI: 10.1016/j.jlumin.2022.119404.

198. Chen, Y.; Shen, Y.; Zhou, L.; Lin, J.; Fu, J.; Fang, Q.; Ye, R.; Shen, Y.; Xu, S.; Lei, L.; Deng, D. Temperature-Dependent Luminescence of Bi^{3+}, Eu^{3+} Co-Activated La_2MgGeO_6 Phosphor for Dual-Mode Optical Thermometry. *J. Lumin.* **2022**, *249*, 118995. DOI: 10.1016/j.jlumin.2022.118995.

199. Xu, S.; Wu, Y.; Xiao, Z.; Lai, F.; Huang, J.; Fu, J.; Ye, X.; You, W. Crucial Processes for Upconversion White Emission and Ultrahigh Sensitivity in Er^{3+}/Tm^{3+}/Yb^{3+} Tri-Doped Double Perovskite Gd_2ZnTiO_6 Phosphors. *Opt. Mater.* **2020**, *110*, 110548. DOI: 10.1016/j.optmat.2020.110548.

200. Bu, Y.; Chen, Y.; Chen, X.; Zhang, Y.; Deng, D.; Shen, Y.; Zhou, L.; Xu, C. A Dual-Mode Self-Referenced Optical Thermometry with High Sensitivity Based on Er^{3+}-Yb^{3+} Co-Doped Sr_2YTaO_6 Thermochromic Phosphor. *J. Lumin.* **2022**, *248*, 118923. DOI: 10.1016/j.jlumin.2022.118923.

201. Jiang, Y. C.; Tong, Y.; Chen, S. Y. Z.; Zhang, W. N.; Hu, F. F.; Wei, R. F.; Guo, H. A Three-Mode Self-Referenced Optical Thermometry Based on up-Conversion Luminescence of Ca_2MgWO_6:Er^{3+}, Yb^{3+} Phosphors. *Chem. Eng. J.* **2021**, *413*, 127470. DOI: 10.1016/j.cej.2020.127470.

202. Piotrowski, W. M.; Maciejewska, K.; Dalipi, L.; Fond, B.; Marciniak, L. Cr^{3+} Ions as an Efficient Antenna for the Sensitization and Brightness Enhancement of Nd^{3+}, Er^{3+}-Based Ratiometric Thermometer in $GdScO_3$ Perovskite Lattice. *J. Alloys Compd.* **2022**, *923*, 166343. DOI: 10.1016/j.jallcom.2022.166343.

203. Pavani, K.; Neves, A. J.; Pinto, R. J. B.; Freire, C. S. R.; Soares, M. J.; Graça, M. P. F.; Kumar, K. U.; Jakka, S. K. $BiLaWO_6$: Er^{3+}/Tm^{3+}/Yb^{3+} Phosphor: Study of Multiple Fluorescence Intensity Ratiometric Thermometry at Cryogenic Temperatures. *Ceram. Int.* **2022**, *48* (21), 31344–31353. DOI: 10.1016/j.ceramint.2022.06.316.

204. Pattnaik, S.; Rai, V. K. Insight into the Spectroscopic and Thermometric Properties of Titanate Phosphors via a Novel Co-Excited Laser System. *Mater. Sci. Eng. B.* **2021**, *272*, 115318. DOI: 10.1016/j.mseb.2021.115318.

205. Wang, C.; Jin, Y.; Yuan, L.; Wu, H.; Ju, G.; Li, Z.; Liu, D.; Lv, Y.; Chen, L.; Hu, Y. A Spatial/Temporal Dual-Mode Optical Thermometry Platform Based on Synergetic Luminescence of Ti^{4+}-Eu^{3+} Embedded Flexible 3D Micro-Rod Arrays: High-Sensitive Temperature Sensing and Multi-Dimensional High-Level Secure Anti-Counterfeiting. *Chem. Eng. J.* **2019**, *374*, 992–1004. DOI: 10.1016/j.cej.2019.06.015.

206. Cheng, H.; Jiang, Z.; Lai, F.; Wang, H.; Xiao, Z.; Sun, J.; You, W. Simultaneous Achievement of Sensitivity Enhancement and Dual-Mode Luminescence through Co-Doping Eu^{3+} in Ca_2MgWO_6:Er^{3+}/Yb^{3+} Phosphor. *J. Lumin.* **2022**, *246*, 118804. DOI: 10.1016/j.jlumin.2022.118804.

207. Eldridge, J. I.; Chambers, M. D. Fiber Optic Thermometer Using Cr-Doped $GdAlO_3$ Broadband Emission Decay. *Meas. Sci. Technol.* **2015**, *26* (9), 095202. DOI: 10.1088/0957-0233/26/9/095202.

Chapter 11

Luminescence temperature sensing based on ratiometric and colorimetric fluorescence in garnet systems

B. Amrithakrishnan, I. N. Jawahar, and Subodh Ganesanpotti

11.1 INTRODUCTION

Optical thermometry based on thermographic phosphors is one of the emerging fields due to its several advantages over conventional contact thermometry [1,2]. Temperature is one of the basic and most important physical quantities that profoundly influence natural, biological, and laboratory processes. Hence, temperature measurement has become an indispensable activity in regular life, which demands accuracy and consistency. In the conventional method, the temperature can be read out using thermocouples, thermistors, and pyrometers via physical contact with the object under study. They cannot be employed for distances beyond 10 μm for sensing temperature in microcircuits and biological samples [1]. Also, they cannot be used in harsh environments such as chemicals, coal mines, corrosive environments, and strong electromagnetic fields [2,3]. In this context, non-contact optical thermometry involves no direct contact with the object under study. It is also a non-invasive technique with a fast response rate, high spatial resolution, short acquisition time, and temperature sensitivity [4–7]. Those phosphor materials that exhibit temperature-responsive optical parameters such as emission intensity, decay or rise time, and emission color with temperature are termed thermographic phosphors (TGPs). Among different classes of phosphors, garnet-based phosphor hosts have a unique place owing to their structural and compositional diversity. Garnets are quite popular for their solid-state lighting applications starting from YAG: Ce [8,9]. Apart from optical purposes, garnet materials are employed for dielectric, magnetic, and Li-ion batteries [8]. The research on garnet-based thermographic phosphors is one of the emerging areas in optical thermometry.

Apart from the variation in the intensity ratio of emission bands, garnet systems are also reported for thermochromic luminescence, which can be applied for safety sign applications in high-temperature environments. Thus, garnet-based thermographic phosphors are reported for ratiometric

DOI: 10.1201/9781032661537-11

and colorimetric temperature sensing. Tables 11.1 and 11.2 list garnet phosphors reported for ratiometric and colorimetric temperature sensing, depicting the increased likelihood of garnet-based systems for optical thermometry over the years.

Table 11.1 List of garnet-based phosphors reported for ratiometric temperature sensing

Garnet	Temperature range (K)	Maximum absolute sensitivity, S_a (K^{-1})	Maximum relative sensitivity, S_r (%K^{-1})	References
$Y_3Al_2Ga_3O_{12}$: Ce^{3+}/Er^{3+}	298–333	–	0.63	[10]
$Ca_3Sc_2Si_3O_{12}$: Bi^{3+}-Tb^{3+}-Eu^{3+}	303–500	0.094	1.0	[11]
$Ca_3Y_2Ge_3O_{12}$: Er^{3+},Yb^{3+}	293–473	0.002	1.29	[12]
$Y_3Al_4GaO_{12}$: Ce^{3+}/Yb^{3+}-Er^{3+}	303–585	1.18×10^{-3}	0.178	[13]
$Y_3Al_5O_{12}$: Dy^{3+}, Cr^{3+}	293–573	0.114	2.32	[14]
$Y_3Al_5O_{12}$: Eu^{3+}, Mn^{4+}	293–573	0.441	4.81	[14]
$Ca_3Al_2Ge_3O_{12}$: Dy^{3+}/Eu^{3+}	303–523	5.51×10^{-4}	0.0359	[15]
$Ca_2TbSn_2Al_3O_{12}$: Sm^{3+}	300–550	0.0027	0.50	[16]
$Gd_3Ga_5O_{12}$:Yb^{3+}/Er^{3+}	300–420	0.0083	1.47	[17]
$Li_3Y_3Te_2O_{12}$: Dy^{3+}	80–300	0.02	1.2	[18]
$Li_3Y_3Te_2O_{12}$: Bi^{3+}/Pr^{3+}	100–300		0.42	[19]
$Lu_3Ga_5O_{12}$: Fe^{3+}, Cr^{3+}, Nd^{3+}	123–573	–	5.9	[20]
$Y_3Ga_5O_{12}$: Fe^{3+}, Cr^{3+}, Nd^{3+}	123–573	–	1.68	[20]
$Gd_3Ga_5O_{12}$: Fe^{3+}, Cr^{3+}, Nd^{3+}	123–573	–	1.57	[20]
$Li_6CaLa_2Nb_2O_{12}$:Yb^{3+}/Er^{3+}	298–573	0.015	$1424.27/T^2$	[21]
$Lu_3Al_5O_{12}$:Ce^{3+}/Mn^{4+}	100–350	0.031	4.37	[22]
$Lu_3Al_5O_{12}$: Eu^{3+}, Mn^{4+}	303–358	0.07	0.7	[23]
$Y_3Al_2Ga_3O_{12}$: Nd^{3+}	100–220		0.45	[24]
$Sr_3Y_2Ge_3O_{12}$: Bi^{3+}, Sm^{3+}	298–498	0.012	0.57	[25]
$Ca_3Ga_2Ge_3O_{12}$: Mn^{4+}	5–75	–	6	[26]
$Ca_2YZr_2Al_3O_{12}$: Bi^{3+}/Eu^{3+}	297–573	0.00826	0.66	[27]
$Y_3(Al,Ga)_5O_{12}$: Pr^{3+}	17–700	–	3.6	[28]
YAG:Yb^{3+}, Mn^{2+}	298–423	0.017	3.4	[29]
$Lu_3Ga_5O_{12}$: Mn^{2+}	78–335	–	5.14	[30]
$Sr_2NaMg_2V_3O_{12}$: Eu^{3+}	300–500	0.0135	1.61	[31]
$Sr_2NaMg_2V_3O_{12}$: Sm^{3+}	300–500	–	2.01	[32]
$Sr_2NaMg_2V_3O_{12}$: Dy^{3+}	200–500	–	0.41	[33]
$LiCa_2Mg_2V_3O_{12}$: Dy^{3+}	80–400	0.0063	0.41	[34]
$Na_2GdMg_2V_3O_{12}$: Sm^{3+}	303–513	0.665	2.12	[35]
$LiCa_3ZnV_3O_{12}$: Sm^{3+}	303–463	0.25	1.8	[36]
$Ca_3LiMgV_3O_{12}$: Sm^{3+}	303–513	0.0911	1.99	[37]
$Ca_2NaMg_2V_3O_{12}$: Sm^{3+}	303–503	0.01649	1.889	[38]
$Ca_2NaMg_2V_3O_{12}$: Eu^{3+}	303–503	0.0156	1.686	[38]
$Na_2YMg(VO_4)_3$: Er^{3+},Yb^{3+}	303–573	0.77	1.104	[39]

Table 11.2 List of garnet-based phosphors reported for colorimetric temperature sensing

Garnet	Temperature range (K)	CIE shift	Thermochromic shift	References
$LiCa_3ZnV_3O_{12}:Sm^{3+}$	303–463	(0.364, 0.388) to (0.569, 0.391)	White to red	[36]
$Lu_3Al_5O_{12}:Ce^{3+}/Mn^{4+}$	298–353	(0.4525, 0.4673) to (0.3665, 0.5247)	Red to green	[22]
$Ca_2NaMg_2V_3O_{12}:Eu^{3+}$	303–503	–	Green to reddish orange	[38]
$Ca_2NaMg_2V_3O_{12}:Sm^{3+}$	303–503	–	Green to orange	[38]
$Ca_2NaZn_2V_3O_{12}:Eu^{3+}$	303–483	(0.421, 0.396) to (0.629, 0.350)	Yellowish red to pure red	[40]

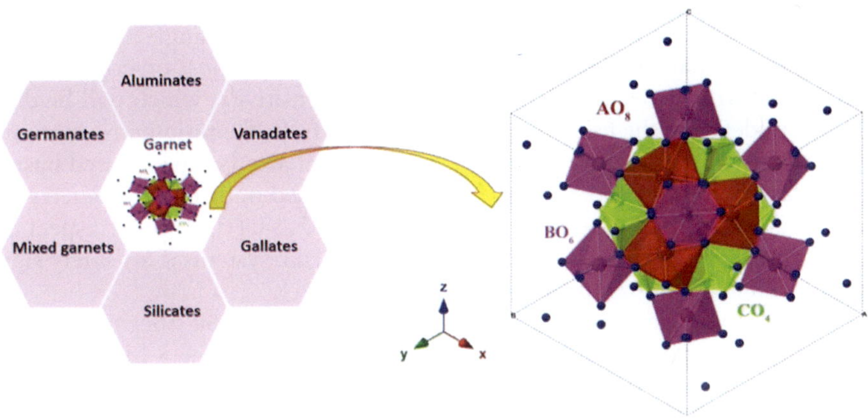

Figure 11.1 The schematic diagram depicting the classification of garnets and the representation of the unit cell along with the polyhedral configuration comprising AO_8, BO_6 and CO_4 tetrahedra [8]. (Reproduced from Ref. [8] with permission from the Taylor & Francis; permission conveyed through Copyright Clearance Center, Inc.)

The attractive feature of garnet-based thermographic phosphors lies in the easier tunability of the luminescence by structural modifications and energy transfer from activator ions. Thanks to the robust interdependence of garnet systems' structural and luminescence properties. Garnet oxides possess a general stoichiometry of $A_3B_2C_3O_{12}$, where A, B, and C correspond to the cations occupying dodecahedral, octahedral, and tetrahedral sites. In short, the garnet structural framework is composed of chains of AO_8, BO_6, and CO_4 polyhedra along the crystallographic directions, offering ample connectivity. Figure 11.1 shows the classification of garnets and representation of the garnet unit cell [8].

This extended connectivity helps to realize structure-dependent luminescence properties. Due to the availability of polyhedral sites of three different coordination, substitution of cations of varied size and charge can be done in garnet structure, which is an added advantage of the garnet structure. The optical centers located in any of the three polyhedral sites will be perturbed owing to the introduction of cations of varied sizes and charges. Consequently, the crystal field strength will be varied, and luminescence can be tuned as per the lighting and optical thermometry demands. Depending on the emission type, phosphors can be classified into self-activated and dopant-activated types. The self-activated phosphors give inherent emission from the host material itself. The presence of self-activated complexes like VO_4^{3-} and WO_4^{2-} in the host composition causes intrinsic broad emission. On the other hand, the introduction of dopant/activator ions like lanthanides-Eu^{3+}, Sm^{3+}, or transition metal ions like Mn^{4+} and main group ions like Bi^{3+} into the non-luminescent host structure can result in the dopant-activated emission. Regarding optical thermometry, it can be noted that both single and multiple activator ions can be introduced into the self-activated host or dopant-activated host to improve the relative temperature sensitivity, which will be discussed in detail in the following sections. Apart from the ratiometric method of emission bands, the temperature sensitivity can also be investigated based on decay time. The reports on the decay time-based optical thermometry in garnet systems are also widely reported [16,19,22,28,41]. Another aspect of temperature sensing that is widely reported in garnet systems is for safety sign applications in high-temperature environments.

In addition, thermographic phosphors, which exhibit thermochromic luminescence, can be used for safety sign applications in a high-temperature environment and as high-temperature alarms. The rapid change in emission color with temperature can be used as an indicator to sense the temperature of the environment.

Thus, garnet systems are one of the exciting classes in phosphors suitable for optical thermometry via thermometric and thermochromic luminescence.

11.2 RATIOMETRIC AND COLORIMETRIC TEMPERATURE SENSING

In ratiometric temperature sensing, the intensity ratio of two emission bands is considered. These emission bands can be from a single or multiple optical center associated with the phosphor. The temperature-dependent variation in the intensity ratio of these emission bands is utilized to evaluate temperature. Hence, the more diverse the thermal responses, the stronger the intensity ratios of these emission bands, and the better the relative temperature sensitivity. The method of dual emission from two optical centers seeks much attention owing to the diverse thermal response, which can reduce errors due to excitation sources or instruments. Thus, the fluorescence intensity ratio (FIR) or ratiometric method is self-calibrating and is unaffected by measurement conditions.

On the other hand, colorimetric temperature sensing is a means of visualizing thermometry. Phosphors with rapidly changing emission color with temperature variation or strong thermochromic luminescence can be employed for colorimetric thermometry. The apparent emission color variation lets one observe the visualized thermometry through the naked eye or by using a digital camera. Hence, the phosphors with enhanced thermochromic luminescence can be effectively applied for safety sign applications in high-temperature environments.

11.3 RATIOMETRIC OPTICAL THERMOMETERS BASED ON GARNET SYSTEMS

11.3.1 Systems based on single activator ions

In this section, the garnet-based phosphors reported for optical thermometry based on the single activator ions will be discussed in detail. Table 11.1 lists garnet phosphors reported for ratiometric and colorimetric temperature sensing. Analyzing the reports, it can be observed that thermally coupled levels (TCLs) of activators are being reported for single activator ions. The commonly found activators are Er^{3+}, known for their prolific energy levels, including TCLs and non-TCLs. Er^{3+} is well-known activator based on upconversion luminescence under 978 nm laser excitation. Owing to the weak absorption cross-section of Er^{3+}, a sensitizer like Yb^{3+} is necessary to enhance the upconversion efficiency; however, intensity of Er^{3+} is alone considered into account. The thermally coupled energy levels (TCELs) of Er^{3+} are $^2H_{11/2}$ and $^4S_{3/2}$, where the intensity ratio of emission bands follows a Boltzmann distribution. Chen *et al.* studied temperature sensing properties of $Ca_3Y_2Ge_3O_{12}$ activated with Er^{3+} and sensitized by Yb^{3+} based on ratiometric method [12]. Maximum relative temperature sensitivity of 1.29%K^{-1} is obtained for $^2H_{11/2}$ and $^4S_{3/2}$ TCELs, as shown in Figure 11.2a and b. On the other hand, Humeyra *et al.* reported a temperature sensitivity of 0.83%K^{-1} in the gallate garnet system, $Gd_3Ga_5O_{12}$: Yb^{3+}/Er^{3+} [17]. However, Du *et al.* reported the sensitivity through the non-TCLs based intensity ratio of upconversion luminescence bands of Er^{3+} in the niobate garnet-$Li_6CaLa_2Nb_2O_{12}$. The non-TCELs $^2H_{11/2}$ and $^4F_{9/2}$ offer better maximum absolute sensitivity of 1.52%K^{-1} at 523 K [21].

However, besides TCLs-based thermometry, Liu *et al.* reported singly activated luminescent thermometers based on the stark splitting of Er^{3+} levels, which is a comparatively new method. In that study, the characteristic NIR-II emission (1450–1700 nm) in the third bioimaging window of Er^{3+} is used. The Boltzmann thermometer with a high relative sensitivity is realized based on the thermal coupling of Stark sub-levels based

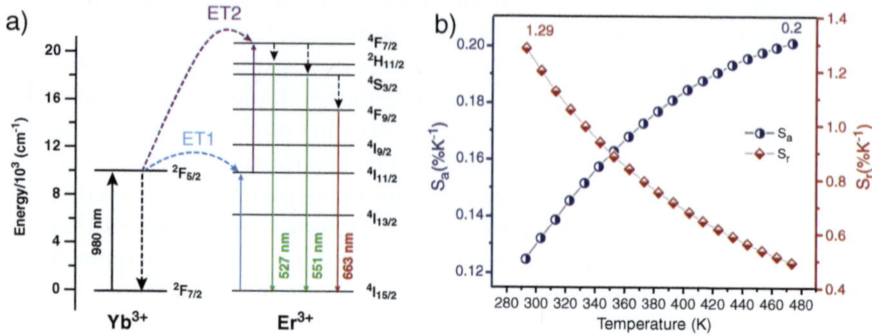

Figure 11.2　(a) Energy level diagram depicting the sensitization of Er^{3+} by Yb^{3+} and (b) Temperature dependence of absolute and relative temperature sensitivities S_a and S_r of $Ca_3Y_2Ge_3O_{12}$: Er^{3+}, Yb^{3+} [12]. (Reprinted from Ref. [12], with permission from Elsevier.)

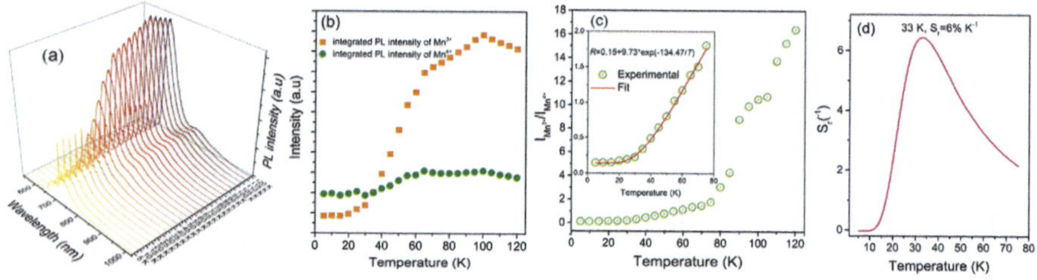

Figure 11.3　Temperature dependence of (a) emission bands, (b) integrated intensities of Mn^{3+} and Mn^{4+}, (c) temperature versus FIR plot and experimental fit, and (d) relative temperature sensitivity of $Ca_3Ga_2Ge_3O_{12}$: Mn phosphor [26]. (Reprinted with permission from Ref. [26], American Chemical Society.)

on the $^4I_{13/2}$ level [10]. A similar exciting investigation on developing two Boltzmann thermometers in the cryogenic and high-temperature range can be found in the Nd^{3+}-activated $Y_3Al_2Ga_3O_{12}$ system [24]. The thermal equilibrium between $^4F_{5/2}$ and $^4F_{3/2}$ states and the Stark sub-levels-Z_1-Z_2 couple of the activator Nd^{3+} is utilized for wide temperature sensing.

Apart from lanthanide ions, transition metal ion-Mn^{3+} is also reported for luminescence ratiometric thermometry. The distinct thermometric response of emissions from Mn^{3+} and Mn^{4+} resulted in an enhanced relative temperature sensitivity of $6\%K^{-1}$ in the cryogenic temperature range of 5–75 K [26]. The inability of Boltzmann thermometers in the cryogenic temperature range is due to the insignificant Boltzmann-based thermal population at extremely low temperatures. Figure 11.3 shows the temperature dependence of emission bands, integrated intensities of Mn^{3+} and Mn^{4+}, experimental fit, and relative temperature sensitivity of $Ca_3Ga_2Ge_3O_{12}$: Mn phosphor [26].

Zhang *et al.* also reported temperature-sensing properties of Mn^{3+} in the $Y_3Ga_5O_{12}$ system. Here, the substitution of Lu^{3+} changes the temperature point of Mn^{3+} from 260 to 290 K, which culminates in the increased sensitivity of 3.78%K^{-1} [41]. Xu *et al.* observed a maximum relative sensitivity of 3.4%K^{-1} at 423 K based on the upconversion (UC) luminescence of Mn^{2+} activated, Yb^{3+} sensitized $Y_3Al_5O_{12}$ system. The orange upconversion emission band at 585 nm corresponds to the transitions-$^4T_1$ (4G) $\rightarrow$ $^6A^1$ (S) in which rapid fall in emission intensity is noted at elevated temperatures which contributes to the increased sensitivity [29].

Dy^{3+} is another activator which is often studied for optical thermometry. Recently, Subodh *et al.* have investigated the temperature sensing properties of Dy^{3+} in tellurate garnet, $Li_3Y_3Te_2O_{12}$ [18]. Based on non-thermally coupled energy levels (NTCELs) of Dy^{3+}, a maximum relative temperature sensitivity of 1.2%K^{-1} is obtained, suitable for low temperature measurements. In addition, Bolek *et al.* investigated the prospects of temperature sensing properties of 0.1% Pr^{3+} activated mixed phosphor systems-$Y_3(Al_3Ga_2)O_{12}$ and $Y_3(Al_1Ga_4)O_{12}$ through which a better relative temperature sensitivity in a wide temperature range of 15–675 K is obtained [42]. Also, the temperature sensitivity can be improved by controlling the cation substitution. In $Y_3(Al, Ga)_5O_{12}$: 0.1% Pr^{3+} system, a maximum relative sensitivity of 3.6%K^{-1} is obtained by controlling Ga:Al composition [28].

11.3.2 Systems based on multi emitting centers

Among garnet phosphors, it can be observed that enhanced sensitivity is obtained by substituting multiple activator ions. The combined contribution to the variation of intensity results in the distinct thermal response favoring improved sensitivity. This can be multiple activators incorporated into the same host or single activators substituted into the self-activated host like vanadate systems.

11.3.2.1 Self-activated systems with activators

In this section, temperature sensing properties of self-activated systems incorporated with activator ions like Eu^{3+} and Sm^{3+} will be discussed in detail. Analyzing garnet-based phosphor systems, it can be noted that vanadate garnets of the stoichiometry $A_3B_2V_3O_{12}$ are an interesting choice among researchers. In vanadate garnets, the emission from self-activated complex-VO_4^{3-} and activator ions like rare-earth ions-RE^{3+} is considered for calculating the FIR. The peculiarity of the vanadate systems is that the broader emission band extends from 350 to 700 nm due to charge transfer from V^{5+} to O^{2-} ions of the VO_4^{3-} complex. The distinct thermal response from the broad emission of vanadate complex and sharp emission from lanthanide ions results in the distinct thermal response. It thus makes vanadate garnet systems, a potential candidate for ratiometric thermometry.

It can be noted that vanadate garnets composed of various cations such as $Ca_3LiMgV_3O_{12}$, $Na_2GdMg_2V_3O_{12}$, $Na_2YMg_2V_3O_{12}$, $Sr_2NaMg_2V_3O_{12}$, and $Ca_2NaMg_2V_3O_{12}$ with notable lanthanide activators like Eu^{3+}, Sm^{3+}, Dy^{3+}, and Er^{3+} are widely reported [31,35,37,38,43]. The excitation energy is transferred from VO_4^{3-} to RE^{3+} ions in vanadate garnets. The relative temperature sensitivity of these phosphors varies with respect to the cation substitution and the energy transfer from the VO_4^{3-} complex to lanthanide ions. Zhou *et al.* investigated ratiometric fluorescence temperature sensing properties of Eu^{3+}/Sm^{3+} activated $Ca_2NaMg_2V_3O_{12}$ system [38]. A higher temperature sensitivity of 1.889%K^{-1} is obtained at 463 K for Sm^{3+} activated $Ca_2NaMg_2V_3O_{12}$ phosphor than the Eu^{3+} counterpart. Among various vanadate garnet compositions, $Sr_2NaMg_2V_3O_{12}$ hosts are quite studied for various lanthanide ions like Eu^{3+}, Sm^{3+}, and Dy^{3+}, as reported by our group [31–33]. The relative temperature sensitivity varied from 0.41, 1.61, and 2.01%K^{-1} for Dy^{3+}, Eu^{3+}, and Sm^{3+}, respectively. The improved sensitivity obtained in the reported vanadate garnets can be attributed to diverse thermal quenching associated with the dual emission from vanadate and rare-earth ions. In such systems, VO_4^{3-} complex follows rapid thermal quenching, whereas RE^{3+} follows a slow thermal quenching, as shown in Figure 11.4 [31,32]. The fast thermal quenching of vanadate garnets might be due to the non-radiative decay via crossover mechanism due to the increased thermal population of higher vibrational levels by excited electrons associated with the VO_4^{3-} complex. On the other hand, RE^{3+} follows a slow thermal quenching owing to the requirement of the increased

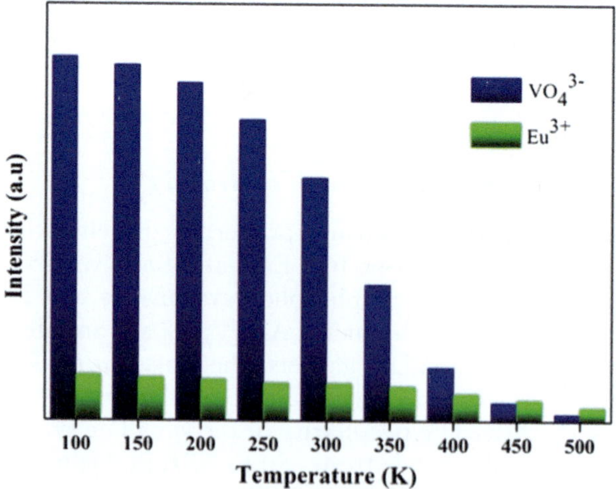

Figure 11.4 The histogram showing the diverse thermal quenching of VO_4^{3-} and Eu^{3+} ions in the Eu^{3+} activated $Sr_2NaMg_2V_3O_{12}$ system [31]. (Reproduced from Ref. [31] under Creative Commons Attribution 4.0 International License.)

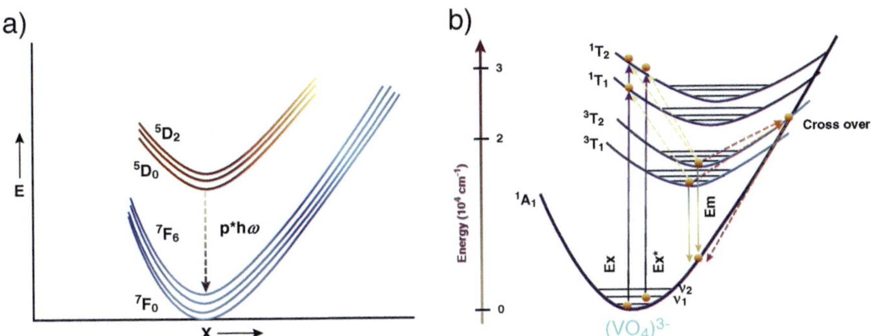

Figure 11.5 (a) Schematic model showing thermal quenching of Eu^{3+} and (b) configurational coordinate diagram for Eu^{3+} activated $Sr_2NaMg_2V_3O_{12}$ system depicting non-radiative relaxation via crossover mechanism [31]. (Reproduced from ref. [31] under Creative Commons Attribution 4.0 International License.)

number of phonons. Figure 11.5 illustrates the diverse thermal quenching behavior of the Eu^{3+}-activated $Sr_2NaMg_2V_3O_{12}$ system via the configurational coordinate diagram [31].

The main decay route of RE^{3+}-based thermal quenching is through non-radiative relaxations, which will be lowered and thus reduce non-radiative decay. In this context, RE^{3+}-activated vanadate garnets are one of the potential candidates for optical thermometry based on the ratiometric approach.

11.3.2.2 Multiple activator-based dopant activated systems

Another interesting strategy for achieving enhanced sensitivity is the introduction of multiple activator ions. In this category, the distinct thermal response of multi-emission from multiple activator ions will be envisaged for calculating the ratiometric variation of intensity for optical thermometry. The common combination found in garnet systems includes combinations of rare earths like Eu^{3+}, Ce^{3+}, and Tb^{3+}, transition metal ions (TM) like Cr^{3+}, Mn^{4+}, and main group element, Bi^{3+} forming multi-emitters of the form RE^{3+}/RE^{3+}, RE^{3+}/TM^{3+}. Sun *et al.* developed co-doped and multi-emitting Bi^{3+}-Tb^{3+}-Eu^{3+} activated $Ca_3Sc_2Si_3O_{12}$ phosphor [11] based on a novel strategy of new TCELs. Based on emission bands, $^3P_1(Bi^{3+})/^5D_0(Eu^{3+})$ and $^5D_4(Tb^{3+})/^5D_0(Eu^{3+})$ construction of new TCELs is made whose relative electron population follows Boltzmann distribution with the relative temperature sensitivity of 1.0 and 0.93%K^{-1} is obtained at 498 K [11].

There are reports where non-TCELs of multiple activators are considered for ratiometric temperature sensing. Mishra *et al.* detailed the temperature sensing properties of $Ce^{3+}/Yb^{3+}/Er^{3+}$ co-doped $Y_3Al_5O_{12}$ system based on the TCELs (524 and 546 nm) and NTCELs (524/546 and 556 nm) of Er^{3+}

ion [13]. Unlike TCELs, which obey the Boltzmann distribution, the intensity ratio of NTCELs is fitted by a polynomial function of temperature. Our group examined the prospects of Bi^{3+}-Pr^{3+} coactivated $Li_3Gd_3Te_2O_{12}$ phosphor in which a relative temperature sensitivity of $0.42\%K^{-1}$ is obtained via the ratiometric approach, which is lower than the decay time method. This can be attributed to the weak energy transfer in the Bi^{3+}-Pr^{3+} pair [19]. In addition, like in the case of vanadate garnets, the intensity ratio of dual emission bands of two activators can also be selected for the ratiometric approach. For example, Zheng et $al.$ followed a ratiometric approach in Bi^{3+}, Eu^{3+} coactivated $Ca_2YZr_2Al_3O_{12}$ system. The intensity ratio is estimated by considering the ratio of emission bands of Bi^{3+} and Eu^{3+} (I_{Bi}/I_{Eu}) [27]. A similar approach can be observed in co-doped $Ca_3Al_2Ge_3O_{12}$: Dy^{3+}, Eu^{3+} system where FIR of Dy^{3+} to Eu^{3+} emission is considered [15]. In most of the dopants, which include sensitizers and activators, energy transfer is the subsequent process we can expect. However, Zhang et $al.$ recently found a way to enlarge the sensitivity of FIR-based thermometers [16]. The efficient energy transfer of around 90% from Tb^{3+} to Sm^{3+} at low temperatures is interrupted at elevated temperatures, which is determined by calculating energy transfer efficiency at each temperature in the $Ca_2TbSn_2Al_3O_{12}$ system. Thus, the novel strategy of thermal interruption of energy transfer of emitters contributes to the enhanced sensitivity of co-doped samples to undoped samples [16]. So far, we have discussed the RE^{3+}/RE^{3+} combination. However, RE^{3+}-TM^{3+}/Bi^{3+} is an interesting combination that can provide improved sensitivity and better signal discriminability required for optical thermometry. One of the added advantages in selecting rare-earths and transition metal ion/main group elements like Bi^{3+} is the presence of distinct thermal quenching behavior, unlike in RE^{3+}/RE^{3+} combinations where both RE^{3+} majorly possess similar quenching behavior. Shen et $al.$ tried out several similar phosphor combinations based on $RE^{3+}=Eu^{3+}$, Tb^{3+}, Dy^{3+}, $TM=Mn^{4+}$, and Cr^{3+} in $Y_3Al_5O_{12}$ to develop self-referencing thermometers [14]. Among the combinations-Tb^{3+}/Mn^{4+}, Dy^{3+}/Mn^{4+}, Eu^{3+}/Cr^{3+}, and Dy^{3+}/Cr^{3+}, Eu^{3+}/Mn^{4+} offers the highest relative temperature sensitivity of $4.81\%K^{-1}$ (Figure 11.6) with Eu^{3+} and Mn^{4+} as the reference signal and temperature signal, respectively, which opens further studies in developing self-referencing ratiometric sensing [14]. The tunable energy difference between RE^{3+} and TM^{3+} is essential for achieving signal discriminability under single-wavelength excitation. Further, the concentration of activators plays a significant role in determining the temperature measurement conditions. Very recently, Sun et $al.$ demonstrated that the sensitivity increases with Sm^{3+} concentration even at $x=0.14$ in $Sr_3Y_2Ge_3O_{12}$ garnet [25]. Another RE^{3+}/TM^{3+} combination is reported in $Lu_3Al_5O_{12}$ garnet, where simultaneous doping of Ce^{3+} and Mn^{4+} is done [22]. Despite the fluctuation associated with the source of excitation and

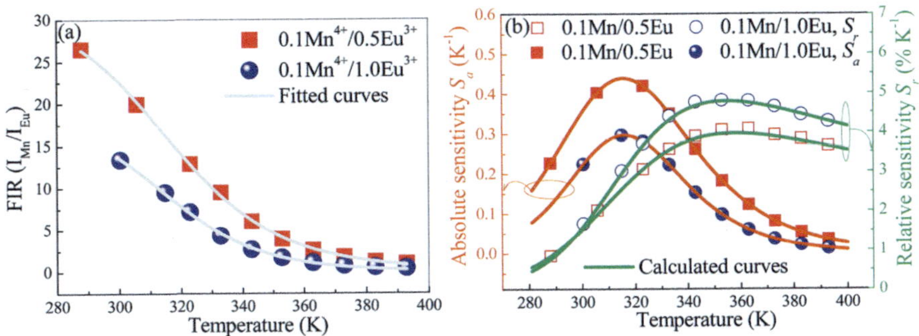

Figure 11.6 Temperature dependence of (a) FIR and (b) absolute and relative temperaturesensitivityofMn/EuactivatedY$_3$Al$_5$O$_{12}$system[14].(ReproducedfromRef.[14] with permission from the Royal Society of Chemistry.)

system errors, Ce^{3+}- and Mn^{4+}-based FIR offers better accuracy. Unlike well-known rare-earth or transition metal ions like Mn^{4+}/Cr^{3+}, Fe^{3+} energy levels can also participate in ratiometric temperature sensing. Kniec *et al.* developed Fe^{3+}, Cr^{3+}, Nd^{3+} co-doped series of phosphors $Y_3Al_5O_{12}$, $Y_3Ga_5O_{12}$, $Lu_3Ga_5O_{12}$, and $Gd_3Ga_5O_{12}$ phosphor. A higher sensitivity of 5.90%/K is obtained for the FIR based on the emission intensities of Fe^{3+} $(^6A_1(^6S) \rightarrow {}^4T_2(^4D))$ transition and Nd^{3+} $(^4F_{3/2} \rightarrow {}^4I_{9/2})$ levels in $Lu_3Ga_5O_{12}$ system [20]. It was found that sensitivity increases with a decrease in crystal field splitting, and thereby, crystal field optimization is another strategy to enhance the temperature sensing properties.

11.4 COLORIMETRIC OPTICAL THERMOMETERS BASED ON GARNET SYSTEMS

Apart from ratiometric sensing of temperature, the change in emission color with an increase in temperature can be used for sensing temperature. Table 11.2 lists garnet phosphors reported for ratiometric and colorimetric temperature sensing. Colorimetric thermometry is a means of visualized thermometry. Thermographic phosphors that exhibit thermochromic luminescence can be mainly used for safety sign applications in high-temperature environments. The search for garnet-based phosphors based on the colorimetric approach is one of the emerging areas. Hence, reports are less prolific than in the case of ratiometric method of temperature sensing. By analyzing reports, it can be observed that the colorimetric approach is often associated with multiple emission centers. This can be due to activators incorporated into the self-activated systems like vanadates or multiple activators substituted into the single host. The emission from multiple optical centers is different. Due to the distinct emission, the thermal response differs, contributing to the lack of similar thermal quenching. A diverse thermal quenching contributes to enhanced thermochromic luminescence. The rapid change in emission color helps to observe through camera or naked eyes. The corresponding drift in CIE coordinates can be expressed using the relation given by,

$$\Delta s = \sqrt{\left(u_T' - u_0'\right)^2 + \left(v_T' - v_0'\right)^2 + \left(w_T' - w_0'\right)^2} \tag{11.1}$$

where $w' = 1 - v' - u'$, $v' = 4y/(3 - 2x + 12y)$, and $u' = 4x/(3 - 2x + 12y)$ and subscripts represent initial temperature and temperature at T, respectively [35]. Depending on the host and tuning strategies, the shift in CIE coordinates can be modulated.

11.4.1 Self-activated systems with activators

Most thermographic phosphors exhibiting thermochromic luminescence are of self-activated vanadate systems doped with activators like Eu^{3+} and Sm^{3+}. It can be noted that red emitting activators like Eu^{3+} and Sm^{3+} ions are usually incorporated into the vanadate garnets so that blue-green emission from vanadate and orange/red emission from activators contribute to the distinct thermal quenching and favors a rapid change in emission color [31,32,35,36]. Hence, as temperature increases, the colorific shift from blue-green region to orange/red emission occurs if Eu^{3+}/Sm^{3+}-like red emitters are introduced. Recently, Chen $et\ al.$ investigated the colorimetric sensing in $Na_2GdMg_2V_3O_{12}$:Sm^{3+} phosphor [35]. A chromaticity shift, Δs of 0.177, is obtained for the temperature range of 303–513 K. Further, the absolute and relative sensitivity of 0.117 and 0.244%/K is obtained with a change in emission color from yellow to red [35]. Zho $et\ al.$ compared the colorific shift of the $Ca_2NaMg_2V_3O_{12}$ system activated with Eu^{3+} and Sm^{3+} in which emission color changes from green to reddish-orange and green to orange for Eu^{3+} and Sm^{3+}, respectively [38]. On the other hand, our group conducted a detailed investigation on the thermochromic luminescence properties of $Sr_2NaMg_2V_3O_{12}$: Eu^{3+} and Sm^{3+} system and the possible mechanism behind the rapid colorimetric shift at elevated temperatures [31,32]. Eu^{3+} activated system exhibits an unprecedented colorific shift from white to the deep red region with temperature rise from 300 to 500 K (Figure 11.7) than Sm^{3+} counterpart, making $Sr_2NaMg_2V_3O_{12}$: Eu^{3+}, a potential candidate for safety sign applications in high-temperature environment.

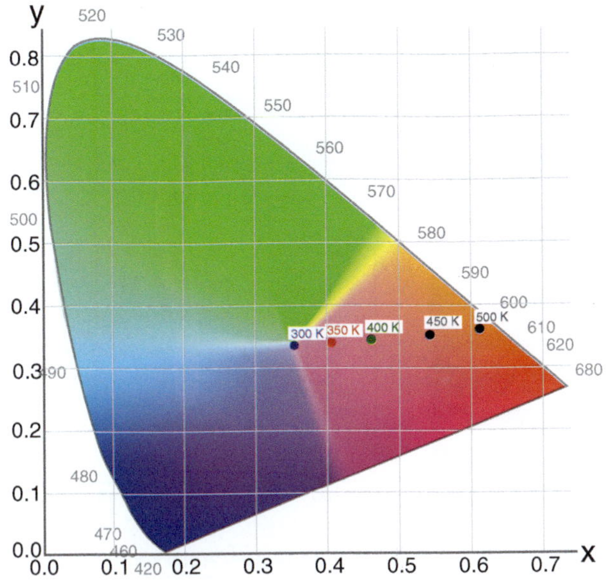

Figure 11.7 CIE diagram depicting the thermochromic luminescence of $Sr_2NaMg_2V_3O_{12}$: Eu^{3+} system [31]. (Reproduced from Ref. [31] under Creative Commons Attribution 4.0 International License.)

11.4.2 Multiple activator-based dopant activated systems

Multiple activators incorporated into the non-luminescent host systems are another method for realizing colorimetric temperature sensing. However, this approach needs more attention. Chen *et al.* [22] studied the thermos-induced chromaticity shift of Mn^{4+}/Ce^{3+} $Lu_3Al_5O_{12}$ garnet.

A strong thermochromic luminescence due to codoping resulted in the chromaticity shift of 0.153, causing emission color to vary from red (298 K) to green (353 K). The visualization of the chromaticity shift is depicted in Figure 11.8. Thus, incorporation of multiple emitters is one of the key strategies to achieve thermochromic luminescence in garnet-based thermographic phosphors.

11.5 CONCLUSIONS AND PROSPECTIVE

This chapter presents a review of garnet-based thermographic phosphors for dual mode of temperature sensing via ratiometric and colorimetric approach. Garnets have attracted research attention as an emerging

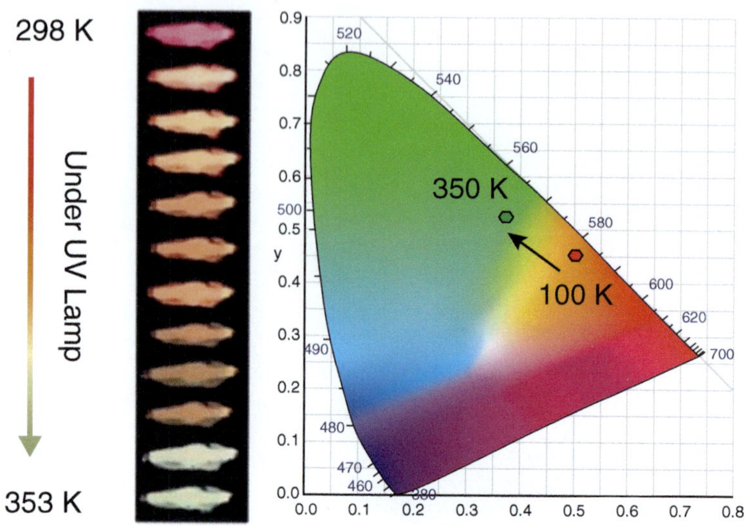

Figure 11.8 The shift of CIE coordinates with temperature and the fluorescent images of the phosphor at various temperatures [22]. (Reprinted from Ref. [22] with permission from the American Chemical Society.)

thermographic phosphor for optical thermometry. In short, the unique feature of the garnet structure is due to its structure-property correlation. Hence, thermographic properties can be tuned via strategies like the introduction of cations of varied size and charge in the polyhedral network and by the energy transfer via the introduction of cations. Apart from achieving high relative temperature sensitivity, thermographic phosphor should possess accuracy in temperature detection and better signal discriminability. Hence, besides TCELs of single activators, novel temperature sensing strategies like NTCELs and dual emission centers are being employed. In particular, at cryogenic temperatures, where the Boltzmann-based thermal population is insignificant, NTCELs of single activator or dual emission centers are a great choice. Apart from rare-earth ions, co-doping with transition metal ions-Mn^{4+}, Mn^{3+}/Fe^{3+} is a better option to contribute to the enhanced sensitivity as it overcomes weak absorption of RE^{3+}. This is due to the tunable energy difference between RE^{3+} and TM^{3+}, which helps to achieve better signal discriminability. Thus, it is high time to focus more on the non-RE^{3+} like TM ions/main group elements.

However, for colorimetric thermometry, distinct thermal responses and subsequent thermal quenching are the requirements, which can be achieved using dual/multiple emission centers. In this context, self-activated systems like vanadate garnets have a long way to be used as safety sign applications in high-temperature environments since the rapid change in emission color is inevitable for this. Introducing multiple emitters/co-doping via energy transfer is another tuning strategy that can be followed in both ratiometric and colorimetric ways of temperature sensing. In particular, the combined strategy of structural modifications and energy transfer from multiple optical centers in the garnet systems helps to develop tailored thermographic phosphors with distinct thermal responses intended for ratiometric and colorimetric temperature sensing. We expect more reports on the strategically tuned garnet thermographic phosphors for efficient and accurate temperature sensing in the future.

Above all, the authors honor the excellent reports on garnet-based thermographic phosphors by various groups for leading garnets as one of the potential candidates for optical thermometry.

REFERENCES

1. W. Liu, B. Yang, Thermography techniques for integrated circuits and semiconductor devices, *Sensor Rev* 27 (2007) 298–309. https://doi.org/10.1108/02602280710821434.
2. S.W. Allison, G.T. Gillies, Remote thermometry with thermographic phosphors: Instrumentation and applications, *Rev Sci Instrum* 68 (1997) 2615–2650. https://doi.org/10.1063/1.1148174.
3. M.G. Nikolić, A.Z. Al-Juboori, V. Đorđević, M.D. Dramićanin, Temperature luminescence properties of Eu^{3+}-doped Gd2O3 phosphors, *Phys Scr* T157 (2013) 014056. https://doi.org/10.1088/0031-8949/2013/T157/014056.
4. F. Vetrone, R. Naccache, A. Zamarrón, A.J. De La Fuente, F. Sanz-Rodríguez, L.M. Maestro, E.M. Rodriguez, D. Jaque, J.G. Sole, J.A. Capobianco, Temperature sensing using fluorescent nanothermometers, *ACS Nano* 4 (2010) 3254–3258. https://doi.org/10.1021/NN100244A.
5. S. Das, S. Som, C.-Y. Yang, S. Chavhan, C.-H. Lu, Structural evaluations and temperature dependent photoluminescence characterizations of Eu^{3+}-activated $SrZrO_3$ hollow spheres for luminescence thermometry applications, *Sci Rep* 6 (2016) 25787. https://doi.org/10.1038/srep25787.
6. C.D.S. Brites, P.P. Lima, N.J.O. Silva, A. Millán, V.S. Amaral, F. Palacio, L.D. Carlos, Thermometry at the nanoscale, *Nanoscale* 4 (2012) 4799–4829. https://doi.org/10.1039/C2NR30663H.
7. L.H. Fischer, G.S. Harms, O.S. Wolfbeis, Upconverting nanoparticles for nanoscale thermometry, *Angew Chem Int Ed* 50 (2011) 4546–4551. https://doi.org/10.1002/ANIE.201006835.
8. A. Bindhu, J.I. Naseemabeevi, S. Ganesanpotti, Distortion and energy transfer assisted tunability in garnet phosphors, *Crit Rev Solid State Mater Sci* 47 (2021) 621–664. https://doi.org/10.1080/10408436.2021.1935211.
9. A. Bindhu, J.I. Naseemabeevi, S. Ganesanpotti, Strategic tuning of photophysical response in the polyhedral framework of the garnet structure toward white light-emitting devices with enhanced color rendering, *Inorg Chem* 62 (2023) 5744–5756. https://doi.org/10.1021/ACS.INORGCHEM.3C00299.
10. S. Liu, J. Ueda, S. Tanabe, Ratiometric luminescent thermometry for the third bioimaging window by Er^{3+} doped garnet with large Stark splitting, *Appl Phys Lett* 123 (2023) 161101. https://doi.org/10.1063/5.0168845.

11. Z. Sun, M. Jia, Y. Wei, J. Cheng, T. Sheng, Z. Fu, Constructing new thermally coupled levels based on different emitting centers for high sensitive optical thermometer, *Chem Eng J* 381 (2020) 122654. https://doi.org/10.1016/J.CEJ.2019.122654.

12. X. Chen, Y. Zhang, Y. Bu, Y. Chen, Y. Chen, J. Fu, J. Li, D. Deng, A multi-mode optical thermometer based on the up-conversion $Ca_3Y_2Ge_3O_{12}$: Er^{3+}, Yb^{3+} phosphor, *J Lumin* 261 (2023) 119907. https://doi.org/10.1016/J.JLUMIN.2023.119907.

13. N. Kumar Mishra, M.M. Upadhyay, S. Kumar, K. Kumar, Efficient dual mode emission in Ce^{3+}/Yb^{3+}/Er^{3+} doped yttrium aluminium gallium garnet for led device and optical thermometry, *Spectrochim Acta A Mol Biomol Spectrosc* 282 (2022) 121664. https://doi.org/10.1016/j.saa.2022.121664.

14. D. Chen, S. Liu, Y. Zhou, Z. Wan, P. Huang, Z. Ji, Dual-activator luminescence of RE/TM: $Y_3Al_5O_{12}$ (RE=Eu^{3+}, Tb^{3+}, Dy^{3+}; TM=Mn^{4+}, Cr^{3+}) phosphors for self-referencing optical thermometry, *J Mater Chem C Mater* 4 (2016) 9044–9051. https://doi.org/10.1039/c6tc02934e.

15. L. Zhou, R. Chen, X. Jiang, T. Zhang, X. Shi, Z. Leng, Y. Yang, Z. Zhang, C. Zuo, C. Li, W. Yang, H. Lin, L. Liu, S. Li, F. Zeng, Z. Su, Garnet structure-activated warm white light phosphors $Ca_3Al_2Ge_3O_{12}$: Dy^{3+}, Eu^{3+} with ultrahigh thermal stability and tunable luminescence, *Inorg Chem* 63 (2024) 1274-1287. https://doi.org/10.1021/acs.inorgchem.3c03739

16. Z. Zhang, J. Yan, Q. Zhang, G. Tian, W. Jiang, J. Huo, H. Ni, L. Li, J. Li, Enlarging sensitivity of fluorescence intensity ratio-type thermometers by the interruption of the energy transfer from a sensitizer to an activator, *Inorg Chem* 61 (2022) 16484–16492. https://doi.org/10.1021/acs.inorgchem.2c02756.

17. H. Örücü, Effect of doping concentration and excitation power on upconversion and temperature sensitivity of $Gd_3Ga_5O_{12}$:Yb^{3+}/Er^{3+} phosphors, *Int J Adv Eng Pure Sci* 35 (2023) 237–245. https://doi.org/10.7240/jeps.1240654.

18. A. Bindhu, J.I. Naseemabeevi, S. Ganesanpotti, Insights into the crystal structure and photophysical response of Dy^{3+} doped $Li_3Y_3Te_2O_{12}$ for ratiometric temperature sensing, *J Sci Adv Mater Devices* 7 (2022) 100444. https://doi.org/10.1016/J.JSAMD.2022.100444.

19. A. Bindhu, A.S. Priya, J.I. Naseemabeevi, S. Ganesanpotti, Deciphering crystal structure and photophysical response of Bi^{3+} and Pr^{3+} co-doped $Li_3Gd_3Te_2O_{12}$ for lighting and ratiometric temperature sensing, *J Alloys Compd* 893 (2022). https://doi.org/10.1016/j.jallcom.2021.162246.

20. K. Kniec, K. Ledwa, K. Maciejewska, L. Marciniak, Intentional modification of the optical spectral response and relative sensitivity of luminescent thermometers based on Fe^{3+}, Cr^{3+}, Nd^{3+} co-doped garnet nanocrystals by crystal field strength optimization, *Mater Chem Front* 4 (2020) 1697–1705. https://doi.org/10.1039/D0QM00097C.

21. P. Du, X. Sun, Q. Zhu, J.G. Li, Garnet-structured $Li_6CaLa_2Nb_2O_{12}$: Yb/Er new phosphor showing superior performance of optical thermometry, *Scr Mater* 185 (2020) 140–145. https://doi.org/10.1016/J.SCRIPTAMAT.2020.04.039.

22. Y. Chen, J. He, X. Zhang, M. Rong, Z. Xia, J. Wang, Z.Q. Liu, Dual-mode optical thermometry design in $Lu_3Al_5O_{12}$:Ce^{3+}/Mn^{4+} phosphor, *Inorg Chem* 59 (2020) 1383–1392. https://doi.org/10.1021/acs.inorgchem.9b03107.

23. B. Yan, Y. Wei, W. Wang, M. Fu, G. Li, Red-tunable LuAG garnet phosphors via $Eu^{3+} \rightarrow Mn^{4+}$ energy transfer for optical thermometry sensor application, *Inorg Chem Front* 8 (2021) 746–757. https://doi.org/10.1039/D0QI01285H.

24. M. Back, J. Xu, J. Ueda, S. Tanabe, Neodymium (III)-doped $Y_3Al_2Ga_3O_{12}$ garnet for multipurpose ratiometric thermometry: From cryogenic to high temperature sensing, *J Ceram Soc Jpn* 131 (2023) 22167. https://doi.org/10.2109/jcersj2.22167.

25. R. Sun, X. Wei, H. Yu, P. Chen, H. Ni, J. Li, J. Zhou, Q. Zhang, Temperature sensing of $Sr_3Y_2Ge_3O_{12}{:}Bi^{3+}$, Sm^{3+} garnet phosphors with tunable sensitivity, *Dalton Trans* 52 (2023) 2825–2832. https://doi.org/10.1039/D2DT03153A.

26. Y. Wang, D. Włodarczyk, M.G. Brik, J. Barzowska, A.N. Shekhovtsov, K.N. Belikov, W. Paszkowicz, L. Li, X. Zhou, A. Suchocki, Effect of temperature and high pressure on luminescence properties of Mn^{3+} ions in $Ca_3Ga_2Ge_3O_{12}$ single crystals, *J Phys Chem C* 125 (2021) 5146–5157. https://doi.org/10.1021/acs.jpcc.0c09845.

27. Z. Zheng, J. Zhang, X. Liu, R. Wei, F. Hu, H. Guo, Luminescence and self-referenced optical temperature sensing performance in $Ca_2YZr_2Al_3O_{12}{:}Bi^{3+}$, Eu^{3+} phosphors, *Ceram Int* 46 (2020) 6154–6159. https://doi.org/10.1016/J.CERAMINT.2019.11.081.

28. P. Bolek, J. Zeler, C.D.S. Brites, J. Trojan-Piegza, L.D. Carlos, E. Zych, Ga-modified YAG: Pr^{3+} dual-mode tunable luminescence thermometers, *Chem Eng J* 421 (2021) 129764. https://doi.org/10.1016/j.cej.2021.129764.

29. Z. Xu, H. Lin, R. Hong, C. Tao, Z. Han, D. Zhang, S. Zhou, Yb^{3+}/Mn^{2+} co-doped $Y_3Al_5O_{12}$ phosphors for optical thermometric application, *Opt Mater (Amst)* 124 (2022) 111949. https://doi.org/10.1016/j.optmat.2021.111949.

30. C. Wang, S. Ye, Q. Zhang, Unraveling the distinct luminescence thermal quenching behaviours of A/B-site Eu ions in double perovskite $Sr_2CaMoO_6{:}Eu^{3+}$, *Opt Mater (Amst)* 75 (2018) 337–346. https://doi.org/10.1016/J.OPTMAT.2017.10.048.

31. A. Bindhu, J.I. Naseemabeevi, S. Ganesanpotti, Vibrationally induced photophysical response of $Sr_2NaMg_2V_3O_{12}$: Eu^{3+} for dual-mode temperature sensing and safety signs, *Adv Photonics Res* 3 (2022) 2100159. https://doi.org/10.1002/ADPR.202100159.

32. A. Bindhu, J.I. Naseemabeevi, S. Ganesanpotti, Delving into the multifunctionality of $Sr_2NaMg_2V_3O_{12}$ via RE^{3+} substitution for dual-mode temperature sensing, latent fingerprint detection and security inks, *Mater Adv* 4 (2023) 3796–3812. https://doi.org/10.1039/D3MA00241A.

33. A. Bindhu, J.I. Naseemabeevi, S. Ganesanpotti, Augmenting cyan emission in vanadate garnets via Dy^{3+} activation for light emitting devices and multi-mode optical thermometry, *Dalton Trans* 52 (2023) 11705–11715. https://doi.org/10.1039/D3DT01895D.

34. R. R, A. P.S, G. N, An insight into Judd-Ofelt analysis and non-contact optical thermometry of $LiCa_2Mg_2V_3O_{12}$: Dy^{3+} phosphors for multifunctional applications, *Opt Mater (Amst)* 145 (2023) 114393. https://doi.org/10.1016/j.optmat.2023.114393.

35. J. Chen, L. Li, T. Pang, H. Guo, L. Chen, $Na_2GdMg_2V_3O_{12}{:}Sm^{3+}$ phosphors for three-mode optical temperature sensing, *J Am Ceram Soc* 106 (2023) 7514–7522. https://doi.org/10.1111/jace.19341.

36. H. Guo, B. Devakumar, R. Vijayakumar, P. Du, X. Huang, A novel Sm^{3+} singly doped $LiCa_3ZnV_3O_{12}$ phosphor: A potential luminescent material for multifunctional applications, *RSC Adv* 8 (2018) 33403–33413. https://doi.org/10.1039/C8RA07329E.

37. J. Q. Chen, J.Y. Chen, W.N. Zhang, S.J. Xu, L.P. Chen, H. Guo, Three-mode optical thermometer based on $Ca_3LiMgV_3O_{12}$:Sm^{3+} phosphors, *Ceram Int* (2023). https://doi.org/10.1016/j.ceramint.2023.01.223.

38. H. Zhou, N. Guo, X. Lü, Y. Ding, L. Wang, R. Ouyang, B. Shao, Ratiometric and colorimetric fluorescence temperature sensing properties of trivalent europium or samarium doped self-activated vanadate dual emitting phosphors, *J Lumin* 217 (2020) 116758. https://doi.org/10.1016/J.JLUMIN.2019.116758.

39. Y. Tong, W.N. Zhang, R.F. Wei, L.P. Chen, H. Guo, $Na_2YMg_2(VO_4)_3$: Er^{3+}, Yb^{3+} phosphors: Up-conversion and optical thermometry, *Ceram Int* 47 (2021) 2600–2606. https://doi.org/10.1016/J.CERAMINT.2020.09.106.

40. P. Du, J.S. Yu, Self-activated multicolor emissions in $Ca_2NaZn_2(VO_4)_3$: Eu^{3+} phosphors for simultaneous warm white light-emitting diodes and safety sign, *Dyes Pigments* 147 (2017) 16–23. https://doi.org/10.1016/J.DYEPIG.2017.07.065.

41. J. Zhang, X. Hou, S. Wang, Z. Ye, Unraveling the valence states of manganese ions and the effects of composition variation and post-processing in YGG_{1-x}-Lu_xGG: Mn garnet optical sensor, *Chem Eng J* 411 (2021) 128448. https://doi.org/10.1016/j.cej.2021.128448.

42. P. Bolek, J. Zeler, L.D. Carlos, E. Zych, Mixing phosphors to improve the temperature measuring quality, *Opt Mater (Amst)* 122 (2021) 111719. https://doi.org/10.1016/j.optmat.2021.111719.

43. W. Zhang, C. He, X. Wu, X. Huang, F. Lin, Y. Liu, M. Fang, X. Min, Z. Huang, Yellow emission obtained by combination of broadband emission and multi-peak emission in garnet structure $Na_2YMg_2V_3O_{12}$: Dy^{3+} phosphor, *Molecules* 25 (2020) 542. https://doi.org/10.3390/MOLECULES25030542.

Chapter 12

Temperature sensing of biological cells and tumours

*Soorya Srinivasan, Madeshwari Ezhilan,
Arockia Jayalatha Kulandaisamy,
John Bosco Balaguru Rayappan,
Girdega Muruganandam, and Noel Nesakumar*

12.1 INTRODUCTION

Temperature fluctuations within cells play a pivotal role in governing a wide spectrum of cellular processes, encompassing vital functions such as cellular metabolism and the intricate process of cell division. To garner a holistic and profound insight into the intricate mechanisms that govern heat generation and its subsequent dispersion across various intracellular compartments, a meticulous approach involving comprehensive thermal mapping becomes imperative. This analytical practice, commonly denoted as intracellular thermometry, holds the key to unravelling the complex interplay within cells [1]. Furthermore, it is universally recognized that pathological states, including cancer and a spectrum of other diseases, are intimately intertwined with the nuanced dynamics of cellular heat generation. Consequently, the application of thermal mapping to analyze the intracellular environment offers not only valuable insights into various cellular processes but also the potential for intracellular disease diagnosis, thereby fostering the development of innovative therapeutic and diagnostic strategies. However, the precise and accurate characterization of temperature distribution within cells is increasingly recognized as a formidable challenge faced by researchers [2]. Notably, commonly utilized temperature measurement methods, such as thermocouples and thermography, suffer from limitations in spatial resolution, rendering them inadequate for investigating cellular dynamics. As a viable alternative, fluorescence microscopy emerges as an appealing option due to its sensitivity to temperature variations relevant to physiological conditions (303.15–323.15 K) [3]. This technique, known for its non-contact nature, combined with advanced cameras featuring hyperspectral technology, offers a powerful tool for studying intracellular temperature with high temporal and spatial resolution.

Traditional quantification of intracellular temperature relies on fluorescence properties of extrinsic and intrinsic fluorescent molecules, including organic dyes and proteins [4]. Although organic dye-based fluorophores

DOI: 10.1201/9781032661537-12

readily integrate into active cells and distribute evenly across organelles, they have limitations such as photostability issues and variable fluorescence properties influenced by microenvironmental factors such as cell viscosity, pH, and cytosolic calcium ion levels. This environmental dependency underscores the need for fluorescence property calibration based on specific microenvironments, emphasizing the impact of fluorophore embedding mediums on thermal sensitivity.

Recent advancements in nanotechnology have yielded light-sensitive and stable nanoscale fluorophores, ranging from 1 to 100 nm. As the nanoscale fluorophores are small in size and can be surface-engineered, these nanoscale probes can be integrated into the living cells, enabling fluorescence cell imaging. Consequently, the potential arises for nanoscale fluorophores emitting fluorescence in response to temperature changes within physiological parameters to function as nano-thermometers, thereby facilitating the intricate task of intracellular temperature mapping [5].

In this chapter, a comprehensive exploration of nano/micro-thermometers is presented, highlighting their pivotal role in the precise measurement of cellular temperatures. These thermometers capitalize on the distinctive traits of nanoscale fluorophores, where the fluorescence attributes are intricately intertwined with both their inherent chemical defects and structural characteristics, thus forming the foundational elements of these sensing probes. To enhance the comprehension of these intricate connections, we have systematically classified nano/micro-thermometers based on their specific chemical defects and structural attributes, which directly influence the observed fluorescence behaviours of the thermometers. Moreover, this chapter takes a progressive stride by delving into the realm of pioneering nanoscale fluorescent probes. These probes are thoughtfully designed to incorporate thermosensitive components derived from green fluorescent proteins, rare earth-doped nanoparticles, molecule beacons, metallic nanoclusters, quantum dots, nanodiamonds, and polymeric nanostructures, significantly expanding the capabilities of temperature-sensitive techniques. This advancement opens up new avenues for temperature-related research and applications.

12.2 LUMINESCENCE-BASED MICRO-NANOTHERMOMETERS

The attributes of non-invasive and non-contact optical imaging techniques have found extensive application across a wide spectrum of domains, including biomedical, clinical, industrial, and scientific research settings. These techniques are effectively employed by integrating luminescent micro-nanothermometers with established imaging methodologies. Notably, these instruments can be operated remotely through wireless connectivity, boasting high spatial and thermal resolutions along with rapid data acquisition capabilities. This renders them particularly well-suited for the comprehensive analysis of biofluids containing intricate biological samples. A decade ago, in 2012, two comprehensive reviews provided concise insights into luminescence-based nano/micro-thermometry when efforts to measure cellular temperature were in their early stages [6,7]. These reviews encompassed diverse perspectives, including theoretical foundations, operational mechanisms, and advantages and disadvantages. The quantification of temperature through luminescence involves assessing factors such as decay of lifetime, alterations in maximum emission wavelength, and changes in fluorescence intensity exhibited by nanoscale fluorescent probes. However, the utilization of luminescence nano/micro-thermometers to monitor cellular temperature has been constrained by limitations related to applicability, reliability, and robustness. Recent years have witnessed a proliferation of luminescence-based methodologies tailored for precise temperature measurement in living cells. In this section, we delve into the most pertinent and consequential approaches, drawing comparisons and considering various factors.

12.2.1 Rare earth-doped nanoparticle

Fluorescence is a phenomenon primarily characterized by the presence of discrete spectral bands within nanomaterials and ion complexes containing trivalent rare earth ions, regardless of the surrounding environment [8]. These rare earth ion complexes and doped nanoparticles have gained significant prominence due to their advantageous characteristics, finding extensive use in the high-contrast imaging of single cells. Notably, their low toxicity levels and excellent optical stability contribute to their appeal in cellular imaging applications [9]. One particularly advantageous aspect is the ability to excite these nanomaterials with infrared light, which enables enhanced penetration depths while minimizing autofluorescence. While rare earth ions remain unaffected by environmental conditions, their fluorescence response exhibits a remarkable sensitivity to fluctuations in temperature. This unique characteristic positions them as invaluable tools for various temperature sensing applications. The domain of rare earth ion-based fluorescence thermal monitoring has generated a plethora of applications. A significant breakthrough occurred in 2007 with the introduction of a prototype fluorescence sensor that employed europium-doped complexes as fluorescent nanothermometers [10]. This sensor demonstrated a linear temperature dependence around 303 K and involved establishing direct contact between a living HeLa cell and a fluorescent rare earth ion complex through a micropipette. Through this direct approach, the sensor enabled the determination of cell membrane temperature, achieving an impressive temperature resolution exceeding 273.65 K. This methodology provided a robust means to delve into the thermal dynamics of individual HeLa cells, especially in scenarios involving externally induced thermogenesis. In parallel, separate research corroborated the efficacy of rare earth ions for single-cell thermal sensing. Ytterbium complexes, which emit green fluorescence upon exposure to near-infrared light, exemplify this capability. These complexes, often referred to as "upconverting nanoparticles", can be uniformly dispersed within cells under specific conditions [11]. This dispersion phenomenon arises from thermally coupled electronic states characterized by a pronounced temperature-dependent behaviour, leading to the emission of green fluorescence.

Achieving a homogeneous dispersion of upconverting nanoparticles within cells without causing adverse effects is indeed feasible under specific conditions. These nanoparticles emit green fluorescence due to the presence of thermally coupled electronic states that exhibit a strong temperature-dependent response. Changes in the nanoparticle temperature lead to shifts in the distribution of populations within these thermally coupled states, resulting in alterations in fluorescence intensity. For example, erbium-doped upconverting nanoparticles can sense temperature through ratiometric methods, overcoming issues associated with intensity measurements stemming from varying nanoparticle concentrations in cells.

In the case of HeLa cells, upconverting nanoparticles were incorporated to detect thermal signals as external heating affected the cells. In an experimental study, the response of a single HeLa cell's optical transmission image to different voltages applied to a metallic plate in direct contact with the cell was examined. The variations in fluorescence band intensity were leveraged to gauge the cell's temperature. Through a thorough analysis of these observed changes, the temperature changes within the cells, ranging from 298 to 318 K in response to varying voltages on the metallic plate, were accurately measured [12].

12.2.2 Biomolecules

12.2.2.1 Green fluorescent proteins

Despite being rare, some researchers have endeavoured to create nanothermometers using biomolecules such as DNA or proteins, capitalizing on their small size to achieve high spatial resolution (Figure 12.1a–d) [13]. In the context of cellular imaging, the green fluorescent protein stands out as a commonly used marker for visualizing cells. Unlike non-specific exogenous probes, appropriately designed green fluorescent proteins can precisely target particular organelles without disrupting cell function. Remarkably, nanothermometers based on the green fluorescent protein have demonstrated the ability to measure intracellular heat production activity. These thermometers utilize the fluorescence polarization anisotropy of the green fluorescent protein, which exhibits a significant variation corresponding to temperature changes within the range of 293–333 K. This property involves

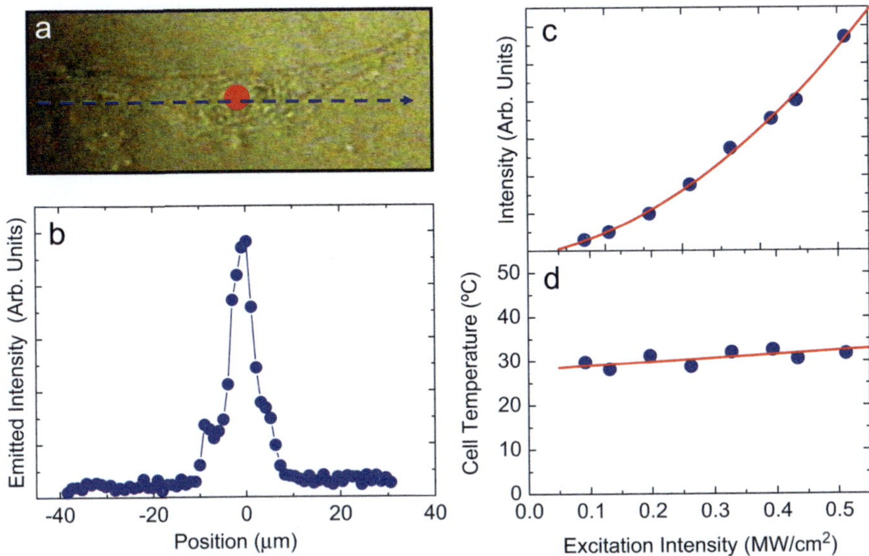

Figure 12.1 (a) Optical transmission image of a HeLa cancer cell, which has been incubated for 3 hours in a PBS solution containing Er/Yb co-doped NaYF$_4$ nanoparticles. (b) Spatial variation of the two-photon emission generated by the Er/Yb co-doped NaYF$_4$ nanoparticles obtained along the scan direction indicated by an arrow in (a). (c) Variation of the two-photon emitted intensity as a function of the 920 nm excitation intensity. (d) Variation of the intracellular temperature as a function of the 920 nm excitation intensity. (Data included in (c) and (d) were obtained at the intracellular location leading to the maximum two-photon luminescence (indicated by a red spot in (a)) Ref. [13] (Copyright received).)

comparing the fluorescence intensity ratios between polarizations parallel and perpendicular to the incident polarization. Importantly, the robustness of the green fluorescent protein's fluorescence polarization anisotropy remains unaffected by external intensity fluctuations, ensuring the accuracy of the results. Interestingly, the calibration curve of fluorescence polarization anisotropy against temperature differs notably when measured within cells compared to buffer solutions, attributed to viscosity variations between the two mediums. This finding underscores the necessity of conducting individual calibrations for each specific cell line. Moreover, the application of a photothermal method for heat delivery assessment, utilizing green fluorescent protein-transfected HeLa cells and U-87 MG cancer cell lines alongside heated gold nanorods, showcases a versatile approach [14]. Furthermore, employing this technique has enabled the real-time measurement of intracellular temperature in *Caenorhabditis elegans*, marking a groundbreaking achievement in temperature monitoring [13].

Recently, there have been advancements in developing thermometers based on green fluorescence protein with increased thermal sensitivity. This progress enables the visualization of thermal activity occurring within specific organelles at the individual cell level [15]. The approach to developing thermosensitive green fluorescent proteins involves the use of coiled-coil proteins. These proteins can detect temperature fluctuations and regulate transcription repression based on temperature changes. In the common configuration, a green fluorescent protein is inserted between consecutive coiled-coil segments of a thermosensitive coiled-coil protein (TIpA). During the process of thermosensing, TlpA undergoes a reversible and swift structural change, resulting in an unfolded monomer state, which triggers a slight change in fluorescence at 310 K. As the temperature increases within the range of 293–323 K, there are distinct changes in the emission peaks. The intensity of the peak at 400 nm decreases, while the intensity of the peak at 480 nm increases. Notably, the thermosensitive green fluorescent protein exhibits 3.5 times less sensitivity than the green fluorescent protein within the linear temperature range of 311–319 K. Moreover, by selecting the appropriate coil region within TlpA, the thermosensitive green fluorescent protein can be tailored to measure intracellular temperature. Moreover, by combining organelle-specific sequences with thermosensitive green fluorescent protein, it becomes possible to observe and track alterations in temperature within distinct cellular compartments, such as the mitochondria and endoplasmic reticulum, while observing living HeLa cells [15]. Furthermore, investigations that encompass cellular treatment with cyanide 3-chlorophenylhydrazone have revealed the presence of diverse mitochondrial thermogenesis patterns. This intriguing thermogenesis phenomenon has also manifested within skeletal muscle myotubes as well as brown adipocytes.

12.2.2.2 Molecule beacons

Molecular beacons, synthetic bioengineered nucleic acid molecules, present a captivating alternative to the conventional green fluorescent protein. Their distinctive capacity to selectively interact with specific RNA or DNA sequences renders them versatile instruments in a range of biosensor applications. To illustrate, the utilization of L-DNA-based beacons, an isomeric variant of natural D-DNA, has been investigated for the creation of an innovative intracellular thermometer (depicted in Figure 12.2a–f) [16]. Structurally, these beacons typically comprise elongated DNA biomolecules intricately folded into a hairpin configuration. Positioned at their opposing ends are a fluorophore and a quencher, strategically attached. The separation distance between these components adjusts in response to temperature fluctuations, forming the basis of their temperature-sensing mechanism. Consequently, this structural arrangement yields minimal fluorescence output at lower temperatures. However, as the temperature increases, a substantial enhancement in fluorescence intensity becomes evident.

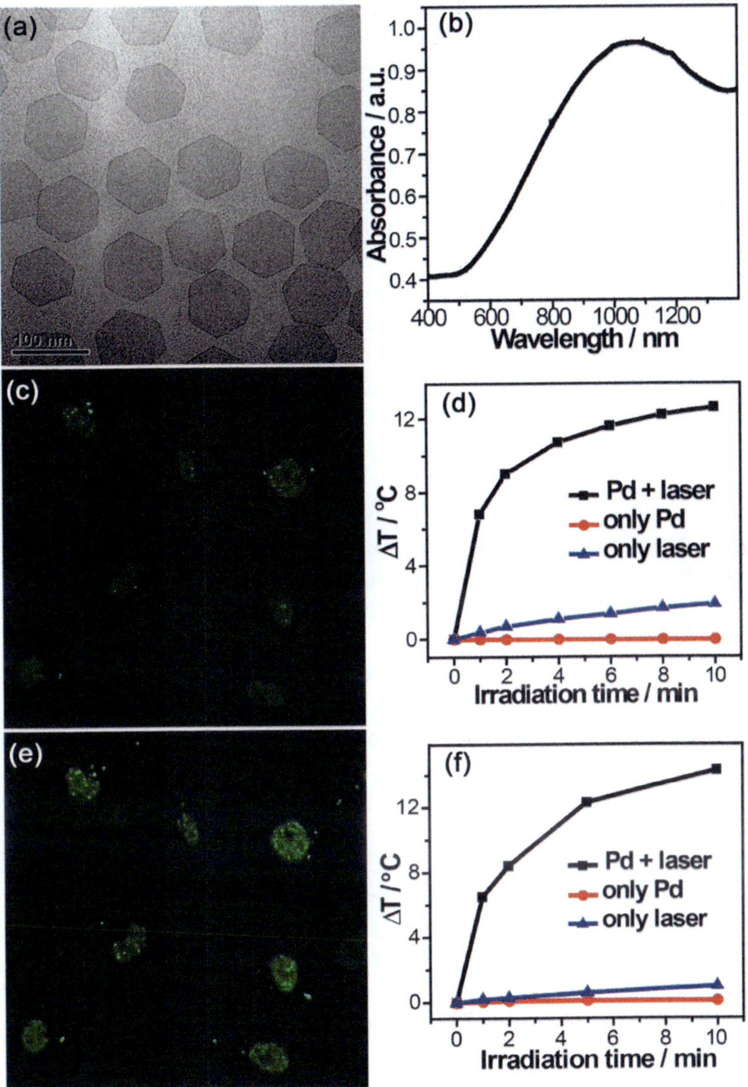

Figure 12.2 Photothermal study using Pd nanosheets for living HeLa cells: TEM (a), absorption spectra (b), and photothermal effect in DMEM medium (d) of Pd nanosheets; confocal images of HeLa cells with Pd nanosheets before (c) and after (e) irradiation by 808 nm laser for 10 minutes; and the photothermal effect of Pd nanosheets on living cells (f). (Reprinted with permission from Ref. [16]. Copyright 2012 American Chemical Society.)

In a different study, L-DNA-based molecular beacons were introduced into HeLa cells through liposome-mediated transfection. Within the cellular nucleus, these beacons were widely dispersed, displaying heightened fluorescence as the temperature was elevated from 293 to 310 K. The utilization of non-natural L-DNA proved crucial in maintaining stability within the complex cellular environment. This stands in contrast to conventional D-DNA-based beacons, which are susceptible to complete metabolism by the cell nucleus, rendering them ineffective in detecting temperature variations. Furthermore, an innovative approach was developed to enhance the utility of molecular beacons. This involved incorporating multiple types of molecular beacons onto gold nanoparticles, which functioned as quenchers. Through this ingenious multiplexing strategy, researchers achieved a comprehensive temperature sensing capability. Gold nanoparticles, with a diameter of 13 nm, were coupled with various thermostable molecular beacons, resulting in an impressive temperature resolution of less than 273 K within a temperature range of 288–333 K [17].

12.2.3 Metallic nanoclusters

Metal clusters are currently recognized for their role as highly sensitive luminescent nanothermometers, primarily due to their extremely small size (less than 1 nm in diameter), ease of preparation, excellent biocompatibility, and stability in colloidal forms [18]. For instance, studies have indicated that the longevity of gold nanoclusters decreases with rising temperatures during physiological conditions [18]. It's worth noting that the impact of temperature on their longevity is significantly influenced by the surrounding environment. This influence of the environment on temperature-induced changes in lifetime is particularly important and varies when comparing the behaviour of metal nanoclusters in biological tissues to those in buffer solutions. Consequently, precise calibration of these gold nanoclusters within living cells is essential to achieve accurate results. In a study involving HeLa cells, fluorescence lifetime imaging microscopy was employed to measure temperature changes after the endocytosis of gold nanoclusters. Similar potential has been observed with copper nanoclusters, which can serve as effective fluorescent probes for both imaging cellular structures and measuring temperature. The design of copper nanocluster-based fluorescent nanothermometers often incorporates glutathione to enhance the stability of the preparation process [19]. Notably, these copper nanoclusters exhibit impressive quantum yields exceeding 5%. The role of glutathione extends beyond stabilization; it also functions as a protective layer and a reducing agent, preventing the aggregation of copper nanoclusters. Interestingly, as the temperature was raised from 288 to 353 K, a noticeable decrease in the emission intensity from the copper clusters was observed. This underscores the intricate relationship between temperature and the luminescent properties of metal nanoclusters.

12.2.4 Quantum dots

In the context of intracellular nanothermometers, quantum dots offer significant appeal due to their widespread accessibility, excellent photostability, tunable fluorescence, and nanoscale size. Once quantum dots are surface-modified with specific molecules, they exhibit thermally sensitive fluorescence, characterized by changes in both fluorescence emission intensity and wavelength [20]. Additionally, quantum dots demonstrate resilience to acidic pH levels within the range of 5–7, as well as the ability to withstand diverse environmental conditions. These attributes make them indispensable for creating thermometers tailored for cellular systems [21]. Numerous articles in the scientific literature delve into the design and fabrication of cellular system thermometers using quantum dots as efficient fluorescent probes. For instance, researchers have successfully synthesized cadmium selenide quantum dots with diameters less than 12 nm, revealing subtle shifts in emission spectra wavelengths as temperature increases [22]. Notably, the calculated thermal sensitivity of nanothermometers based on these quantum dots is 3.66×10^{-4} nm/K, measured between 297.55 and 316.75 K, closely aligning with established bulk measurements (3.406×10^{-4} nm/K). However, a word of caution is warranted against relying solely on individual cadmium selenide-based quantum dots for absolute temperature determination. Generally, quantum dot-based nanothermometers yield more reproducible and accurate results compared to their conventional counterparts. An interesting recent development involves the utilization of core-shell quantum dots composed of zinc sulphide and cadmium selenide to quantify the thermal sensitivity of the NIH3T3 cell environment, resulting in a determined value of 2.087 nm/K [23]. In contrast, introducing streptavidin to the surface of core-shell quantum dots led to a temperature increase of 274.65 K within NIH3T3 cells following calcium ion infusion. Notably, the observable shifts in photoluminescence spectra caused by endocytosed streptavidin-modified core-shell quantum dots enable the measurement of cellular heat generation in response to hypothermic conditions and Ca^{2+} stress, while also allowing direct observation of temperature variations within cells.

Surface modifications of quantum dots have been shown to have a significant impact on both their photostability and thermal sensitivity. For example, the introduction of thiol-modified cyclodextrin to aqueous cadmium telluride quantum dots resulted in a remarkable increase in thermal sensitivity, reaching a value of 1.025×10^{-3} nm/K. This enhancement was approximately 2.4 times greater compared to quantum dots functionalized with monothiol ligands [24]. Furthermore, this modification played a crucial role in minimizing the variability in quantum dot size when exposed to high temperatures and hydrazine treatment. Another noteworthy study focused on a novel approach involving the concurrent copolymerization of two distinct monomers, namely methacrylic acid and methyl methacrylate. This polymerization strategy yielded a unique polymeric matrix known as poly(methyl methacrylate-comethacrylic acid) [25]. Within this matrix, cadmium selenide-zinc sulphide core-shell quantum dots were effectively embedded. Notably, this composite system exhibited intriguingly reversible behaviour and demonstrated impressive resistance against changes in both ionic strength and pH. In addition to these advancements, cadmium selenide quantum dots were utilized as highly effective nanothermometers, enabling the monitoring of temperature fluctuations within HeLa cells through the implementation of two-photon fluorescence microscopy. Notably, the utilization of two-photon excitation proved to be considerably more sensitive to temperature changes in comparison to single-photon excitation (Figure 12.3) [26]. This innovative approach took full advantage of the unique capabilities of a two-photon fluorescence microscope operating within the near-infrared spectrum. This not only led to a substantial enhancement in spatial resolution but also contributed to a reduction in photodamage to specimens, simultaneously allowing for increased penetration depth.

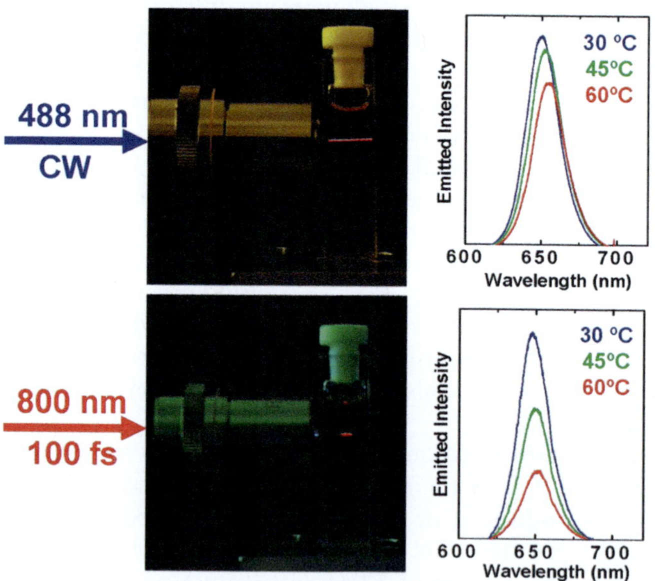

Figure 12.3 (Left side) Digital images of the CdSe-QDs/PBS solution obtained at room temperature under one-photon (top) and two-photon (bottom) optical excitation, respectively. (Right side) Emission spectra of the CdSe-QDs/PBS solution at three different temperatures (30°C, 45°C, and 60°C) as obtained under one-photon (top) and two-photon (bottom) excitation, respectively. (Reprinted with permission from Ref. [26]. Copyright 2010 American Chemical Society.)

12.2.5 Nanodiamonds

A recent study has demonstrated that nanocrystalline materials can accurately measure intracellular temperature [27]. This precision is attributed to nitrogen vacancy colour centers, which are point defects resulting from the displacement of a lattice carbon atom by a nitrogen atom. These colour centres confer thermal sensitivity to nanodiamonds. Fundamentally, a nitrogen vacancy centre possesses a quantum spin state as its electronic ground state, wherein two energy sublevels respond markedly to temperature changes [28]. To exemplify, the utilization of nanothermometers composed of gold nanoparticles and nanodiamonds enabled the monitoring of intracellular temperature in WS1 (embryonic fibroblast) cells. In this innovative approach, gold nanoparticles acted as photo-induced nanoheaters, while nanodiamonds functioned as nanothermometers [27]. Both nanoparticle types exhibit fluorescence and can be excited by a 532 nm YAG laser, allowing for simultaneous imaging through fluorescence microscopy. By applying laser illumination, thermal modulation was induced within the intracellular environment at the gold nanoparticle location. Subsequently, employing a spatial resolution finer than 1 µm facilitated precise temperature quantification in the vicinity of nanodiamonds. In a separate investigation, the impact of laser power on cell viability was examined. Remarkably, under low laser power conditions, heat diffusion from gold nanoparticles to the nanodiamond site led to a cell surface temperature increase of approximately 273.65 K, while cell viability was maintained. Conversely, elevating the laser power by a factor of 10 resulted in a temperature elevation of 277.05 K within cells at the nanodiamond site. Notably, a temperature rise of around 353.15 K near the gold nanoparticles led to cell death, evident from the penetration of a fluorescent marker containing ethidium into the cell membrane. In addition to their enhanced thermal sensitivity, nanodiamonds hold a distinct advantage for nanothermometry. This is attributed to the prevalence of nitrogen vacancy centres as predominantly inner defects, rendering them less susceptible to significant environmental effects. However, despite these favourable characteristics, the lack of documented cellular thermal distribution images can be attributed to the limited fluorescence intensity arising from the relatively small number of nitrogen vacancy centres present in the nanodiamonds.

12.2.6 Polymer-based nanothermometers

In 2009, Gota and his research team published a study that showcased the practical application of fluorescent nanothermometers based on polymeric nanostructures [29]. These nanothermometers had the unique capability to visualize and track thermal changes within cells. The study specifically utilized complex fluorescent nanogels, composed of either synthetic polymers or naturally derived polymers with a strong thermal response. Notably, these nanogels exhibited intriguing behaviour: at temperatures below 303 K, they expanded due to water absorption, resulting in a gradual reduction in fluorescence intensity. However, this effect tapered off at higher temperatures (above 303 K) as the nanogels contracted and released encapsulated molecules. One of the pivotal advantages of these fluorescent nanogels was their proficiency in detecting variations in fluorescence intensity within the critical temperature range of 303–315 K. This range corresponds to the temperature realm in which a multitude of crucial biophysical processes take place. Beyond their performance in test environments, the researchers successfully injected fluorescent nanogels into COS-7 cells, demonstrating their precision in measuring intracellular temperatures (Figure 12.4a–c) [29]. In parallel, another avenue of research emerged involving polymer nanostructures derived from block copolymers, presenting a comparable approach to intracellular temperature assessment. These nanostructures encompassed both a fluorescent complex and a thermoresponsive unit, housed within the block copolymer's architecture. The 303–308 K temperature range induced remarkable fluorescence intensity shifts, mirroring the block copolymer's expansion and contraction behaviour in response to temperature variations [30]. This property was effectively harnessed to determine temperatures within canine kidney cells, showcasing the applicability of these polymer nanostructures. Nonetheless, the application of polymer nanostructures as nanothermometers is marred by a significant drawback – their relatively large size, coupled with a limited affinity for aqueous environments. This combination of attributes fosters intracellular aggregation, leading to an uneven distribution within cells. Consequently, this restricts their effectiveness for comprehensive intracellular thermal imaging.

To address these inherent limitations, researchers devised an innovative strategy: the development of a fluorescent polymer-based nanothermometer designed to facilitate enhanced diffusion within living cells [31]. This novel nanothermometer is constructed from three key components: a fluorescent module, a hydrophilic unit, and a thermosensitive element. Its behaviour is characterized by structural elongation at lower temperatures, causing a reduction in fluorescence lifespan due to quenching by water molecules. In contrast, higher temperatures trigger structural contraction, expelling water molecules and resulting in an extended fluorescence lifespan [31]. Remarkably, the fluorescence lifespan of these polymer-based nanothermometers demonstrates a linear correlation with temperatures within the range of 301–313 K. One notable advantage of these nanothermometers lies in their ability to function in a lifespan mode. This mode effectively circumvents potential inaccuracies stemming from the non-uniform distribution within cells, which can be a concern for other types of nanothermometers. However, it is important to note that the fluorescence lifespan of polymer-based fluorescent nanothermometers is susceptible to environmental factors. This susceptibility to external influences has provided an opportunity for researchers to calibrate fluorescence lifespan images originating

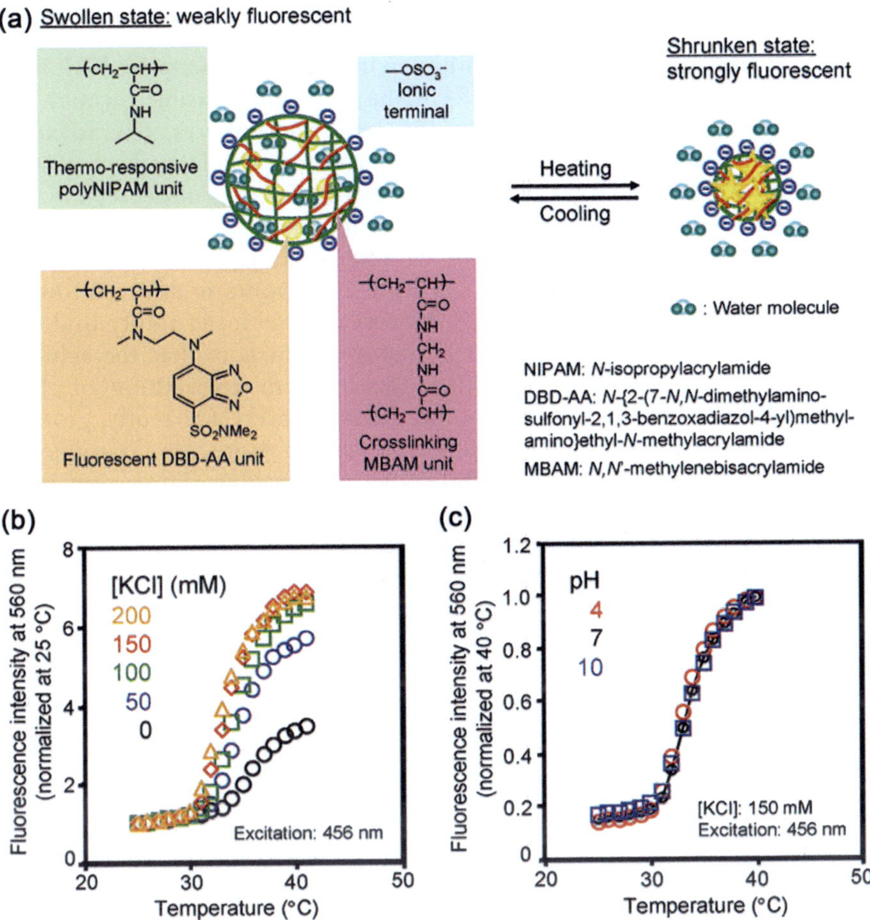

Figure 12.4 Fluorescent nanogel thermometer 1. (a) Schematic diagram and chemical structures of the components. At a lower temperature, 1 swells by absorbing water into its interior, where the water-sensitive DBD-AA units are quenched by the neighbouring water molecules. When heated, 1 shrinks with the release of water molecules, resulting in fluorescence from the DBD-AA units. (b) Fluorescence responses to temperature variation in water and in KCl solutions. (c) Insensitivity to pH variation. HCl and KOH were used for the pH adjustments. The fluorescence quantum yield was 0.37 at 40°C. (Reprinted with permission from Ref. [29]. Copyright 2009 American Chemical Society.)

from tumour cells that are equipped with fluorescent polymeric nanothermometers. This calibration process enables the creation of detailed intracellular thermal maps, contributing to a deeper understanding of cellular thermal dynamics.

12.3 CONCLUSION AND FUTURE PERSPECTIVES

Thermal imaging and sensing within living cells, characterized by temperature resolutions lower than 274 K, have become feasible through the utilization of organic molecules and nanoscale fluorophores. This advancement enables the monitoring of cellular processes involving heat generation and transfer between the cell and its environment. Luminescent nano/micro-thermometers, when introduced into living cells, enable non-invasive remote temperature measurements without significantly disrupting cell dynamics. The ongoing focus of scientific endeavours is directed toward the development of novel nanoscale fluorophores with the ability to detect sub-nanometer thermal signals. It is important to note that the achievement of thermal imaging hinges on the uniform distribution of these nanoscale fluorophores throughout the cellular matrix. Currently, proteins and polymers have been employed as sources for nanoscale fluorophores, contributing to successful intracellular thermal imaging. These constructs have demonstrated a remarkable degree of thermal resolution within living cells. However, challenges persist in terms of stability and thermal sensitivity during the integration of nanoscale fluorophores into cellular environments, impeding their broader application. In contrast, fluorescent nanoparticles present an appealing alternative due to their resilience to environmental conditions and sustained stability over time. Nevertheless, achieving uniform distribution of these nanoparticles within cells remains an ongoing challenge, currently restricting their application to localized intracellular temperature measurements and estimation of mean cell temperature. The trajectory of future research in cellular thermal imaging is expected to centre around the development of novel materials with intrinsic thermal responsiveness independent of external conditions. Furthermore, the exploration of innovative methodologies and scientific approaches for intracellular nanoparticle manipulation is poised to attract considerable interest. These advancements could enable real-time, three-dimensional temperature profiling of living cells using fluorescent nanothermometers. Notably, the achievement of time-dependent temperature sensing for effective photothermal therapy within living cells may necessitate systems capable of concurrent heating and temperature measurement, a concept under exploration involving rare earth-doped nanoparticles, nanodiamonds, and metallic nanoparticles. Nanothermometry systems hold significant promise in addressing unresolved biological inquiries. However, despite the emerging nature of nanothermometry utilizing fluorescent molecules, substantial work remains to enhance our current understanding. Many researchers are actively engaged in pioneering approaches to serve as intracellular fluorescent nanothermometers, with the anticipation that their efforts will culminate in a system capable of achieving temperature measurements within cells at a precision of 273.25 K, resistant to potential physical or chemical interference. While these aspirations hold potential, the present state of research remains in an early stage. Complexities surrounding the reciprocal interactions between cells and fluorescent thermometers present challenges in confidently addressing biological concerns. As research progresses, it becomes imperative to establish the optimal operational range for nanothermometers, considering the minimal temperature fluctuations inherent to most biochemical processes within cells. Unregulated heating resources utilized during measurements may obscure temperature variations arising from physiological processes. A significant source of unregulated heating arises from the persistent absorption of cellular components at excitation or emission wavelengths. To mitigate this issue, the application of fluorescent nanothermometers with a spectral operating range corresponding to wavelengths where water absorption is reduced is recommended. In practice, fluorescent nanothermometers operating within the first biological window of 750–920 nm offer a viable solution to counteract unwanted heating, enhancing the resolution of fluorescent thermal imaging. Conducting experiments within this spectral window necessitates the development of innovative instruments, including advanced microscopes and enhanced infrared-sensitive fluorescent nanothermometers. Such advancements hold the potential to facilitate real-time thermal monitoring and imaging, particularly in the context of livestock research.

REFERENCES

1. K. Okabe, R. Sakaguchi, B. Shi, S. Kiyonaka, Intracellular thermometry with fluorescent sensors for thermal biology, *Pflügers Arch. Eur. J. Physiol.* 470 (2018) 717–731. https://doi.org/10.1007/s00424-018-2113-4.
2. C. Immerz, M. Schweins, P. Trinke, B. Bensmann, M. Paidar, T. Bystroň, K. Bouzek, R. Hanke-Rauschenbach, Experimental characterization of inhomogeneity in current density and temperature distribution along a single-channel PEM water electrolysis cell, *Electrochim. Acta.* 260 (2018) 582–588. https://doi.org/10.1016/j.electacta.2017.12.087.
3. T. Kondo, W.J. Chen, G.S. Schlau-Cohen, Single-molecule fluorescence spectroscopy of photosynthetic systems, *Chem. Rev.* 117 (2017) 860–898. https://doi.org/10.1021/acs.chemrev.6b00195.
4. S. Howell, M. Dakanali, E.A. Theodorakis, M.A. Haidekker, Intrinsic and extrinsic temperature-dependency of viscosity-sensitive fluorescent molecular rotors, *J. Fluoresc.* 22 (2012) 457–465. https://doi.org/10.1007/s10895-011-0979-z.
5. H. Gao, C. Kam, T.Y. Chou, M.-Y. Wu, X. Zhao, S. Chen, A simple yet effective AIE-based fluorescent nano-thermometer for temperature mapping in living cells using fluorescence lifetime imaging microscopy, *Nanoscale Horizons.* 5 (2020) 488–494. https://doi.org/10.1039/C9NH00693A.
6. C.D.S. Brites, P.P. Lima, N.J.O. Silva, A. Millán, V.S. Amaral, F. Palacio, L.D. Carlos, Thermometry at the nanoscale, *Nanoscale.* 4 (2012) 4799–4829. https://doi.org/10.1039/C2NR30663H.

7. D. Jaque, F. Vetrone, Luminescence nanothermometry, *Nanoscale.* 4 (2012) 4301–4326. https://doi.org/10.1039/C2NR30764B.

8. J. Solé, L. Bausa, D. Jaque, *An introduction to the optical spectroscopy of inorganic solids*, John Wiley & Sons, 2005.

9. L.M. Maestro, E.M. Rodriguez, F. Vetrone, R. Naccache, H.L. Ramirez, D. Jaque, J.A. Capobianco, J.G. Solé, Nanoparticles for highly efficient multi-photon fluorescence bioimaging, *Opt. Express.* 18 (2010) 23544–23553. https://doi.org/10.1364/OE.18.023544.

10. M. Suzuki, V. Tseeb, K. Oyama, S. Ishiwata, Microscopic detection of thermogenesis in a single HeLa cell, *Biophys. J.* 92 (2007) L46–L48.

11. M. Haase, H. Schäfer, Upconverting nanoparticles, *Angew. Chemie Int. Ed.* 50 (2011) 5808–5829. https://doi.org/10.1002/anie.201005159.

12. D. Jaque, L.M. Maestro, E. Escudero, E.M. Rodríguez, J.A. Capobianco, F. Vetrone, A. Juarranz de la Fuente, F. Sanz-Rodríguez, M.C. Iglesias-de la Cruz, C. Jacinto, U. Rocha, J. García Solé, Fluorescent nano-particles for multi-photon thermal sensing, *J. Lumin.* 133 (2013) 249–253. https://doi.org/10.1016/j.jlumin.2011.12.022.

13. J.S. Donner, S.A. Thompson, C. Alonso-Ortega, J. Morales, L.G. Rico, S.I.C.O. Santos, R. Quidant, Imaging of plasmonic heating in a living organism, *ACS Nano.* 7 (2013) 8666–8672. https://doi.org/10.1021/nn403659n.

14. J.S. Donner, S.A. Thompson, M.P. Kreuzer, G. Baffou, R. Quidant, Mapping intracellular temperature using green fluorescent protein, *Nano Lett.* 12 (2012) 2107–2111. https://doi.org/10.1021/nl300389y.

15. S. Kiyonaka, T. Kajimoto, R. Sakaguchi, D. Shinmi, M. Omatsu-Kanbe, H. Matsuura, H. Imamura, T. Yoshizaki, I. Hamachi, T. Morii, Y. Mori, Genetically encoded fluorescent thermosensors visualize subcellular thermo-regulation in living cells, *Nat. Methods.* 10 (2013) 1232–1238. https://doi.org/10.1038/nmeth.2690.

16. G. Ke, C. Wang, Y. Ge, N. Zheng, Z. Zhu, C.J. Yang, l-DNA molecular beacon: A safe, stable, and accurate intracellular nano-thermometer for temperature sensing in living cells, *J. Am. Chem. Soc.* 134 (2012) 18908–18911. https://doi.org/10.1021/ja3082439.

17. S. Ebrahimi, Y. Akhlaghi, M. Kompany-Zareh, Å. Rinnan, Nucleic acid based fluorescent nanothermometers, *ACS Nano.* 8 (2014) 10372–10382. https://doi.org/10.1021/nn5036944.

18. L. Shang, F. Stockmar, N. Azadfar, G.U. Nienhaus, Intracellular thermometry by using fluorescent gold nanoclusters, *Angew. Chemie Int. Ed.* 52 (2013) 11154–11157. https://doi.org/10.1002/anie.201306366.

19. C. Wang, L. Ling, Y. Yao, Q. Song, One-step synthesis of fluorescent smart thermo-responsive copper clusters: A potential nanothermometer in living cells, *Nano Res.* 8 (2015) 1975–1986. https://doi.org/10.1007/s12274-015-0707-0.

20. J. Lee, N.A. Kotov, Thermometer design at the nanoscale, *Nano Today.* 2 (2007) 48–51. https://doi.org/10.1016/S1748-0132(07)70019-1.

21. A.M. Smith, H. Duan, A.M. Mohs, S. Nie, Bioconjugated quantum dots for in vivo molecular and cellular imaging, *Adv. Drug Deliv. Rev.* 60 (2008) 1226–1240. https://doi.org/10.1016/j.addr.2008.03.015.

22. S. Li, K. Zhang, J.-M. Yang, L. Lin, H. Yang, Single quantum dots as local temperature markers, *Nano Lett.* 7 (2007) 3102–3105. https://doi.org/10.1021/nl071606p.

23. J.-M. Yang, H. Yang, L. Lin, Quantum dot nano thermometers reveal heterogeneous local thermogenesis in living cells, *ACS Nano.* 5 (2011) 5067–5071. https://doi.org/10.1021/nn201142f.

24. D. Zhou, M. Lin, X. Liu, J. Li, Z. Chen, D. Yao, H. Sun, H. Zhang, B. Yang, Conducting the temperature-dependent conformational change of macrocyclic compounds to the lattice dilation of quantum dots for achieving an ultrasensitive nanothermometer, *ACS Nano.* 7 (2013) 2273–2283. https://doi.org/10.1021/nn305423p.

25. H. Liu, Y. Fan, J. Wang, Z. Song, H. Shi, R. Han, Y. Sha, Y. Jiang, Intracellular temperature sensing: An ultra-bright luminescent nanothermometer with non-sensitivity to pH and ionic strength, *Sci. Rep.* 5 (2015) 14879. https://doi.org/10.1038/srep14879.

26. L.M. Maestro, E.M. Rodríguez, F.S. Rodríguez, M.C.I. la Cruz, A. Juarranz, R. Naccache, F. Vetrone, D. Jaque, J.A. Capobianco, J.G. Solé, CdSe quantum dots for two-photon fluorescence thermal imaging, *Nano Lett.* 10 (2010) 5109–5115. https://doi.org/10.1021/nl1036098.

27. G. Kucsko, P.C. Maurer, N.Y. Yao, M. Kubo, H.J. Noh, P.K. Lo, H. Park, M.D. Lukin, Nanometre-scale thermometry in a living cell, *Nature.* 500 (2013) 54–58. https://doi.org/10.1038/nature12373.

28. Y. Sonnefraud, A. Cuche, O. Faklaris, J.-P. Boudou, T. Sauvage, J.-F. Roch, F. Treussart, S. Huant, Diamond nanocrystals hosting single nitrogen-vacancy color centers sorted by photon-correlation near-field microscopy, *Opt. Lett.* 33 (2008) 611–613. https://doi.org/10.1364/OL.33.000611.

29. C. Gota, K. Okabe, T. Funatsu, Y. Harada, S. Uchiyama, Hydrophilic fluorescent nanogel thermometer for intracellular thermometry, *J. Am. Chem. Soc.* 131 (2009) 2766–2767. https://doi.org/10.1021/ja807714j.

30. J. Qiao, L. Qi, Y. Shen, L. Zhao, C. Qi, D. Shangguan, L. Mao, Y. Chen, Thermal responsive fluorescent block copolymer for intracellular temperature sensing, *J. Mater. Chem.* 22 (2012) 11543–11549. https://doi.org/10.1039/C2JM31093G.

31. K. Okabe, N. Inada, C. Gota, Y. Harada, T. Funatsu, S. Uchiyama, Intracellular temperature mapping with a fluorescent polymeric thermometer and fluorescence lifetime imaging microscopy, *Nat. Commun.* 3 (2012) 705. https://doi.org/10.1038/ncomms1714.

Chapter 13

Flexible Temperature Sensor

*Saurav Kumar Maity, Uplabdhi Tyagi,
and Gulshan Kumar*

13.1 INTRODUCTION

Temperature is regarded as one of the most fundamental physical parameters in daily life and all branches of natural science [1]. The thermometers evolved in parallel to the technological demands of industry and society. The very first temperature measurement device, i.e. thermoscope, was developed during the 16th and 17th centuries in Europe, essentially by Galileo Galilei, Santorio, Amontons, Fahrenheit, and Celsius [2]. Nowadays, the domains of metrology, aerodynamics, meteorology, oceanography, chemistry, biology, pharmacy, and many more fields require the monitoring of temperature changes (Figure 13.1) [3]. Thus, the development of a temperature sensor plays a very crucial role in scientific research and technological

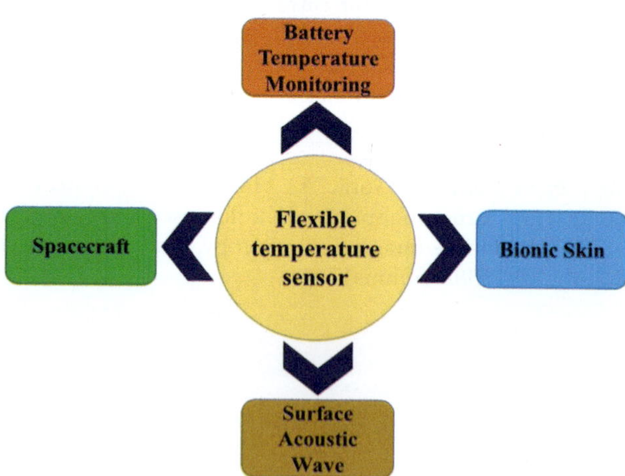

Figure 13.1 Applications of flexible temperature sensor in diverse areas. (Adopted from Ref. [3], Copyright © 2021 Liu, He, Cao, Sun, Zhu, and Li, an open-access article under the Creative Commons Attribution License (CC BY).)

DOI: 10.1201/9781032661537-13

advancements. In contrast, recently, flexible and stretchable devices have drawn tremendous interest owing to its compressibility, bendability, and compatibility with irregular and curvilinear surfaces [4,5]. These flexible and stretchable devices offer several advantages including mechanical resilience, biocompatibility, multifunctionality, and comfort compared to traditional rigid sensors. Thus, next-generation sensors can be prepared by combining the advantageous property of flexible and stretchable with a soft technology. Depending on the type of application, different flexible sensors offer different properties. An ideal flexible temperature sensor must be quick and effective in identifying target signals required in various applications such as in wearable health devices, medical monitoring, industrial and manufacturing, HVAC systems (heating, ventilation, and air conditioning system), food processing and storage, environmental monitoring, energy management [6,7]. At present, temperature sensors account for approximately 80% of the worldwide sensor market, which is expected to reach around $6.86 billion by the end of 2023 (Figure 13.2). The ability of a human skin to sense a particular temperature is very crucial for aiding in the diagnosis of illnesses, preventing mishaps, and providing information about the surroundings. Any deviation from the normal temperature range of the body has been recognized as a sign of disease. When the typical

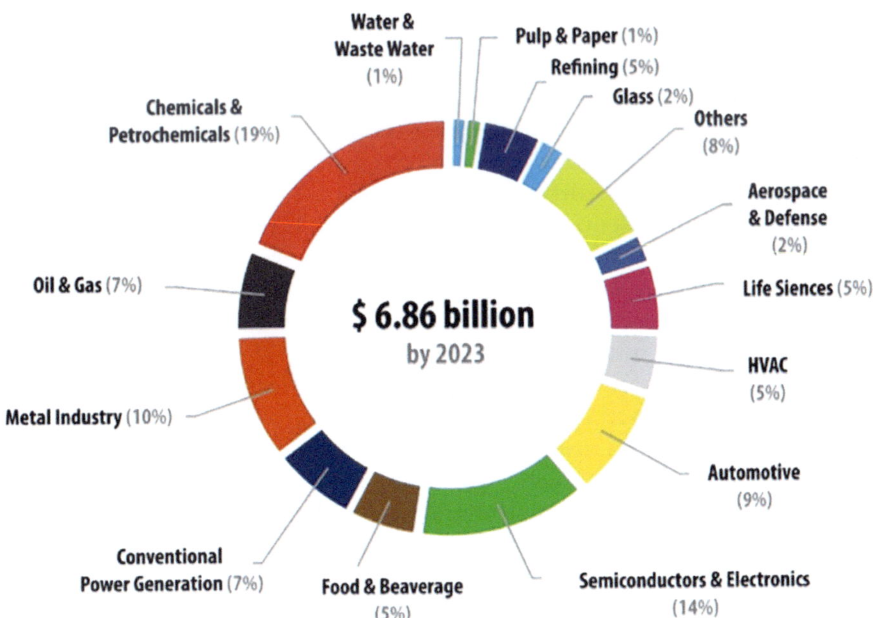

Figure 13.2 Revenue generation for the global temperature sensors market. (Reproduced with permission from Ref. [17], © 2018, John Willey & Sons.)

temperature range is exceeded or decreased, a person may not be in a completely healthy state. The steady body temperature of an adult typically falls between 37°C and 37.5°C [8]. The temperature sensors must be sensitive to the actual temperature of human skin and temperature variations (within 0.1°C from 37°C to 39°C and 0.2°C below 37°C and above 39°C). Several temperature-dependent physical features such as electric conductance, electric potential, and volume, have led to the invention of numerous types of thermometers. These thermometers consist of thermocouples, thermistors, bimetallic thermometers, and glass thermometers filled with liquid [9]. Typically, hard temperature detectors like thermometers can be effectively used to determine the temperature of the skin. These conventional temperature sensors are referred to as contact thermometers because the heat transfer process requires equilibrium between the sample and the thermal probe in order to measure temperature. However, a key issue is often encountered in generating pressure points against the skin using conventional stiff sensor in order to monitor conditions in an enclosed setting (such as within the garment or prosthetic parts). Additionally, this method is not suited for kids because thermometers cannot be conveniently placed under the arm as desired. Since a direct body contact is highly preferred for accurate temperature measurements with high precision. Thus, traditional rigid temperature detectors are unable to make conformal contact with irregular surfaces of skin and body [10]. Additionally, conventional thermometers are insufficient for measuring temperatures in fast-moving objects or at micro scale (microfluidics, microcircuits, or inside the cells) because of their contact nature.

Furthermore, the conventional thermocouples and thermistors-based temperature sensors need an electrical connection in the sensor system which prevents their application in high electromagnetic noise environment, dangerous sparks, and corrosive environments. Therefore, the solution to this key issue is to essentially develop advanced flexible and wearable temperature sensors. To meet the wearable criteria, sensors with distinctive properties such as soft, flexible, biocompatible, lightweight, strong, and pain-free must be created. Recently, researchers are motivated to prepare novel, flexible, non-invasive thermometers that provide accurate and precise measurements and can function even in powerful electromagnetic fields due to the above-mentioned inherent limitations of conventional thermometers.

Thus, recently, a number of optical techniques have been investigated for application in thermometry approaches, including Raman scattering, thermal reflection, infrared thermography, and luminescence. Out of all the optical methods, luminescence-based thermometry has been recognized as a possible alternative owing to its quick response time, and great sensitivity [11,12]. Recently, an active field has been dedicated to the manufacture and development of highly accurate and sensitive light thermometers.

Luminescent-based flexible thermometers are temperature sensors that utilize luminescent materials to measure temperature and can be incorporated into flexible or bendable substrates. These thermometers are designed to provide a visual indication of temperature through the variation in the emitted light. They offer several advantages, including flexibility, robustness, and the potential for integration into various applications such as wearable devices, medical sensors, and in environmental sensing. Several luminescent materials such as organic dyes [12], lanthanide-doped phosphors [13], quantum dots (QDs) [14], metal complexes or clusters [15], and metal-organic frameworks (MOFs) [16] are widely utilized for the preparation of luminescent thermometer.

Thus, we have presented a detailed and dedicated chapter on luminescent-based flexible temperature sensors. Firstly, we have listed several types of flexible luminescent temperature sensors. Different materials used in the fabrication of flexible luminescent temperature sensors are subsequently summarized in Section 13.2. In Section 13.3, the general fabrication methods for different types of flexible luminescent thermometer are discussed. Lastly, Section 13.6 includes the conclusion and details on the flexible reliability and challenges of luminescent-based temperature sensor.

13.2 LUMINESCENT THERMOMETRY TECHNIQUE

Luminescence is the term used to describe the ability of a substance to emit light. It occurs in electronically stimulated states that are supplied with excitation from an outside source. The relationship between luminescence properties and temperature is used in luminescence thermometry. This aids in the temperature sensing and mapping process by doing a spectrum and spatial analysis of the luminescence produced by the object. The Boltzmann distribution of electrons inside the excited state level causes the luminescence of all materials which tends to be temperature-dependent. If a material in which the only low-temperature transitions are radiative from the excited state's lowest level to the ground state, Nonradiative transitions are however possible at higher temperature because thermal energy will excite the electrons inside the excited state to several vibrational modes that are overlapping at different energy levels. The difference in energy between the bottom of the excited state and the overlap point to a potential nonradiative decay state (E) is proportional to $\exp(-\Delta E/kT)$, where k is the Boltzmann constant and T represents temperature [18]. This relationship states that as temperature rises, the nonradiative decay rate increases, which reduces luminescence duration and intensity. Nevertheless, most materials exhibit more nuanced temperature-dependent luminescence due to their more complex electronic band structure, which consists of several excited bands with different overlap energies.

13.2.1 Types of Luminescent thermometry

Temperature-dependent luminescence features include shifts in the positions and forms of the spectrum, adjustments to the luminescence intensity or decay duration, and adjustments to anisotropy. Luminous thermometry is frequently divided into four categories viz. intensity-based, lifetime-based, wavelength-based, and ratiometric method-based depending on these variable parameters.

13.2.1.1 Luminescence thermometry based on intensity

This technique detects the temperature by analyzing the luminescence intensity. The heat activation of nonradiative deactivation mechanisms is generally responsible for the temperature-induced variations in luminescence intensity. For example, a number of organic dyes, including fluorescein, isothiocyanates, rhodamine, and their derivatives, have become very popular in luminescence thermometry in microfluidic devices or cells because of the severe thermal quenching at high temperatures. Although fluctuations in probe concentration, excitation power, or detection efficacy can often lead to mistakes in intensity-based luminescence thermometry, temperature-related changes in luminescence intensity are easier to detect. Indeed, even in cases where all experimental parameters, including the concentration of luminescence centers, the excitation wavelength, and the strength of the light source, are held constant throughout the tests, the absorption and scatter cross-section could differ between samples. These shortcomings decrease the precision, making this approach unsuitable for accurate temperature sensing [19].

13.2.1.2 Lifetime-based luminescence thermometry

The fundamentals behind lifetime-based thermometry are that lifespan is temperature-dependent. The time period during which the luminescence intensity decreases to 1/e of its starting value following a pulsed excitation is known as the lifetime of luminescence. Lifetime value is unaffected by probe concentration, size, and geometry in contrast to luminescence intensity. Additionally, it is unaffected by light reflection, light scattering, or the excitation source's intensity variations. The time-consuming computational treatment and more complex read-out are needed for lifetime determination, though. Additionally, lifetime-based luminescence thermometry is not much practical for measuring large-area gradients in temperature and is ineffective in obtaining dynamic measurements of temperature variations that take place at intervals shorter than or equal to the light lifespan [20].

13.2.1.3 Wavelength-based luminescence thermometry

In this particular method, a correlation between the temperature values and the wavelength (or frequency) shift of luminescence spectra is usually done to detect the temperature. An illustration of a triaryl boron-based fluorescent molecule (dipyren-1-yl(2,4,6-triisopropylphenyl) borane (DPTB)) shows a temperature-dependent luminescence wavelength shift within the temperature range of −50°C to 100°C. This is because the temperature

Table 13.1 Quantum yields Φ of DPTB in MOE at different temperatures

T (°C)	−50	−30	−10	10	30	50	70	90	100
Φ	0.64	0.68	0.73	0.78	0.80	0.83	0.81	0.71	0.67

Source: Reproduced with permission from Ref. [21] © 2011, John Willey & Sons.

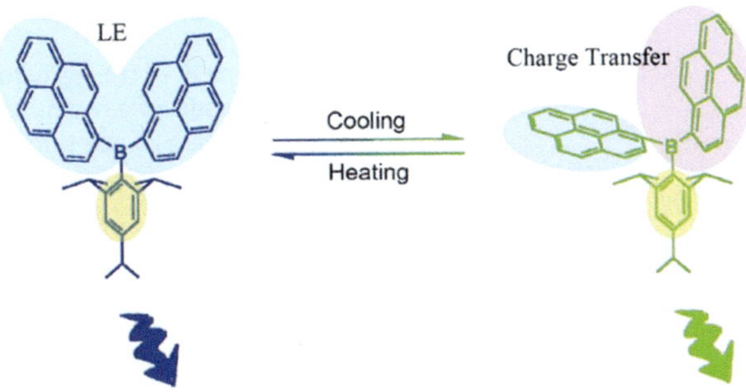

Figure 13.3 Mechanism of emission changes with temperature. (Reproduced with permission from Ref. [21] © 2011, John Willey & Sons.)

has a significant impact on the dynamic equilibrium between the twisted intramolecular charge transfer (TICT) excited state emission and the local excited state emission (LE). DPTB shows temperature-dependent green to blue luminescence with a very high quantum yield (Table 13.1) over a wide temperature range (−50°C to +100°C) [21]. Figure 13.3 shows the mechanism of emission changes with temperature. The major drawback of this method is the broad emission band which makes it very challenging to precisely identify the maximum luminescence wavelength [22].

13.2.1.4 Ratiometric luminescence thermometry

To measure the temperature, this approach uses the intensity ratio of two distinct transitions from the same materials, each of which has a luminescence spectrum with a number of bands. One major limitation of intensity-based measurements of a single transition is that temperature can be sensed independently of the concentration and heterogeneity of luminous centers as well as the optoelectronic drifts of the excitation source and detectors. This can be achieved through ratiometric thermometry. The major advantage of this method is that this self-referenced thermometry technique can ensure an exact and absolute temperature measurement because there is no requirement for calibration in this method. Thus, it is obvious that the ratiometric luminescence thermometry method (as opposed to the other approaches mentioned earlier) is the best way to measure temperature [23].

Table 13.2 Parameters used to evaluate the performance of a thermometer [24–26]

Parameters	Definition	Unit	Physical meaning
Relative thermal sensitivity	$S_r = 1/\Delta * \left\lvert \dfrac{\partial \Delta}{\partial T} \right\rvert$	% K^{-1}	Relative change on the thermometric parameter per unit of temperature change
Temperature uncertainty	$\partial T = \dfrac{1}{S_r}\dfrac{\partial \Delta}{\Delta}$ or SD	K	Minimum temperature value resolved by the thermometer
Accuracy	MSD or MUD	K	Closeness of the agreement between the result of a measurement and a true value of the temperature
Spatial resolution	$\partial X = \dfrac{\partial T}{\lvert \nabla_x T \rvert}$	M	Distance between points whose temperature difference is lower than δT or SD
Temporal resolution	$\partial t = \dfrac{\partial T}{\lvert \partial T / \partial t \rvert}$	S	Time interval between instants whose temperature difference is lower than δT or SD
Repeatability	$R = 1 - \dfrac{\max(\Delta_c - \Delta)}{\Delta_c}$	%	Maximum difference on Δ relatively to the mean value (Δc) at given conditions
Reproducibility	NA	NA	Ability to reproduce the same calibration curve under distinct conditions (equipment, operator, etc.)

MSD, mean signed deviations; MUD, mean unsigned (absolute) deviations; SD = standard deviation.

The performance of ratiometric luminescent thermometers is quantified using the relative thermal sensitivity (S_r), the temperature uncertainty (δT), or thermal precision, accuracy, spatial and temporal resolution, repeatability or test-retest reliability, and in terms of reproducibility. The physical meaning and mathematical expression of these parameters are mentioned in Table 13.2.

13.3 TYPES OF MATERIALS USED IN LUMINESCENT THERMOMETER

Luminescent-based flexible thermometers are advanced temperature sensors that rely on the luminescence properties of certain materials to indicate temperature changes. These materials exhibit changes in their luminescent characteristics, such as fluorescence or phosphorescence, in response to temperature fluctuations. These materials can be integrated into flexible substrates or coatings to create luminescent-based thermometers. The choice of material depends on several factors such as the desired temperature range, sensitivity, and the specific application requirements. These thermometers are particularly useful in applications where conventional thermometers might not be practical due to their rigidity or size limitations. Some of the materials commonly used in luminescent-based flexible thermometers are discussed below:

13.3.1 Organic dyes-based flexible fluorescent thermometer

An organic dye-based luminescent thermometer is a temperature-sensing device that utilizes the temperature-dependent luminescence properties of certain organic dyes to measure and indicate temperature changes [27]. These types of thermometers are designed to be flexible for various purposes where conventional thermometer might not be practical such as in wearable health devices, environmental sensing, and biomedical applications. They are also used in research and industrial applications where accuracy and sensitivity are highly important. The working principle of organic dye-based flexible luminescent thermometer is mentioned below:

- **Organic Dye**: The key component of this thermometer is an organic dye that exhibits temperature-dependent luminescence. These dyes emit light when excited by a light source, and the intensity or wavelength of the emitted light changes with temperature.
- **Temperature Sensing**: When the organic dye is exposed to a temperature change, its luminescent properties change. This change can be measured and correlated with temperature. Typically, the dye's emission spectrum or intensity shifts in a predictable manner with temperature variation.
- **Flexible Substrate**: The thermometer is constructed on a flexible substrate, often made of materials like polymers or elastomers. This flexibility allows the device to conform to curved or irregular surfaces, making it suitable for various applications, including wearable technology and medical devices.
- **Readout System**: The luminescent signals from the organic dye are detected and processed by a readout system. This can be a photodetector or a camera equipped with the necessary filters to capture the specific wavelength of light emitted by the dye.
- **Calibration**: To ensure accuracy, the thermometer must be calibrated to account for any variations in the luminescent properties of the organic dye and the flexible substrate. Calibration is typically done by subjecting the device to known temperatures and recording its luminescent response.

Advantages of organic dye-based luminescent thermometers include their high sensitivity and potential for accuracy, especially in a narrow temperature range. They are also non-contact and can be used in situations where direct contact with the object being measured is not possible or desirable. However, they can be sensitive to factors such as environmental conditions and the stability of the dye used, which may require regular calibration and maintenance. The choice of the organic dye and the specific luminescent properties can be tailored to the application's temperature range and sensitivity requirements.

13.3.2 Lanthanide-doped phosphors-based flexible luminescent thermometer

Lanthanide-doped phosphors are common materials used in luminescent thermometers. These materials are valued for their ability to emit visible or near-infrared light when exposed to specific excitation sources, and the temperature can be determined by measuring the intensity or spectral characteristics of the emitted light [28]. This property makes them useful in various temperature-sensing applications. The working process of lanthanide-doped phosphors flexible luminescent thermometers is as follows:

- **Doping with Lanthanide Ions**: Lanthanide ions, such as Europium (Eu), Dysprosium (Dy), and Terbium (Tb) are incorporated into the crystal structure of the phosphor material. After that, the lanthanide-doped phosphor is embedded in or coated on a flexible substrate (e.g., polymer film) to create a flexible thermometer. These ions have energy levels with distinct electronic transitions, which are responsible for luminescent properties.
- **Excitation**: To measure temperature, you need to excite the phosphor material by providing it with energy. This energy can be in the form of heat, light, or electrical stimulation. Typically, these phosphors are excited by shining a light source, like a laser or LED, onto them.
- **Luminescent Emission**: When the phosphor absorbs energy, electrons in the lanthanide ions become excited and move to higher energy levels. As they return to their ground state, they emit photons of characteristic wavelengths. These photons can be in the visible or near-infrared spectrum.
- **Temperature Dependence**: The intensity or spectral characteristics of the emitted light are temperature-dependent. In other words, as the temperature of the phosphor changes, so does the luminescent emission. This temperature dependence is due to the temperature-sensitive energy levels and interactions within the crystal lattice.
- **Temperature Measurement**: By measuring the intensity or spectrum of the emitted light, you can determine the temperature of the phosphor. This is often done using a photodetector or a spectrometer to analyze the emitted light.
- **Calibration**: To use a lanthanide-doped phosphor as a thermometer, you must calibrate it. This involves determining the relationship between the luminescent output and temperature. Calibration is necessary because the temperature sensitivity can vary between different materials and dopant concentrations.

Lanthanide-doped phosphors are advantageous in luminescent thermometry because they offer several benefits such as they can operate over a wide temperature range, have good stability and repeatability, are resistant to electromagnetic interference, can be used in harsh environments, and provide a non-contact and non-invasive way to measure temperature.

These materials find applications in various fields, including materials science, biology, and engineering, where precise temperature monitoring is essential. The choice of the specific lanthanide-doped phosphor and the calibration process may vary depending on the application's requirements.

13.3.3 QDs-based flexible luminescent thermometer

A QDs-based flexible luminescent thermometer is a type of temperature sensor that uses QDs as the sensing element to measure temperature. QDs are nanoscale semiconductor particles that exhibit unique optical properties, including size-tunable emission wavelengths [29]. These properties make them suitable for various applications such as flexible electronics and environmental sensing. The working principle of QDs-based flexible luminescent thermometer is explained below:

- **Quantum Dots**: QDs are tiny semiconductor particles typically made of materials like cadmium selenide (CdSe), lead sulfide (PbS), or indium arsenide (InAs). These materials are chosen because the energy gap between their valence and conduction bands determines the emission wavelength, which changes with temperature.
- **Emission Spectrum**: When QDs are excited with light (e.g., ultraviolet or visible light), they emit light in a specific wavelength range. This emission spectrum is sensitive to temperature changes. As temperature increases, the energy gap between the QD's valence and conduction bands changes, causing a shift in the emitted wavelength.
- **Temperature Calibration**: To use QDs as a thermometer, they need to be calibrated. This calibration involves exposing the QDs to a known temperature and measuring the resulting emission wavelength or intensity. A calibration curve is then generated to correlate the temperature with the QD emission.
- **Temperature Sensing**: In practical applications, the QDs are incorporated into a material or placed at the location where temperature needs to be measured. When the temperature at that location changes, the emission wavelength or intensity of the QD changes accordingly. By monitoring this change, the thermometer can determine the temperature.

Some of the advantages of QDs-based luminescent thermometers include high sensitivity since QDs can exhibit high sensitivity to temperature changes, making them suitable for precise measurements. They can be operated across wide temperature range, depending on the choice of QDs.

They can be easily miniaturized because QDs are nanoscale materials, allowing for the development of compact and miniature temperature sensors. These flexible thermometers provide non-contact sensing as QDs can be integrated into non-contact temperature measurement devices, which is useful in various applications.

However, there are also some challenges associated with QDs-based flexible luminescent thermometers which include the potential toxicity of certain QD materials (e.g., CdSe) which can limit their use in biological applications. Researchers are actively working on addressing these issues and developing safer and more efficient QD-based sensors.

13.3.4 Metal cluster-based flexible luminescent thermometer

A metal cluster-based luminescent thermometer is a type of temperature-sensing device that utilizes the temperature-dependent luminescence properties of metal clusters or nanoparticles to measure temperature [30]. These thermometers are commonly used in various scientific and industrial applications, where precise temperature monitoring is essential. Some of the key features and working of metal cluster-based flexible luminescent thermometer are as follows:

- **Luminescence Properties:** Metal clusters or nanoparticles exhibit luminescent properties, which means they can emit light when exposed to certain external stimuli, such as heat or light. In the context of thermometry, these clusters are engineered to emit light in a temperature-dependent manner.
- **Temperature-Dependent Luminescence:** The luminescence properties of metal clusters are influenced by temperature. As the temperature changes, the energy levels within the metal clusters also change, leading to shifts in their emission spectra. These changes in luminescence can be used to determine the temperature.
- **Calibration:** To use metal cluster-based luminescent thermometers, they need to be calibrated. This involves establishing a relationship between the temperature and the observed changes in luminescence, typically through a calibration curve or equation. The calibration process ensures accurate temperature measurements.
- **Integration into Flexible Substrate:** Metal clusters are incorporated into a flexible substrate or material, such as a thin film, polymer, or even a flexible microfabricated structure. This flexibility allows the thermometer to conform to various surfaces and environments.
- **Luminescence Measurement:** An energy source, such as a laser or LED, is used to excite the metal clusters, causing them to emit light. A detector or sensor captures this emitted light, and its wavelength, intensity, or other luminescent properties are measured.

- **Temperature Calculation:** The measured luminescent properties are compared to the calibration curve to determine the temperature of the metal clusters, which in turn reflects the temperature of the environment.
- **Sensitivity:** The sensitivity of a metal cluster-based luminescent thermometer depends on the specific type of metal cluster or nanoparticles used. Some metal clusters are highly sensitive to temperature changes, making them suitable for precise temperature measurements.

Metal cluster-based luminescent thermometers have applications in various fields, including materials science, chemistry, and biology. They are particularly useful in situations where traditional thermocouples or resistance thermometers may not be practical or accurate enough. The advantages of metal cluster-based luminescent thermometer such as non-contact temperature measurement, which can be useful in situations where physical contact with the object is not possible or is undesirable. They can also provide high spatial resolution. While metal cluster-based luminescent thermometers offer unique advantages, they may have limitations, such as a limited temperature range or susceptibility to interference from external factors that can affect luminescence, like ambient light or impurities in the material being measured.

In summary, metal cluster-based luminescent thermometers are a specialized type of thermometer that relies on the temperature-dependent luminescence properties of metal clusters or nanoparticles to measure temperature. They are used in various scientific and industrial applications where non-contact and high-sensitivity temperature measurements are required. Calibration is essential to ensure accurate temperature readings, and the choice of metal clusters or nanoparticles can affect the sensitivity and temperature range of the thermometer.

13.3.5 Metal-organic framework-based flexible luminescent thermometer

Metal-organic frameworks (MOFs) have gained significant attention in recent years due to their diverse range of applications, including their use in luminescent thermometry. MOFs are a class of crystalline materials composed of metal ions or clusters coordinated to organic ligands [31]. They offer a high degree of tunability, which makes them suitable for various sensing and imaging applications, including luminescent thermometry. The working principle of MOF-based flexible luminescent thermometer is as follows:

- **Choice of MOF:** The first step is selecting an appropriate MOF with luminescent properties. The choice of metal ions and organic ligands determines the MOF's luminescent characteristics. The structure and composition of the MOF can be tailored to optimize its performance as a luminescent thermometer.

- **Sensing Mechanism:** The luminescent properties of the MOF are sensitive to temperature changes. This sensitivity can be attributed to several factors, including changes in the energy levels of the electronic states of the MOF components or alterations in the MOF's structure due to temperature variations. For instance, the fluorescence intensity or emission wavelength of the MOF may change with temperature.
- **Calibration:** To use the MOF as a thermometer, it's necessary to calibrate the luminescent response with temperature. This is typically done by measuring the luminescence at various known temperatures to establish a calibration curve.
- **Integration into Flexible Substrate:** MOFs are incorporated into a flexible substrate or material, such as a thin film, polymer, or other flexible structures. This flexibility enables the thermometer to conform to various surfaces and environments.
- **Luminescence Measurement:** An excitation source, typically a light-emitting diode (LED) or laser, is used to excite the MOFs, causing them to emit light. A detector or sensor captures the emitted light, and its luminescent properties are measured.
- **Temperature Sensing:** When the MOF is exposed to a particular environment, the luminescent response can be monitored using appropriate optical equipment (e.g., spectrofluorometers or imaging systems). The temperature can be inferred from the luminescent response by referencing the calibration curve.

Some of the advantages of MOF-based luminescent thermometers include high sensitivity because MOFs can exhibit a high sensitivity to temperature changes, making them suitable for precise temperature measurements. They have excellent versatility because MOFs can be customized to target specific temperature ranges or environmental conditions. This flexible luminescent thermometry using MOFs is a non-invasive and non-contact technique, which is particularly useful in applications where direct temperature measurement is not feasible. They can also be used for remote sensing especially hard-to-reach areas or in hazardous environments. These flexible thermometers can also be miniaturized because MOFs can be integrated into small, portable, or even microscale devices for temperature sensing in various applications such as biomedical or environmental monitoring. The sensing manners of luminescent MOFs with some examples of luminescent MOFs sensing and detection toward temperature by different classifications are mentioned in Table 13.3.

Thus, MOF-based flexible luminescent thermometers have shown promise in a range of fields, including materials science, chemistry, and biology. They offer a versatile and adaptable solution for temperature monitoring in diverse settings. Researchers continue to explore new MOF compositions and designs to improve their sensitivity and applicability in different environments.

Table 13.3 List of some MOFs sensing sensitivity toward temperature [32–34]

Metals	MOFs	Sensitivity
Eu^{+3}	Eu (III) L^{1}	283–333 K, sensing resolution of 0.3 K
Tb^{+3}	Tb-THBA	288–338 K
Zn^{+2}	Zn$_2$(TCPE)	Fluorescence intensity decreased
Zn^{+2}	Zn$_3$(TDPAT) (H$_2$O)$_3$	164–276 K
Cd^{+2}	(H$_2$NMe$_2$) [(Cd (TTAA)]·2H$_2$O	77–298 K
Ag^{+}	[(HMTA)Ag$_4$I$_4$]·(HMTA)	120–180 K and 200–300 K
Eu^{+3}/Tb^{+3}	[Eu(btfa)$_3$(MeOH)(bpeta)] [Tb(btfa)$_3$(MeOH)(bpeta)]	10–350 K, sensitivity of 4.9%/K

btfa, 4,4,4-trifluoro-1-phenyl-1,3-butanedione; bpeta, 1,2-bis(4- pyridyl)ethane); H$_3$TTAA, *N,N′, N″*-1,3,-5-triazine-2,4,6-triyltris(4-aminomethylbenzoicacid); HMTA, hexamethylenetetramine); TCPE, tetrakis (4-carboxyphenyl) ethylene); TDPAT, 2,4,6-tris(3,5-dicarboxylphenylamino)-1,3,5-triazine; THBA, tris[(2-hydroxy-benzoyl)-2-aminoethyl] amine).

13.4 FABRICATION TECHNIQUE FOR DIFFERENT FLEXIBLE LUMINESCENT THERMOMETERS

The fabrication techniques for different types of flexible luminescent thermometers can vary depending on the specific luminescent material used (e.g., QDs, metal clusters, MOFs) and the intended application. A suitable flexible substrate, such as a polymer film, that can be easily integrated depending on the application. The flexibility of the substrate is essential for conforming to various surfaces. An appropriate deposition technique can also be used. Depending on the specific luminescent and substrate, some of the common fabrication methods include spin-coating, dip-coating, drop-casting, chemical vapor deposition (CVD), and inkjet printing [8,35].

13.4.1 Organic dye-based luminescent thermometer

Preparation of an organic dye-based flexible luminescent thermometer involves using organic dyes that exhibit temperature-dependent luminescence. The typical fabrication process for organic dye-based flexible substrate involves the choice of an organic luminescent dye (Rhodamine B [36], Rhodamine 6G [37], or others that exhibit temperature-sensitive luminescence [38]), transparent polymer matrix (Polydimethylsiloxane, PMMA), solvent (ethanol, toluene), temperature calibration equipment (thermocouples), an UV–vis spectrophotometer, fluorescence spectrometer, cuvettes or glass slides, syringes and needles, UV light source (UV LED), and spectrometer or photodetector.

The fabrication steps are as follows:

- **Dye Selection:** Firstly, an organic dye is selected that exhibits temperature-dependent luminescence. The choice of dye is greatly depending on the desired temperature range and sensitivity of the thermometer.
- **Matrix Preparation:**
  - In this step, the selected dye is then dissolved in a suitable solvent to make a dye solution. The concentration of the dye should be chosen carefully to ensure a good balance between sensitivity and linearity of the luminescence response to temperature.
  - The dye solution is then mixed with a transparent polymer matrix. This matrix will serve as a host material to disperse the dye molecules evenly.
- **Casting the Thermometer:**
  - Pour the dye-matrix mixture into a mold or onto a glass slide, depending on the desired thermometer shape and size.
  - The mixture is allowed to solidify or dry (depending on the choice of the matrix material).
- **Calibration:**
  - The luminescence emission spectra of the fabricated thermometer at different temperatures using a fluorescence spectrometer are then measured. The thermometer must be in contact with a controlled heat source.
  - Simultaneously, the temperature can be measured using a thermocouple or another accurate temperature sensor.
- **Data Analysis:**
  - The luminescence intensity (or wavelength) against temperature can be plotted. This would result in a calibration curve for the thermometer.
- **Testing:** The thermometer can be tested at various temperatures within the desired range to ensure its accuracy and reliability.
- **Application:** Afterwards, the calibrated thermometer can be used for the specific application. The luminescence of the organic dye will change as the temperature changes, allowing to monitor temperature variations.
- **Optional Encapsulation:** If the thermometer is intended for the use in harsh environments, then proper encapsulation is necessary in a protective, transparent coating to prevent damage or contamination.

The sensitivity and accuracy of the thermometer will depend on several factors such as the choice of dye and its concentration, polymer matrix, and the calibration method [39]. Careful experimentation and calibration are essential to ensure accurate temperature measurements.

13.4.2 Lanthanide-doped phosphors-based luminescent thermometer

The fabrication of lanthanide-doped phosphors-based luminescent thermometers involves the preparation of a temperature-sensitive material that emits different levels of luminescence depending on the temperature. Lanthanide ions, such as Europium (Eu), Terbium (Tb), or Dysprosium (Dy) are commonly used for this purpose due to their unique luminescent properties [40–42]. A general overview of the fabrication process involves Lanthanide salt such as europium chloride (EuCl3), terbium chloride (TbCl$_3$), or dysprosium chloride (DyCl$_3$) as the luminescent dopants, a phosphor host material that will absorb energy from the lanthanide ions and emit luminescence (common host materials include yttrium aluminum garnet (YAG), yttrium orthovanadate (YVO$_4$), or lanthanide-activated phosphors) and a flux material such as lithium fluoride (LiF) may be used to facilitate the synthesis process.

Fabrication steps are mentioned below:

- **Dopant Selection:** Firstly, the lanthanide ion (Eu, Tbf, and Dy) is selected which exhibits the desired luminescent properties for the measured temperature range.
- **Phosphor Synthesis:** The phosphor host material is synthesized using a solid-state reaction, co-precipitation, or sol-gel method. The lanthanide salts are usually mixed with the host material during synthesis.
- **Doping:** The lanthanide ion as a dopant is introduced into the phosphor host during synthesis. The concentration of the dopant can be adjusted to tune the sensitivity to temperature.
- **Annealing:** The synthesized material is heated at a specific temperature for a certain duration to optimize its luminescent properties and to remove any defects.
- **Particle Size Control:** Depending on the application, the particle size of the phosphor particles can be controlled. This can be achieved through various methods such as milling or sieving.
- **Luminescence Characterization:** The luminescence properties of the doped phosphor as a function of temperature is measured. This involves exciting the material with a suitable light source and recording the emitted luminescence.
- **Calibration:** Preparation of a calibration curve that relates luminescence intensity to temperature for the specific material and dopant combination.
- **Thermometer Assembly:** Depending on the application, integration of the phosphor-based thermometer into a suitable device or system is done. This may involve encapsulation or coating to protect the phosphor material.

- **Temperature Measurement:** The fabricated luminescent thermometer can be used to measure temperature by exciting the material with the appropriate light source and measuring the luminescence emitted. The intensity of the emitted luminescence is then correlated with the temperature using the calibration curve.
- **Application:** Lastly, implementing the luminescent thermometer in the desired applications, such as temperature monitoring in scientific experiments, industrial processes, or medical devices.

It's important to note that the specific details of the fabrication process may vary depending on the chosen phosphor host, lanthanide dopant, and the desired temperature range and sensitivity. Extensive characterization and calibration are essential to ensure accurate temperature measurements using lanthanide-doped phosphors [43].

13.4.3 QDs-based luminescent thermometer

Fabricating a QDs-based luminescent thermometer involves creating a temperature-sensitive material that emits different levels of luminescence in response to changes in temperature. QDs are nanoscale semiconductor particles that exhibit unique optical properties and can be tailored to emit specific colors of light. A general overview of the fabrication process includes QDs that emit a color of light suitable for luminescent thermometer (common QD materials include cadmium selenide (CdSe), indium phosphide (InP), or other semiconductor nanocrystals), surface ligands include organic molecules like oleic acid, oleylamine, or other ligands are often used to stabilize and functionalize the QDs, Thermosensitive ligands that exhibit temperature-dependent properties, such as phase transitions, for temperature sensitivity, appropriate solvents for QD synthesis and dispersion and any additional chemicals and materials for surface modification and functionalization [44–46].

Fabrication steps involve are as follows:

- **QD Synthesis:** The QDs with the desired emission properties are firstly synthesized. This is typically done using a solution-phase method such as the hot injection method or the colloidal synthesis approach.
- **Surface Functionalization:** Modifying the QD surfaces with suitable ligands to stabilize and disperse them in solution. This is crucial for controlling their properties.
- **Temperature-Sensitive Ligands:** Introduction of temperature-sensitive ligands that undergo changes in their properties (e.g., phase transitions) with temperature. These ligands will be used to modify the QD surfaces and impart temperature sensitivity to the QDs.

- **Ligand Exchange:** Exchange of the initial surface ligands with the thermosensitive ligands. This may involve a ligand exchange process where the QDs are treated with the thermosensitive ligands.
- **Characterization:** Characterization of the QDs to ensure they exhibit the desired luminescence properties and temperature sensitivity. This includes measuring their emission spectra and assessing their response to temperature changes.
- **Calibration:** Creating a calibration curve that relates the luminescence intensity of the QDs to temperature for the specific material and ligand combination.
- **Thermometer Assembly:** Depending on the application, integrate the QD-based thermometer into a suitable device or system. This may involve encapsulation or coating to protect the QDs.
- **Temperature Measurement:** The fabricated luminescent thermometer can be used to measure temperature by exciting the QDs with an appropriate light source and measuring the emitted luminescence. The intensity of the emitted luminescence is then correlated with the temperature using the calibration curve.
- **Application:** Lastly, implementing the luminescent thermometer in the desired applications, such as temperature monitoring in scientific experiments, industrial processes, or medical devices.

It's important to note that the specific details of the fabrication process may vary depending on the chosen QD material, temperature-sensitive ligands, and the desired temperature range and sensitivity [47]. Careful characterization and calibration are essential to ensure accurate temperature measurements using QDs-based luminescent thermometers.

13.4.4 Metal cluster-based luminescent thermometer

The fabrication of a metal cluster-based luminescent thermometer involves creating a temperature-sensitive material that emits luminescence (light) with intensity or wavelength that changes with temperature. Metal clusters, particularly noble metal nanoparticles, have unique optical properties that can be harnessed for this purpose. The general steps to fabricate a metal cluster-based luminescent thermometer involve noble metal nanoparticles (gold or silver nanoparticles) [48,49], organic ligands or stabilizers, temperature-sensitive organic dye, solvent (water, ethanol), UV–vis spectrophotometer, fluorimeter, glass vials or cuvettes and thermocouples or a temperature control system.

Fabrication steps involve the following steps:

- **Synthesis of Metal Nanoparticles:**
  - Preparation of noble metal nanoparticles using established methods like the Turkevich method or citrate reduction method. These nanoparticles should be small and well-dispersed in a solvent.

- **Surface Functionalization:**
  - Modifying the surface of the metal nanoparticles with temperature-sensitive organic ligands or dyes. These ligands should have luminescent properties that change with temperature. Covalent or non-covalent interactions can be used to attach the ligands to the nanoparticle surfaces.
- **Characterization:**
  - Characterization of the modified nanoparticles using techniques such as UV–vis spectroscopy, transmission electron microscopy (TEM), and dynamic light scattering (DLS) to confirm their size, stability, and luminescent properties.
- **Temperature Calibration:**
  - To establish a temperature-luminescence relationship, measurement of the luminescence of the nanoparticles at various known temperatures using a fluorimeter or similar equipment is required. These data points will be used to calibrate the thermometer.
- **Construction of the Thermometer:**
  - Combining the functionalized metal nanoparticles with a suitable solvent to create a stable dispersion.
  - The dispersion is then placed in a glass vial or cuvette.
- **Temperature Measurement:**
  - Measurement of the temperature of the sample using a thermocouple or a temperature control system.
- **Luminescence Measurement:**
  - Measurement of the luminescence of the nanoparticle solution using a fluorimeter. The luminescence will vary with temperature based on the temperature-sensitive ligands attached to the nanoparticles.
- **Calibration Curve:**
  - A calibration curve is then plotted by recording luminescence intensity or wavelength against temperature for a range of temperatures. This curve will allow to determine the temperature based on the luminescence of the nanoparticles.
- **Temperature Sensing:**
  - To use the thermometer, measurement of the luminescence of the nanoparticle solution is required followed by referring to the calibration curve to determine the temperature of the system.

It's important to note that the success of a metal cluster-based luminescent thermometer depends on the selection of appropriate temperature-sensitive ligands or dyes, as well as careful calibration. The choice of noble metal nanoparticles and ligands can be tailored to specific temperature ranges and sensitivity requirements [50]. This type of thermometer can find applications in various fields, including biotechnology, materials science, and environmental monitoring.

13.4.5 Metal-organic framework-based luminescent thermometer

Creating a metal-organic framework (MOF)-based luminescent thermometer involves designing and synthesizing a MOF that exhibits temperature-dependent luminescent properties. Compared with the traditional luminescent organic molecules and coordination compounds, the MOFs also have the advantage of collaboratively interplaying/interacting with each other among the periodically located luminescent centers to direct new luminescent functionalities. Such attractive characteristics can allow to design and construct luminescent MOFs with significantly temperature-dependent emissions, which is crucial for luminescence temperature sensing. Figure 13.4 shows the three types of possible strategies to design dual-emitting MOFs for ratiometric temperature sensing [51–53]. However, a general outline of the fabrication process involves the selection of metal ions that can be incorporated into the MOF structure (common choices include Zn, Cd, and Cu), organic ligands that can coordinate with metal ions to form the MOF (Terephthalic acid and 2-aminoterephthalic acid are some of the common choices), solvents for the synthesis

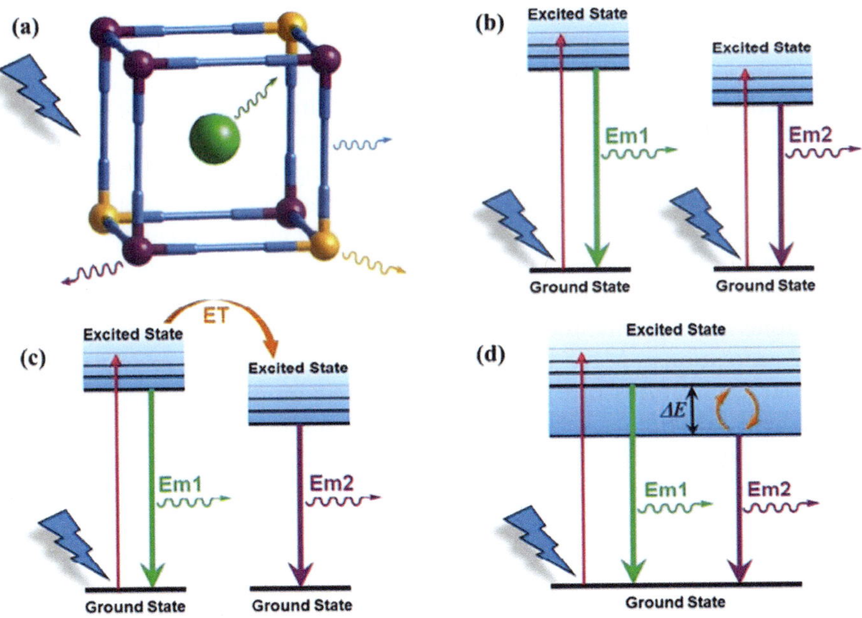

Figure 13.4 Schematic representation of possible luminescence centers in MOFs (a), and three different energy level structures of dual-emitting MOFs for ratiometric luminescence temperature sensing (b–d). ET: energy transfer, Em: emission. (Reproduced with permission from Ref. [53] © 1998 Royal Society of Chemistry.)

(such as *N,N*-dimethylformamide (DMF) or dimethyl sulfoxide (DMSO)), temperature-sensitive luminescent molecule that emits light in a temperature-dependent manner (examples include europium or terbium complexes), fluorescence spectrophotometer or spectrofluorometer to measure luminescent properties.

Synthesis and fabrication steps are as follows:

- **Preparation of a Metal-Organic Framework (MOF):** Firstly, dissolve the chosen metal ions and organic ligands in a suitable solvent (e.g., DMF). Secondly, mixing them under controlled conditions (temperature and time) to allow the MOF to form. Precise conditions will depend on the specific MOF and ligands chosen. Lastly, isolation of the MOF crystals using filtration and wash with appropriate solvents. The resulting MOF must be a solid powder.

- **Incorporation of Luminescent Molecule:** Preparation of a solution of the temperature-sensitive luminescent molecule is required. Afterward, the addition of the MOF powder into the solution of the luminescent molecule and allow them to react. The luminescent molecule should be incorporated into the MOF structure.

- **Characterization:** Characterization of the luminescent properties of the MOF-based thermometer using a fluorescence spectrophotometer is required. Measurement of the temperature-dependent luminescence emission spectrum of the material is necessary.

- **Calibration:** Generation of a calibration curve is needed by measuring the luminescence intensity at various known temperatures. This will help establish the relationship between temperature and luminescence.

- **Temperature Sensing:** To use the MOF-based luminescent thermometer, exposure of thermometer to the target environment is necessary. Measurement of the luminescent intensity is done using the fluorescence spectrophotometer. Lastly, referring to the calibration curve is done to determine the temperature of the environment.

- **Applications:** MOF-based luminescent thermometers have potential applications in various fields, such as monitoring temperature in chemical processes, biological systems, and environmental studies.

- **Further Development:** Further, research and development may be needed to optimize the MOF structure and luminescent properties for specific applications.

The choice of metal ions, ligands, and luminescent molecules will significantly impact the performance and sensitivity of the MOF-based luminescent thermometer. Additionally, ensure that the MOF is stable and doesn't degrade under the target temperature and environmental conditions [54]. Collaborating with researchers experienced in MOF synthesis and luminescence is often beneficial for successful fabrication.

13.5 CHALLENGES ASSOCIATED WITH LUMINESCENT-BASED TEMPERATURE SENSOR

Achieving accurate and consistent temperature measurements with flexible luminescent thermometers can be challenging. These devices often require precise calibration, and the calibration process may need to be repeated periodically to maintain accuracy. Also, the sensitivity of flexible luminescent thermometers may vary with temperature, and this can complicate temperature measurement in extreme temperature ranges or when high precision is required. Like many temperature sensors, flexible luminescent thermometers may also experience drift over time, which can result in a decrease in accuracy if not properly checked [55]. Several environmental conditions, such as humidity, light exposure, and chemical exposure can also affect the luminescent materials and their response, leading to errors in temperature measurements. Sometimes flexibility can make these thermometers vulnerable to mechanical damage, and they may need to be protected or shielded in certain applications to ensure their durability. The response time of flexible luminescent thermometers can vary depending on the materials used and the design. In applications where rapid temperature changes occur, the response time may be a limiting factor [56]. The choice of luminescent materials is critical to the performance of these thermometers. Selecting the right materials for the desired temperature range and application can be a challenge. Some luminescent materials may require excitation sources, such as light or electrical energy, which can increase power consumption in the sensing system. Managing power efficiently is thus, may be necessary. The fabrication of flexible luminescent thermometers can be complex and require specialized techniques, making them more challenging to produce compared to traditional temperature sensors. Integrating these thermometers into specific applications, such as wearable devices or flexible electronics, can be challenging, as it requires special attention to design, encapsulation, and connectivity [57]. Developing and manufacturing flexible luminescent thermometers can be more expensive than conventional temperature sensors, which may affect their widespread adoption in cost-sensitive applications. Extracting temperature data from luminescent materials often involves sophisticated readout and signal processing techniques, which can add complexity to the temperature measurement system. The response of luminescent materials to temperature changes may not always be linear, and compensation or correction methods may be required to achieve accurate readings.

Thus, overcoming these challenges involves a combination of materials science, calibration techniques, sensor design, and environmental control. The suitability of flexible luminescent thermometers for a specific application depends on the specific requirements, budget constraints, and the feasibility of addressing these challenges effectively.

13.6 CONCLUSION

The flexible luminescent thermometer represents a remarkable advancement in temperature measurement technology. Its unique design and capabilities have opened up new avenues for non-invasive and versatile temperature monitoring in various fields from medical applications to industrial processes. The ability to provide real-time, contactless temperature readings, and its flexibility to conform to complex shapes and surfaces make it a valuable tool in situations where traditional thermometers fall short. Furthermore, the luminescent properties of these flexible thermometers offer several advantages such as high sensitivity, accuracy, and rapid response time. Its potential for remote and continuous monitoring makes it especially beneficial in critical environments where maintaining a stable temperature is essential. This innovation has the potential to revolutionize temperature measurement, offering increased precision and adaptability. As the technology behind flexible luminescent thermometers continues to advance, we can anticipate further improvements in their design and applications. These devices have the potential to transform the way we monitor temperature in various settings, offering more accurate, reliable, and adaptable solutions for a wide range of industries. Thus, the flexible luminescent thermometer is a promising tool that will undoubtedly play a significant role in temperature measurement in the future.

REFERENCES

1. Su, Y., Ma, C., Chen, J., Wu, H., Luo, W., Peng, Y., & Li, H. (2020). Printable, highly sensitive flexible temperature sensors for human body temperature monitoring: a review. *Nanoscale Research Letters*, 15, 1–34.
2. Martins, J. C., Brites, C. D., Neto, A. N. C., Ferreira, R. A., & Carlos, L. D. (2023). An overview of luminescent primary thermometers. *Luminescent Thermometry: Applications and Uses*, 105–152.
3. Liu, R., He, L., Cao, M., Sun, Z., Zhu, R., & Li, Y. (2021). Flexible temperature sensors. *Frontiers in Chemistry*, 9, 539678.
4. Wang, H., Li, S., Lu, H., Zhu, M., Liang, H., Wu, X., & Zhang, Y. (2023). Carbon-based flexible devices for comprehensive health monitoring. *Small Methods*, 7(2), 2201340.
5. Khayrudinov, V., Koskinen, T., Grodecki, K., Murawski, K., Kopytko, M., Yao, L., & Haggren, T. (2022). InSb nanowire direct growth on plastic for monolithic flexible device fabrication. *ACS Applied Electronic Materials*, 4(1), 539–545.
6. Yao, P., Bao, Q., Yao, Y., Xiao, M., Xu, Z., Yang, J., & Liu, W. (2023). Environmentally stable, robust, adhesive, and conductive supramolecular deep eutectic gels as ultrasensitive flexible temperature sensor. *Advanced Materials*, 35, 2300114.

7. Yin, R., Yang, S., Li, Q., Zhang, S., Liu, H., Han, J., & Shen, C. (2020). Flexible conductive Ag nanowire/cellulose nanofibril hybrid nanopaper for strain and temperature sensing applications. *Science Bulletin, 65*(11), 899–908.

8. Kuzubasoglu, B. A., & Bahadir, S. K. (2020). Flexible temperature sensors: a review. *Sensors and Actuators A: Physical, 315*, 112282.

9. Zhang, Z., Yan, J., Zhang, Q., Tian, G., Jiang, W., Huo, J., & Li, J. (2022). Enlarging sensitivity of fluorescence intensity ratio-type thermometers by the interruption of the energy transfer from a sensitizer to an activator. *Inorganic Chemistry, 61*(41), 16484–16492.

10. Sukul, P. P., & Swart, H. C. (2023). Synergistic red dominancy over green upconversion studies in $PbZrTiO_3$: Er^{3+}/Yb^{3+} phosphor synthesized via two different modified technique and flexible thin-film thermometer demonstration on C1: $PbZrTiO_3$@ PDMS substrates. *Journal of Alloys and Compounds, 966*, 171656.

11. Wang, Q., Liao, M., Lin, Q., Xiong, M., Mu, Z., & Wu, F. (2021). A review on fluorescence intensity ratio thermometer based on rare-earth and transition metal ions doped inorganic luminescent materials. *Journal of Alloys and Compounds, 850*, 156744.

12. Runowski, M., Woźny, P., & Martín, I. R. (2021). Optical pressure sensing in vacuum and high-pressure ranges using lanthanide-based luminescent thermometer–manometer. *Journal of Materials Chemistry C, 9*(13), 4643–4651.

13. Peng, M., Kaczmarek, A. M., & Van Hecke, K. (2022). Ratiometric thermometers based on rhodamine B and fluorescein dye-incorporated (nano) cyclodextrin metal–organic frameworks. *ACS Applied Materials & Interfaces, 14*(12), 14367–14379.

14. Abbas, M. T., Khan, N. Z., Mao, J., Qiu, L., Wei, X., Chen, Y., & Khan, S. A. (2022). Lanthanide and transition metals doped materials for non-contact optical thermometry with promising approaches. *Materials Today Chemistry, 24*, 100903.

15. Xue, Z. Z., Li, X. Y., Xu, L., Han, S. D., Pan, J., & Wang, G. M. (2021). Novel silver (I) cluster-based coordination polymers as efficient luminescent thermometers. *CrystEngComm, 23*(1), 56–63.

16. Li, Y. (2020). Temperature and humidity sensors based on luminescent metal-organic frameworks. *Polyhedron, 179*, 114413.

17. Brites, C. D., Balabhadra, S., & Carlos, L. D. (2019). Lanthanide-based thermometers: at the cutting-edge of luminescence thermometry. *Advanced Optical Materials, 7*(5), 1801239.

18. Mishra, K. C., & Collins, J. (2022). Formulation of radiative and nonradiative transitions of a polyatomic system within the crude adiabatic approximation. *Optical Materials: X, 15*, 100190.

19. Ćirić, A., Periša, J., Zeković, I., Antić, Ž., & Dramićanin, M. D. (2022). Multilevel-cascade intensity ratio temperature read-out of Dy^{3+} luminescence thermometers. *Journal of Luminescence, 245*, 118795.

20. Kniec, K., Kochanowska, A., Li, L., Suta, M., & Marciniak, L. (2022). A ratiometric and lifetime-based luminescent thermometer exploiting the Co^{3+} luminescence in $CaAl_2O_4$: Co^{3+} and $CaAl_2O_4$: Co^{3+}, Nd^{3+}. *Journal of Materials Chemistry C, 10*(24), 9278–9286.

21. Feng, J., Tian, K., Hu, D., Wang, S., Li, S., Zeng, Y., & Yang, G. (2011). A tri-arylboron-based fluorescent thermometer: sensitive over a wide temperature range. *Angewandte Chemie International Edition, 50*(35), 8072–8076.

22. Song, D. N., Zhang, D. J., Wang, Y. L., Wang, J. J., Xing, X. S., Lv, Z. Y., & Zhang, R. (2020). Luminescent thermochromic silver iodides as wavelength-dependent thermometers. *Inorganic Chemistry, 59*(18), 13067–13077.

23. Wang, C., Jin, Y., Zhang, R., Yao, Q., & Hu, Y. (2022). A review and outlook of ratiometric optical thermometer based on thermally coupled levels and non-thermally coupled levels. *Journal of Alloys and Compounds, 894,* 162494.

24. Li, P., Jia, M., Liu, G., Zhang, A., Sun, Z., & Fu, Z. (2019). Investigation on the fluorescence intensity ratio sensing thermometry based on nonthermally coupled levels. *ACS Applied Bio Materials, 2*(4), 1732–1739.

25. Wu, M., Deng, D., Ruan, F., Chen, B., & Xu, S. (2020). A spatial/temporal dual-mode optical thermometry based on double-sites dependent luminescence of $Li_4SrCa(SiO_4)_2$: Eu^{2+} phosphors with highly sensitive luminescent thermometer. *Chemical Engineering Journal, 396,* 125178.

26. Jia, M., Sun, Z., Zhang, M., Xu, H., & Fu, Z. (2020). What determines the performance of lanthanide-based ratiometric nanothermometers? *Nanoscale, 12*(40), 20776–20785

27. Wan, Y., Yu, L., & Xia, T. (2022). A dye-loaded nonlinear metal-organic framework as self-calibrated optical thermometer. *Dyes and Pigments, 202,* 110234.

28. Piotrowski, W., Kuchowicz, M., Dramićanin, M., & Marciniak, L. (2022). Lanthanide dopant stabilized Ti^{3+} state and supersensitive Ti^{3+}-based multiparametric luminescent thermometer in $SrTiO_3$: Ln^{3+} ($Ln^{3+=}Lu^{3+}$, La^{3+}, Tb^{3+}) nanocrystals. *Chemical Engineering Journal, 428,* 131165.

29. Li, Y., Xiao, X., Wei, Z., & Chen, Y. (2022). A ratio fluorescence thermometer based on carbon dots & lanthanide functionalized metal-organic frameworks. *Zeitschrift für Anorganische und Allgemeine Chemie, 648*(9), e202100323.

30. Xue, Z. Z., Meng, X. D., Li, X. Y., Han, S. D., Pan, J., & Wang, G. M. (2021). Luminescent thermochromism and white-light emission of a 3D $[Ag_4Br_6]$ cluster-based coordination framework with both adamantane-like node and linker. *Inorganic Chemistry, 60*(7), 4375–4379.

31. Cheng, Y., Wu, M., Du, Z., Chen, Y., Zhao, L., Zhu, Z., & Zeng, C. (2023). Tetra-nuclear cluster-based lanthanide metal–organic frameworks as white phosphor, information encryption, self-calibrating thermometers, and Fe^{2+} sensors. *ACS Applied Materials & Interfaces, 15,* 24570–24582.

32. Li, Y. (2020). Temperature and humidity sensors based on luminescent metal-organic frameworks. *Polyhedron, 179,* 114413.

33. Zhou, Y., Zhang, D., Zeng, J., Gan, N., & Cuan, J. (2018). A luminescent lanthanide-free MOF nanohybrid for highly sensitive ratiometric temperature sensing in physiological range. *Talanta, 181,* 410–415.

34. Wang, S., Jiang, J., Lu, Y., Liu, J., Han, X., Zhao, D., & Li, C. (2020). Ratiometric fluorescence temperature sensing based on single-and dual-lanthanide metal-organic frameworks. *Journal of Luminescence, 226,* 117418.

35. Zhang, G., Zhou, S., Jiang, T., Pu, Z., Li, J., Shi, J., & Yu, B. (2022). Quantum dots micro-probe based on selective etching and fiber end-face inkjet printing for temperature sensing. *Measurement, 188,* 110615.

36. Zhu, Z., Sun, Z., Guo, Z., Zhang, X., & Wu, Z. C. (2019). A high-sensitive ratiometric luminescent thermometer based on dual-emission of carbon dots/Rhodamine B nanocomposite. *Journal of Colloid and Interface Science*, *552*, 572–582.

37. Senapati, S., & Naik, R. (2022). Ratiometric optical temperature sensing based on greenish-white light emitting ZnO nanosheets and rhodamine 6G composite. *Materials Research Bulletin*, *153*, 111904.

38. Ma, H., Zhou, T., Dai, Z., Hu, J., & Liu, J. (2019). In situ self-assembly of ultrastable crosslinked luminescent gold nanoparticle and organic dye nano-hybrids toward ultrasensitive and reversible ratiometric thermal imaging. *Advanced Optical* Materials, 7(12), 1900326.

39. Wu, Y., Liu, J., Ma, J., Liu, Y., Wang, Y., & Wu, D. (2016). Ratiometric nanother-mometer based on rhodamine dye-incorporated F127-melamine-formaldehyde polymer nanoparticle: preparation, characterization, wide-range temperature sensing, and precise intracellular thermometry. *ACS Applied Materials & Interfaces*, 8(23), 14396–14405.

40. Yu, H., Su, W., Chen, L., Deng, D., & Xu, S. (2019). Excellent temperature sensing characteristics of europium ions self-reduction $Sr_3P_4O_{13}$ phosphors for ratiometric luminescence thermometer. *Journal of Alloys and Compounds*, *806*, 833–840.

41. Xue, J., Yu, Z., Noh, H. M., Lee, B. R., Choi, B. C., Park, S. H., & Song, M. (2021). Designing multi-mode optical thermometers via the thermochromic $LaNbO_4$: Bi^{3+}/Ln^{3+} (Ln=Eu, Tb, Dy, Sm) phosphors. *Chemical Engineering Journal*, *415*, 128977.

42. Liu, Q., Wu, M., Chen, B., Huang, X., Liu, M., Liu, Y., & Huang, Z. (2023). Optical thermometry based on fluorescence intensity ratio of Dy^{3+}-Doped oxysilicate apatite warm white phosphor. *Ceramics International*, *49*(3), 4971–4978.

43. Nikolić, M. G., Antić, Ž., Ćulubrk, S., Nedeljković, J. M., & Dramićanin, M. D. (2014). Temperature sensing with Eu^{3+} doped TiO_2 nanoparticles. *Sensors and Actuators B: Chemical*, *201*, 46–50.

44. Jorge, P. A. S., Mayeh, M., Benrashid, R., Caldas, P., Santos, J. L., & Farahi, F. (2006). Quantum dots as self-referenced optical fibre temperature probes for luminescent chemical sensors. *Measurement Science and Technology*, *17*(5), 1032.

45. Li, S., Zhang, K., Yang, J. M., Lin, L., & Yang, H. (2007). Single quantum dots as local temperature markers. *Nano Letters*, 7(10), 3102–3105.

46. Saxena, N., Kumar, P., & Gupta, V. (2015). CdS: SiO_2 nanocomposite as a luminescence-based wide range temperature sensor. *RSC Advances*, *5*(90), 73545–73551.

47. Maestro, L. M., Rodriguez, E. M., Rodríguez, F. S., la Cruz, M. I. D., Juarranz, A., Naccache, R., & Solé, J. G. (2010). CdSe quantum dots for two-photon fluorescence thermal imaging. *Nano Letters*, *10*(12), 5109–5115.

48. Xue, Z. Z., Li, X. Y., Xu, L., Han, S. D., Pan, J., & Wang, G. M. (2021). Novel silver (I) cluster-based coordination polymers as efficient luminescent thermometers. *CrystEngComm*, 23(1), 56–63.

49. Maity, S., Bain, D., & Patra, A. (2019). An overview on the current under-standing of the photophysical properties of metal nanoclusters and their potential applications. *Nanoscale*, *11*(47), 22685–22723.

50. Zhang, W., Lin, D., Wang, H., Li, J., Nienhaus, G. U., Su, Z., & Shang, L. (2017). Supramolecular self-assembly bioinspired synthesis of luminescent gold nanocluster-embedded peptide nanofibers for temperature sensing and cellular imaging. *Bioconjugate Chemistry, 28*(9), 2224–2229.

51. Cheng, Y., Wu, M., Du, Z., Chen, Y., Zhao, L., Zhu, Z., & Zeng, C. (2023). Tetra-nuclear cluster-based lanthanide metal–organic frameworks as white phosphor, information encryption, self-calibrating thermometers, and Fe^{2+} sensors. *ACS Applied Materials & Interfaces, 15*(20), 24570–24582.

52. Cui, Y., Xu, H., Yue, Y., Guo, Z., Yu, J., Chen, Z., & Chen, B. (2012). A luminescent mixed-lanthanide metal–organic framework thermometer. *Journal of the American Chemical Society, 134*(9), 3979–3982.

53. Cui, Y., Zhu, F., Chen, B., & Qian, G. (2015). Metal–organic frameworks for luminescence thermometry. *Chemical communications, 51*(3), 7420–7431.

54. Han, Y. H., Tian, C. B., Li, Q. H., & Du, S. W. (2014). Highly chemical and thermally stable luminescent Eu_xTb_{1-x} MOF materials for broad-range pH and temperature sensors. *Journal of Materials Chemistry C, 2*(38), 8065–8070.

55. Yi, F. Y., Chen, D., Wu, M. K., Han, L., & Jiang, H. L. (2016). Chemical sensors based on metal–organic frameworks. *ChemPlusChem, 81*(8), 675–690.

56. Brites, C. D., Marin, R., Suta, M., Carneiro Neto, A. N., Ximendes, E., Jaque, D., & Carlos, L. D. (2023). Spotlight on luminescence thermometry: basics, challenges, and cutting-edge applications. *Advanced Materials, 35*(36), 2302749.

57. Ramalho, J. F., Correia, S. F., Fu, L., António, L. L., Brites, C. D., André, P. S., & Carlos, L. D. (2019). Luminescence thermometry on the route of the mobile-based Internet of Things (IoT): how smart QR codes make it real. *Advanced Science, 6*(19), 1900950.

Chapter 14

Health monitoring with wearable thermometers

*Manju Bala, Manoj Verma,
Renuka Bokolia, and Vivek Chopra*

14.1 WEARABLE HEALTH MONITORING SYSTEMS

Wearable health devices started way back in 1980 when the first electro-cardiogram (ECG) was invented. The number of wearable devices and their usage has increased many fold in the last four decades. Nearly 320 million health wearable devices were used worldwide in 2022, and it is estimated to reach around 600 million by 2025.

Smart wearable medical sensors have drawn a lot of attention from researchers in the past ten years because of their ability to precisely identify patient pathophysiological factors and, as a result, track patients' physiological states in real time. Generally, wearable sensors are used to identify target analytes in a range of biological bodily fluids. A great deal of information about the human body has been made available by the wearable sensors' continual monitoring of vital indicators.

Numerous flexible sensor types can be attached directly to the human body or integrated into textile fibers, clothes, and elastic bands. Heart rate, body temperature, electrodermal activity, oxygen saturation, blood pressure, respiration rate, electrocardiogram, electromyogram, and heart rate are among the physiological indications that the sensors may measure. Furthermore, activity-related data are frequently evaluated by tiny motion sensors built on micro-electro-mechanical systems (MEMS), such as gyroscopes, accelerometers, and magnetic field sensors.

One of the most important applications of wearable technology in healthcare is patient monitoring. Wearable technology can assist medical professionals in better understanding their patients' daily routines and emotional states by monitoring patient activities and data.

14.2 TYPES OF WEARABLE SENSORS

There are four main types of wearable sensors: chemical, electromechanical, optical, and electrical sensors (Bhelkar & Shedge, 2017). Chemical sensors quantify the amount of chemicals present in perspiration or

DOI: 10.1201/9781032661537-14

other body secretions. Examples are sensors that assess stress chemicals in perspiration and glucose monitors for diabetics. Electromechanical sensors track mechanical movement by measuring electrical signals. Electromechanical sensors are useful for measuring and gathering data related to physical activity. They can also be used as an accelerometer. The most popular wearable health gadgets are optical sensors, which are capable of measuring a wide range of health indicators. Optical sensors measure variations in various biological signals, such as blood pressure, oxygen saturation, and heart rate, using light. Optical sensors record and quantify the amount of light that is absorbed, reflected, or transmitted through the body. Light sources (LEDs) in these sensors release light into the body. Electrical sensors use electrical signals to measure heart rate or brain activity. An example of an electrical sensor is the ECG, which captures the electrical activity of the heart and can be used to diagnose abnormalities like arrhythmias. Another kind of electrical sensor is an electroencephalogram (EEG), which records electrical activity in the human brain.

14.3 BEHAVIOR OF BODY TEMPERATURE

Body temperature can be divided into two broad categories: core temperature and skin surface temperature. As mammals need a nearly constant internal body temperature, they are homoeothermic. The temperature of the hypothalamus, which is the body's regulating center, is known as core temperature. Rectal temperature is a more precise way of measuring hypothalamus temperature than measures taken in the stomach, esophagus, or auditory canal, which researchers use to estimate core temperature. The hypothalamus serves as the center of the thermoregulatory system, which controls body temperature. This system can keep the core temperature within a certain range by using perspiration and vasomotor. The core temperature fluctuates according to the circadian rhythm, with a difference of approximately 1°C between the lowest value in the early morning and the peak in the afternoon. The skin on the trunk typically has a temperature that ranges from 33.5°C to 36.9°C (Taylor et al., 2014). Figure 14.1 shows the core and skin temperature distribution and the symptoms seen if the temperature varies by ±2°. Compared to superficial arteries, the skin's surface temperature is lower above superficial veins. Additionally, it is lower than prominently curved body features like the nose, ears, fingers, and toes. The convenience of having easy access to a thermometer is one benefit of taking skin temperature. On the other hand, the skin lies on the edge of media that have various temperatures. Moreover, airflow, heat radiation, perspiration, cutaneous blood flow, and covering by clothing can all cause rather noticeable

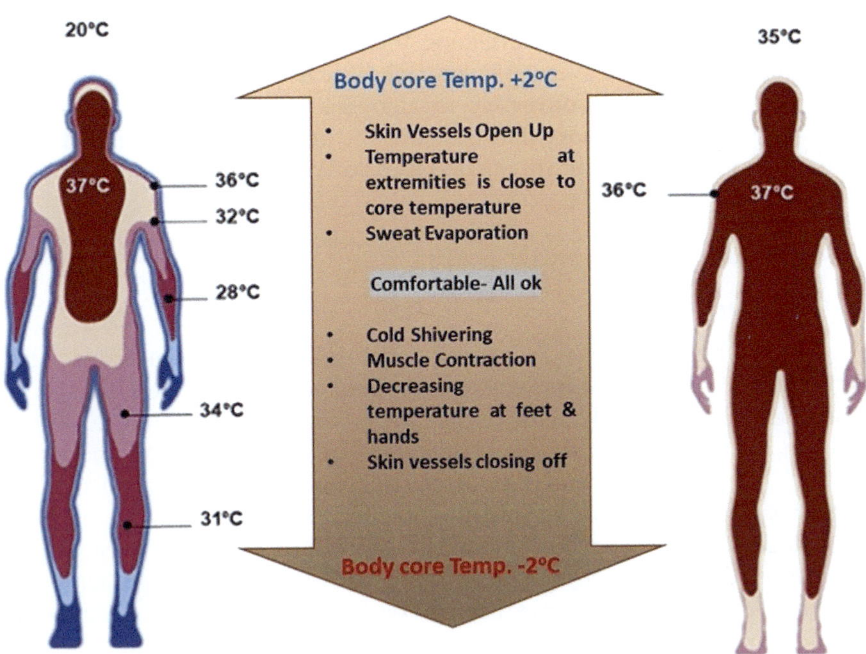

Figure 14.1 Core and skin surface temperature distribution. (Adapted from Tamura et al. (2018b), Open Access.)

temperature gradients from the tissue to the surrounding air. Even while skin temperature varies and is susceptible to outside influences, it may nevertheless be a reliable indicator of how core temperature changes based on activity level and measurement location (Lenhardt & Sessler, 2006). The main mechanism for transferring heat from the skin to the environment is the thermoregulatory system, which also connects the skin and core temperatures. Therefore, in thermoneutral conditions, the skin temperature at places such as the axilla or upper thorax, which is comparatively less affected by the surrounding environment, may be able to reject changes in core temperature.

14.4 PRINCIPLE OF TEMPERATURE DETECTION

For clinical applications, a wide variety of temperature sensors are available; these are usually built into surface probes, catheters, and needles. However, there have been few advancements in this area, and only a small number of sensors are embedded in wearable devices. Two categories exist for temperature detection: non-contact/indirect and contact/direct measures. Contact thermometers include thermistors and integrated circuits (ICs) that measure the average temperature differential between the skin's surface and the sensor. Conversely, thermopiles and pyro sensors are examples of non-contact sensors that measure the amount of thermal radiation received from a known or anticipated location on the surface.

14.4.1 Contact thermometers

14.4.1.1 Thermistors

A resistor whose electrical resistance fluctuates with temperature is called a thermistor, sometimes known as a thermal resistor. While all resistors' resistance varies slightly with temperature, a thermistor is more susceptible to changes in temperature. Resistors are a passive component in a circuit. They are a dependable, affordable, and precise method of measuring temperature [4]. Thermistors are the best type of sensor for many applications, even though they don't work well in excessively hot or cold environments. Thermistors work by using resistance that changes with temperature. A thermistor's resistance can be measured with an ohmmeter. If the exact relationship between the change in resistance and temperature is known, temperature can be determined by monitoring the resistance. The thermistor's material type affects how much the resistance varies. There are two types of thermistors:

- NTC thermistor having negative temperature coefficient
- PTC thermistor having positive temperature coefficient

An NTC thermistor's resistance drops as its temperature rises. Consequently, in an NTC thermistor, resistance and temperature have an inverse relationship. These are the most popular varieties of thermistor. The relationship between resistance and temperature in an NTC thermistor is determined by the following expression:

$$R_T = R_0 \exp\left[\beta\left(\frac{1}{T} - \frac{1}{T_0}\right)\right]$$

where:

R_T is the resistance at temperature T (K)
R_0 is the resistance at temperature T_0 (K)
T_0 is the reference temperature (normally 25°C)
β is a constant, its value is dependent on the characteristics of the material. The nominal value is taken as 4000.

A high value of β indicates a very good resistor–temperature relationship. A larger value of β indicates a greater variation in resistance for the identical temperature rise, increasing the thermistor's sensitivity and accuracy. The resistance temperature coefficient can be determined by the following expression.

$$\alpha_T = \frac{1}{R_T}\left(\frac{dR_T}{dt}\right) = \frac{-\beta}{T^2}$$

The expression above makes it clear that the α_T has a negative sign. $\alpha_T=-0.0045/°K$ is the result if $T=298$ K and $\beta=4000$ K. This is significantly more than platinum RTD's sensitivity. This device has the capability to detect even the smallest variations in temperature. Nevertheless, more expensive versions of highly doped thermistors with a positive temperature coefficient are currently accessible. The thermistors are unquestionably nonlinear sensors because the expression makes it impossible to approximate the curve linearly over even a modest temperature range. Temperature and resistance have the opposite relationship in a PTC thermistor. Therefore, temperature and resistance in a PTC thermistor are inversely proportional. Figure 14.2 shows the various kinds of thermistors that are available in the market.

14.4.1.2 IC thermometers

The output across a P-N junction with a constant forward-bias current shows linear temperature dependency over a wide temperature range. When the current bias is constant, the voltage drop across a silicon P-N diode junction shows a temperature coefficient of around –2 mV/°C. Since P-N junctions are often found in diodes, transistors, and integrated circuits, temperature sensing can be applied to a wide range of devices at a low cost. This offers a chance to develop wearable temperature sensors using microprocessors.

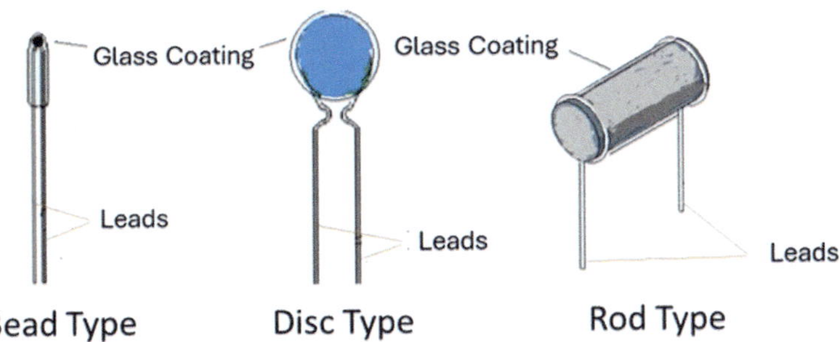

Figure 14.2 Type of thermistors.

14.4.2 Non-contact thermometers

A non-contact device for taking body temperature readings is an infrared thermometer. Infrared radiation thermometry is appealing in a number of difficult temperature measurement scenarios since it is a quick, non-intrusive, and non-contact method of measuring temperature, (Fletcher et al., 2018). The basic physical rules of thermal radiation were developed more than a century ago by Kirchhoff, Stefan, Boltzmann, Wien, and Planck. The blackbody radiation released is directly correlated with the ambient temperature of the emitting source, whether it is resolved spectrally or not. However, there are a number of uncertainties in real-world radiation thermometry readings, including surface emissivity as well as environmental elements such as surrounding temperature and reflected solar radiation. There is a clear correlation between the amount of electromagnetic radiation produced by humans and the temperature of the Earth's surface. Without coming into touch with skin, one can measure the intensity of an object's radiation to establish its temperature. The majority of heat radiation released by the human body falls into the infrared part of the electromagnetic spectrum, which has wavelengths between 0.8 and 100 µm.

An object's total radiation power, P at temperature T_{obj} and ambient temperature, T_{amb} is equal to $P = \sigma\varepsilon(T_{obj}^4 - T_{amb}^4)$, where ε is the object's emission factor and σ is the Stefan-Boltzmann constant. A voltage V, corresponding to the power of incident radiation Prad, is produced by the sensor. Several studies have shown that the body's core temperature is influenced by a wide range of factors, including ambient temperature, exercise, sweat, blood pressure, blood flow, heat arising from internal tissues (muscle), and heat escaping through the skin's surface. Environmental factors including water vapor, skin temperature, and ambient temperature all have a significant impact on how much heat transmission happens between the human body and its surroundings.

The radiation temperature sensors that are frequently seen in wearable thermometers are based on thermopiles or pyroelectric sensors.

14.4.2.1 Thermopiles

A thermopile is an electronic device that transforms thermal energy into electrical energy. Thermopile is made up of many thermocouples connected in series as shown in Figure 14.3. The thermoelectric effect is the process by which this type of device produces a voltage in response to a temperature differential between its thermocouples – dissimilar metals. The output voltage of the series is determined by the temperature differential between the active layer and the reference layer. As more thermocouple couples are joined in series, the voltage output increases in magnitude. When a closed circuit is made of several metal junctions and

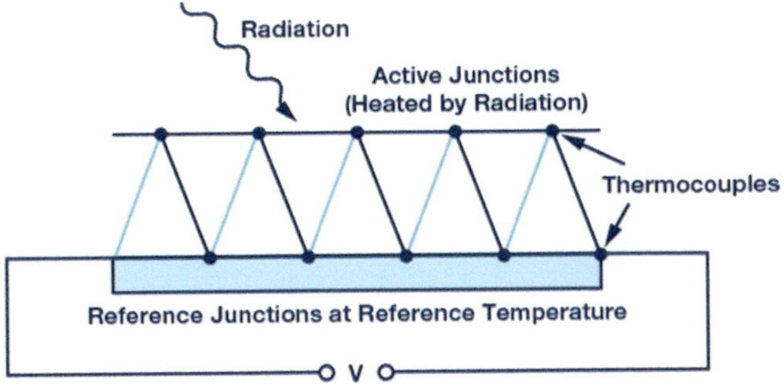

Figure 14.3 Schematic of a thermopile detector. (Adapted from Tamura et al. (2018b), Open Access.)

is placed into differential temperature, a current is produced because of a difference in potential between the hot and cool junction. The output of a thermopile usually varies from tens to hundreds of millivolts (Itoigawa et al., 2005). If both junctions have the same temperature, the voltages cancel out to zero.

A thermopile's output voltage (V) is determined by multiplying the Seebeck coefficient (S), thermocouple count (N), and temperature differential (ΔT).

$$V = S \times N \times \Delta T$$

14.4.2.2 Pyroelectric sensors

These types of temperature sensors are based on the pyroelectric phenomenon and use infrared radiations. A pyroelectric crystal's surface produces a detectable voltage when heated. It is possible to infer the crystal's temperature change from this voltage. There are just a few crystals with low enough crystal symmetry to show pyroelectric action. Because of their temperature-dependent electrical polarization, they produce pyroelectric charges in response to temperature variations. A pyroelectric sensor is essentially a capacitor made of a piece of pyroelectric crystal material with electrodes on two sides (Lee et al., 1998). A black covering covers one of those electrodes that is in direct contact with the incident radiation. When heat is transferred through the electrode and into the crystal, incoming light is absorbed on the coating and partially heats the crystal. As a result, the crystal generates some pyroelectric voltage, which can be electronically detected (Figure 14.4). That pyroelectric signal would eventually disappear for a constant optical power, making the apparatus unsuitable for detecting continuous-wave radiation intensity. Typically, light pulses are utilized with

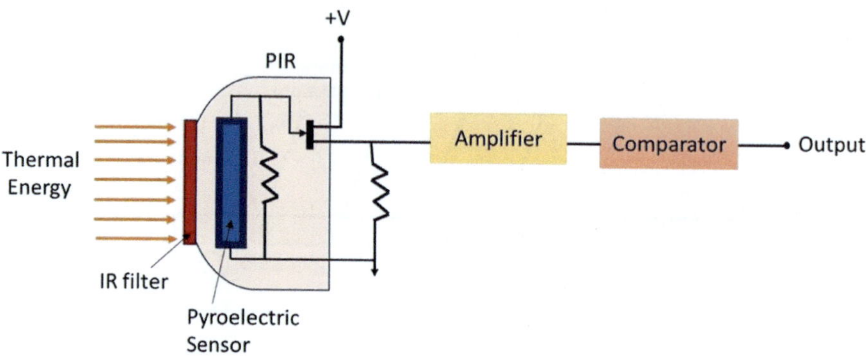

Figure 14.4 Schematic of a pyroelectric sensor to measure the dynamic temperature.

this type of detector to produce a bipolar pulse structure. This results in an initial voltage in one direction followed by a voltage in the opposite direction, which makes it challenging to miniaturize the device.

14.5 WEARABLE TEMPERATURE SENSORS

Accurate body temperature monitoring is achieved using wearable temperature sensors. They can be applied to the skin's surface or placed inside the body to detect and measure the heat emitted by the human body. Numerous varieties of wearable temperature sensors exist, each with unique benefits and uses. The next section goes into further detail on a few of them.

The basal temperature of an infant or newborn is commonly measured with wearable or attachable thermometers. The purpose of these thermometers is to measure the temperature of the skin's surface. Simple gadgets are attached to the skull or axilla. On the other hand, basal temperature monitoring systems allow for continuous nighttime temperature recording.

It is customary to continuously measure body temperature during the menstrual cycle in order to track the prime time for conception. The luteal phase ends with a drop in basal body temperature (BBT). A smartphone can become a sophisticated fertility tracker with the help of some smart sensors. The wearer only needs to apply the sensor straight to their skin or wear it with an armband while they sleep. The basal temperature is automatically transferred to a tablet or smartphone when they take it off after waking up.

In addition, the temperature sensors are built inside pacifiers, which have Bluetooth and GPS capabilities and may be linked to a smartphone app. In addition, it may monitor a child's medication history and historical medical data.

14.6 DESIGNS OF WEARABLE TEMPERATURE THERMOMETERS

14.6.1 Wearable flexible sensors

Because the materials used to make these sensors are flexible, they may adapt to the shape of the body. These sensors typically use wireless transmission to send temperature data to a linked device, like a smartwatch or smartphone. They are popular in consumer wearable devices because they are convenient and simple to use. The basal temperature of an infant or newborn is commonly measured with wearable or attachable thermometers (Nag et al., 2017). The purpose of these thermometers is to measure the temperature of the skin's surface. Simple gadgets are attached to the skull or axilla. On the other hand, basal temperature monitoring systems allow for continuous nighttime temperature recording.

It is customary to continuously measure body temperature during the menstrual cycle in order to track the prime time for conception. With the help of the smart sensor Tempdrop TM, a smartphone may become an advanced fertility monitor (Tamura et al., 2018b). The luteal phase ends with a drop in BBT. The wearer only needs to apply the sensor straight to their skin or wear it with an armband while they sleep. The basal temperature is automatically transferred to a tablet or smartphone when they take it off after waking up.

"Ran's Night" (Chen et al., 2009) is a touchable skin thermometer that has been invented and used to measure the temperature of skin beneath garments at night. With a temperature range of 32°C–40°C, the device's accuracy is ±0.5°C. When you sleep, the sensor – which is fastened around your abdomen – measures your skin's temperature every ten minutes. The device's LCD displays the measured temperature data as a two-dimensional picture. These 2D coded images are captured by a mobile phone, which then decodes the data and sends it to a database server.

A Velcro bandage is used to wear the iSense thermometer (Tamura et al., 2018a). It features a 0.1°C relative precision, an operating range of 30°C–45°C, and a rechargeable battery that enables 40 hours of continuous use. Figure 14.5 shows some of the wearable thermometers commonly used.

The temperature readings from the BBT thermometer earbud are reliable and independent of the surrounding air temperature. Numerous studies on taking children's temperatures have been conducted in both inpatient and outpatient settings, using various thermometers and techniques.

Additionally, pacifiers have temperature sensors implanted in them. The Pacif-I (Figure 14.5) is a pacifier that is connected to a smartphone app and has Bluetooth and GPS capabilities. In addition, it may monitor a child's medication history and historical medical data (Memon et al., 2020).

The choice of materials is essential for creating a flexible, wearable temperature sensor. It is preferred that the materials utilized to produce

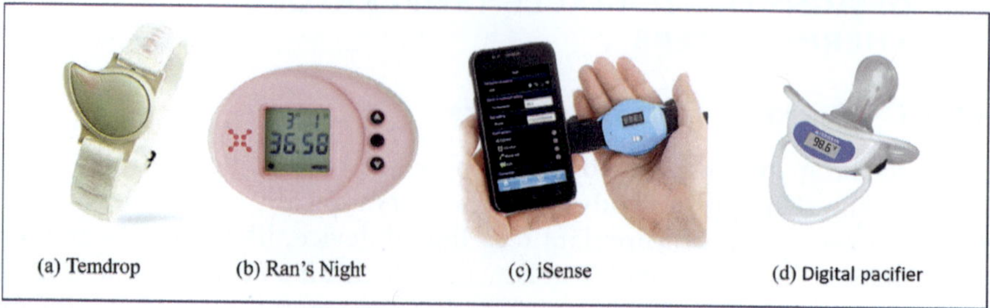

Figure 14.5 Various types of wearable thermometers.

wearable biosensors have qualities that match those of the underlying substrate and the other materials that make up the sensor's structure. In general, the materials must adapt to the non-planar surface without experiencing appreciable physical or sensory response degradation. Moreover, they must also be biocompatible, portable, lightweight, soft, flexible, and stretchable.

14.6.2 Printed temperature sensors

The qualities required by the temperature sensors generated by printing technologies vary depending on the type of fabrication. Colloidal or nano-composite materials are typically used in printing, with the rheological properties of the solution modified to fit the specific printing technique. When selecting a printing technique, some of the key factors taken into account are the concentration of nanoparticles in the solution, spreading coefficient, work of adhesion, surface tension, and solution viscosity. Since printed wearable temperature sensors are designed as either thermistors or RTDs, the sensor structure is constructed using both intrinsic conducting materials and temperature sensing semiconductors. For various printing methods, a large range of materials with the required solution properties are produced (Khan et al., 2023).

The thermal coefficient of resistance (TCR) is utilized to calculate the associated temperature fluctuations since printed resistor thermodynamics (RTDs) operate on the premise of altering resistance as temperature rises (Davaji et al., 2017; Mehmood et al., 2018). For RTDs, printing the required structure with conducting nanoparticle ink alone completes the sensor in a single step. As a result, the process of creating RTD is simple, and using printing technology to create it all at once has several benefits. However, since thermistor is composed of two distinct materials, the production process requires two printing cycles to be completed. The device's initial layer, known as interdigital electrodes (IDEs), is printed with metallic ink (Dan & Elias, 2020). Thermally sensing material is applied using printing or thin film coating technology to the active area covered by IDEs

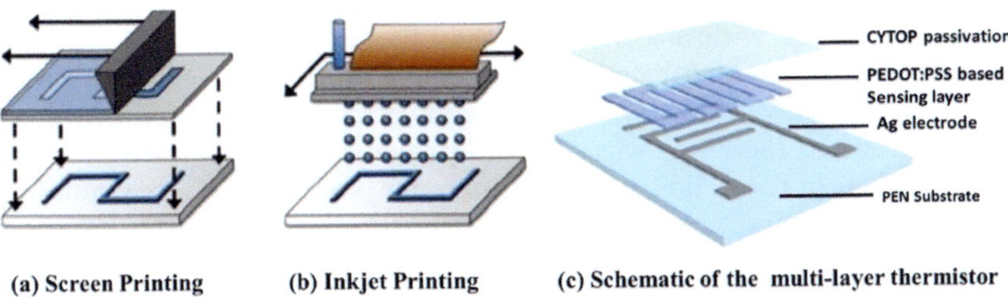

Figure 14.6 (a) Screen printing technology (b) inkjet printing technology and (c) schematic of the multi-layer thermistor. (Adapted from Khan et al. (2023), Open Access Under a Creative Commons License.)

in the second printing cycle. The investigation of thermistors and RTDS has advanced significantly in recent years thanks to the examination of many materials and production methods. The temperature sensing material of a unique type of thermistor, shown schematically and in a printed model in Figure 14.6, is patterned instead of occupying the entire effective zone surrounded by the IDEs. This type of sensor is claimed to be far more effective than ordinary thermistors, in which the entire sensing layer is coated as a thin film. High-resolution Ag patterning is achieved in meander on a cellulose substrate that is biocompatible by the application of inkjet printing technique.

A whetstone bridge is designed using an all-printing method for the patterned deposition of PEDOT-PSS and carbon ink (Wang et al., 2020). Similarly, the TCR value of inkjet printed sensors made of a carbon and PEDOT:PSS mixture is approximately 0.25%/°C. For the creation of a skin-based temperature sensor, a 3:1 wt% mixture of CNT ink and PEDOT:PSS was used to create an inkjet printable ink. There have been reports of a wearable temperature sensor that resembles a human wristband. This temperature sensor is made using PEDOT:PSS inkjet printing. Stretchable fabric is impregnated with a nanocomposite of Ag nanoparticles, PEDOT:PSS, and graphene ink by screen printing in order to create a wearable, self-powered temperature sensor. This kind of sensor is said to have many appealing qualities, including stretchability, high flexibility, lightweight, and ultra-thinness, which allow the sensor to make conformal contact with human skin (Jung et al., 2018).

14.6.3 Adhesive thermometers

Adhesive thermometers come in three varieties: tattoo temperature sensors, throwaway sensors, and a wearable sensor mounted on adhesive plaster. Owing to their ease of handling, adhesive thermometers plays an important role in the field of wearable thermometers.

Earlier, thermometer patches were made of rubber and utilized analog electronics along with a low-power microprocessor. They were made to transmit the data with tablet or smartphone with the help of an integrated radio. Generally, very low power (in microwatts) was dissipated by microprocessor when the radio was off, and power of few milliwatts was dissipated when in use. Thus, a lithium-ion coin battery with a few hundred milliampere-hours of capacity can power the patch for reasonable duration.

Later models of these thermometers begin to sense temperature via thick film technology. Using planar circuit board technology, thick film electrodes are easily produced on textile materials. Steel powder is used in the synthesis process to lower contact resistance and enable textiles and sensor films to move freely. A customizable gain and bandwidth amplifier can handle a wide range of dynamic signal conditions. For the sensor to make skin contact, it must have enough flexibility. The flexible printed circuit board is sensitive enough to monitor vital signs even if it is too rigid to adhere directly to the skin. As a result, the skin and circuit board are separated by a tiny, sticky area. Many temperature sensor patches that are sold commercially use this kind of fabrication. Figure 14.7 shows various disposable electrical temperature sensor patches.

FeverFrida TM thermometer (Tamura et al., 2018b) is a good example of adhesive thermometers. The Fever-FridaTM thermometer satisfies ASTME1112-00 standards for accuracy. Similar specifications apply to Feversmart TM (Perego, 2019), yet no particulars about the device are known. The STEMP sensor, which measures body temperature continuously and instantly using medical-grade adhesives, integrates smoothly with a smartphone application (Figure 14.7a). The TempTraqTM, a wearable temperature device with Bluetooth capability, is a cushiony and cozy patch (Parent et al., 2017).

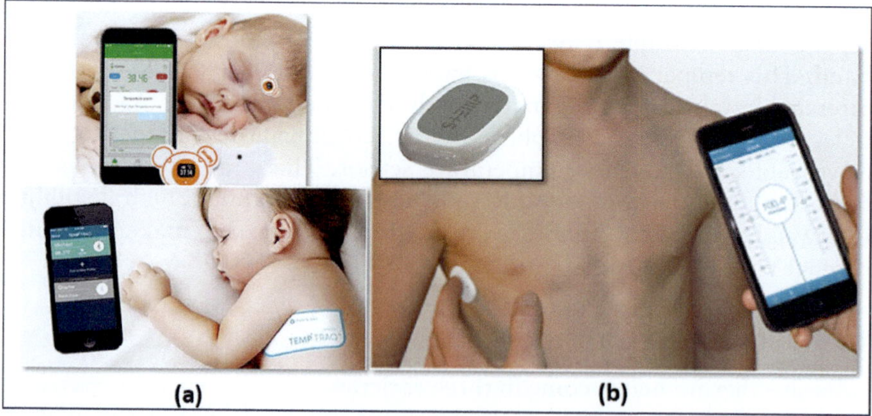

(a)   (b)

Figure 14.7 (a) Disposable patch type wearable thermometer and (b) rechargeable STEMP sensor.

The STEMP sensor is an adhesive patch that is placed on the skin behind the armpit and contains a rechargeable temperature sensor (Figure 14.7b). It gathers information, spots patterns, and makes deft choices over the course of several hours, days, or even weeks. Once fully charged it can be used up to 30 days. Additionally, children with sensitive skin can safely use the medical-grade adhesive, which is similar to that found in high-end bandages.

14.6.3.1 Temporary transfer tattoo temperature monitor

Rogers et al.'s (Kim et al., 2011) groundbreaking work on electronic materials and hybrid fabrication processes has paved the way for the creation of sophisticated electronic devices that can assess physiological factors, such as skin temperature, by contacting the epidermis. An artificial skin or other biocompatible material is used to install the sensor circuit. Platinum resistors, which are register-based temperature sensors, are employed. These incredibly thin rubbery sheets are perforated to let skin ease and breathe in its own way. The skin-mounted sensors measure temperature with millikelvin accuracy. These devices' prototypes demonstrated their ability to collect incredibly sensitive data on therapy-relevant parameters like skin moisture and blood flow (Bandodkar et al., 2015).

Continuous, precise heat measurements can be obtained by pliably adhering an ultrathin, skin-like sensor/actuator to the epidermis. Prototypes of these devices showed that they could gather extremely sensitive data on aspects related to therapy, such as blood flow and skin moisture. The fundamental concepts of operation were established by theoretical and experimental research, which also set engineering standards for device design. Two notable instances of functionality are the accurate evaluation of skin moisture using thermal conductivity measures and the observation of minute fluctuations in skin temperature related to physical activity, mental activity, and vasoconstriction/dilation.

There are now two varieties of temperature sensors available. Initially, sets of temperature sensors are built and set up according to the temperature coefficient of resistance (TCR) in thin, narrow gold sheets made using microlithographic methods. In the second, PIN diodes are created by patterning Si nanomembrane doping, and these are then used to construct multiplexed sensor arrays. When the temperature changes, there are discernible variations in the voltage levels. The devices can function as temperature sensors, local microscale heaters, or both at the same time in each scenario. Moreover, the sensor circuit monitors temperature, humidity, and touch simultaneously when it is put into a biocompatible material, such as artificial skin. The sensitivity of mono-layer-capped nanoparticle (MCNP) films to temperature, humidity, and pressure is high. These sensors have 1°C of resolution, and good accuracy of 95%.

14.6.4 Radiation thermometer

Wearable thermometers can use microwave and infrared radiation. An integrated silicon sensor-based skin-contact thermography wearable has been created. After simulating the responses of three different sensor types – thermocouple, passive temperature coefficient, and integrated silicon – integrated silicon sensors were chosen.

The scientists next tested the device's effectiveness in a medical context by subjecting subjects to a maximum rate of skin temperature variation of 3.1°C/0.25 hours. We contrasted the results with those from a digital video camera that uses infrared. The biggest difference was less than 0.14°C. An infrared camera integrated into a smartphone was used to detect fever (Figure 14.8). The camera can read the infrared camera output signal, detect the surface of the forehead automatically in less than a second, compute the internal body temperature with clinical accuracy, and display the result on the screen. The working range of 35°C – 42°C, with an accuracy of 0.1°C, meets the ISO criterion for clinical thermometers.

An appealing approach to internal thermometry is microwave radiometry, which is used in a wearable gadget that can record, store, and send temperature readings to a computerized medical record while continuously monitoring the temperature of various body areas. The probe is conformal and is very light in mass because it is made on a flexible substrate. Relative temperature differences can be measured with a sensitivity of 0.2 K and an inaccuracy of less than 0.5 K.

Withings Thermo uses the temporal artery at the side of the skull (Mah et al., 2021), a non-invasive body temperature monitoring device, to collect data from 16 separate infrared sensors. Withings claims that this "HotSpot Sensor Technology" can take thousands of measurements in two seconds and can measure infrared signals produced by heat very quickly. After a carefully formulated algorithm accounts for biases like ambient temperature

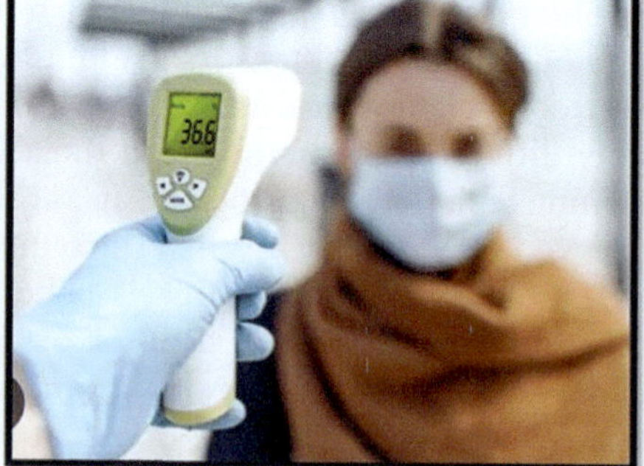

Figure 14.8 (a) Infrared (IR) camera embedded in a smartphone and (b) a forehead IR thermometer.

and heat loss, the temporal artery's corresponding hottest spot is identified to produce an accurate reading. Placing the gadget on the forehead allows for the detection of temperature. The device's button must be clicked to begin a temperature measurement, and vibration alerts users when it's finished. The device's LED shows the temperature along with feedback that is color-coded: green indicates a normal temperature, orange indicates a moderate temperature, and red indicates an excessive temperature (Ring et al., 2009).

Thermo temperature readings are immediately synchronized via Bluetooth or Wi-Fi with an iOS smartphone, where they are saved in the companion app. This makes it possible for parents to monitor long-term trends in their child's temperature. Additionally, a doctor can access the gathered data anytime anywhere. The thermo device runs on a single pair of two AAA batteries, which can last up to two years. The US Food and Drug Administration (FDA) has authorized it as a Class IIa medical device.

14.6.4.1 Deep body thermometers

Deep body thermometers (DBT) are used to monitor core temperature non-invasively. Based on zero heat flow theory, Fox and Solman invented first DBT in the 1970s. A cutaneous probe is heated until it loses all internal temperature gradients, which prevents heat from the skin and, consequently, from the inside of the body from reaching the skin and then will coincide with the deep body temperature measured by the probe as shown in Figure 14.9.

Later on, Terumo Co. (Tokyo, Japan) upgraded this technique. It has been demonstrated that the thermometer accurately measures blood temperature during cardiac surgery. This thermometer compensates for heat flow with a heater, but its high power requirements make it unsuitable for long-term usage as a wearable thermometer (Yamakage et al., 2002).

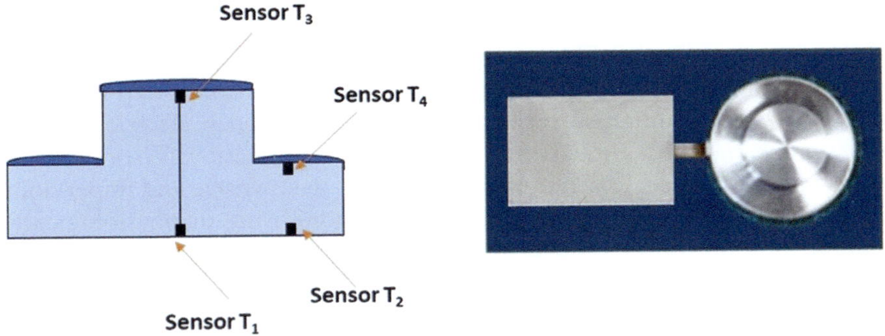

Figure 14.9 (a) Schematic and (b) protype of deep body thermometer monitor.

A wearable DBT is based on dual heat-flux method (DHFM). Kitamura et al. (2010) introduced DHFM, a relatively novel technology that determines the DBT based on the heat flux inside a probe. The temperature sensors can determine the DBT because of the double heat path inside the probe. The substrate material that comprises the core of the probe has four embedded temperature sensors. The majority of the heat transported from the body's core via the skin-to-DBT temperature differential will be absorbed by the substrate material when the DBT is in close proximity to the skin, as it shares physical features with the skin. Heat will also flow longitudinally if there is a peripheral border condition for heat isolation.

With reference to a DBT, a comparative analysis was conducted. Core Temp is a standard thermometer that can be used to measure tissue temperature which is 10 mm deep inside the skin surface, which allows for the monitoring of DBT (Matsukawa et al., 1996). When matched with the reference thermometer, the measurements from the DHFM device prototype varied by less than 0.1°C. Huang et al. (2015) refined this technique through experimental validation and theoretical modeling. Eliminating the external heater significantly lowers the device's power consumption, making it good candidate for wearable temperature sensor.

14.7 OPTICAL SENSORS

The development of medical diagnostic technologies is increasingly reliant on optical sensors. Optical sensors do not always need to make direct contact with the human body or demand a high degree of contact quality. Nowadays, optical sensors are getting a lot of attention and are starting to replace more conventional electrical or mechanical sensors. They have outstanding metrological properties such as low sensitivity and low zero drift, sensitivity, good accuracy, and large usable bandwidth. They can accomplish exceptional miniaturization, are protected from electromagnetic interference, and are able to capture nanoscale volumes, which enables the non-invasive study of biological matter at quite deep penetrations. They are highly useful for measuring the human body's physiology. Photoplethysmography (PPG), radiation, biochemical, and optical fibre sensors are examples of optical approaches. Optical sensors provide simple miniaturization, immunity to electromagnetic interference, electrical safety, great metrological qualities, and non-invasive examination. They can also collect nanometer-scale volumes. They are also inexpensive and impervious to corrosion and water. With the constant monitoring of physiologically significant indicators provided by optical sensors, healthcare expenditures can be reduced. An athlete's performance, for instance, has improved. Developing remote biomedical imaging will be critical in the future.

Optical sensors can be used in many different kinds of devices to measure a broad variety of physiological data. They are the telemedicine of the future and are highly sought after in wearable electronics because of

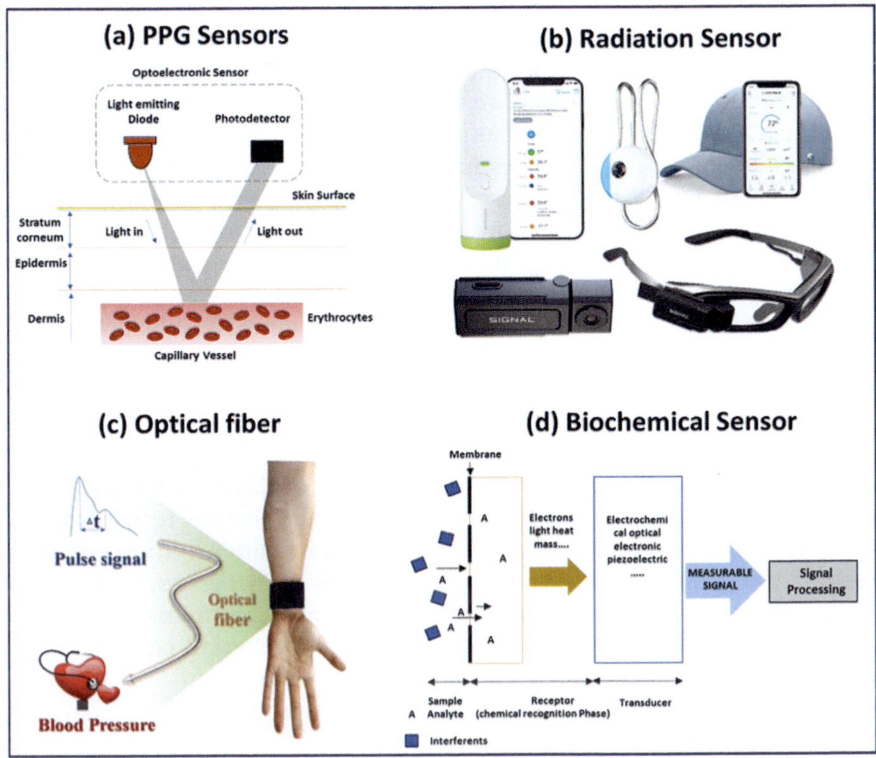

Figure 14.10 Wearable optical sensors for measuring human physiology: (a) PPG sensor (b) radiation sensor (c) optical fiber and (d) biochemical sensor. (Adapted from Vavrinsky et al. (2022), Open Access.)

their many advantages. Optical bio-sensors have some limitations that prevent them from being widely used in the biosensing field. These include the small penetration depth into the skin and other bioliquids, as well as the interference of ambient light with signal readings. Particularly with electrochemical sensors, passivation and biological contamination of the electrode can cause an inaccurate response, making sensor stability a potential issue. However, a common issue with optical fiber sensors is the possibility of system damage during field placement or operation, which needs to be managed and calls for more study in this area (Figure 14.10).

14.8 CHALLENGES AND FUTURE OUTLOOK

Most of the ideas and methods presented here are based on either directly adhering the polymeric substrate to human skin or employing a bi-adhesive tape as a temporary solution for installing a wearable module. However, there are issues with biocompatibility; it may irritate a big portion of the skin or be toxic to some metallic conductors, among other things. Long-term exposure to potentially harmful and cancer-causing substances on or near human skin requires careful research, as does the search for alternative power sources that are safe for use on human skin. Signal display: Another difficult issue is making connections with the printed devices and circuits. Conducting epoxies have been used extensively in practice to connect printed pads; nevertheless, these epoxies cannot be applied because biocompatible substrates have lower glass transition temperatures.

For laboratory-level testing, it would be possible to establish a regular physical contact; nevertheless, trustworthy connections must be investigated for real-time applications. The contact resistance must be low to prevent small fluctuations caused by improper interactions with the pads from producing inaccurate data, particularly when the data is generated by the temperature of the human body.

Most sensors come into direct contact with the human forehead, chest, or forearm/wrist. Due to the constant exposure of these body parts to the outside environment, it is highly desirable for the sensors to be properly encapsulated and operate under appropriate conditions. Biocompatible materials and breathable substrates present further challenges, particularly in the context of wearable sensing applications. Since the human body goes through a number of biological processes and the materials can interact in different ways, it is challenging to predict how one's skin or tissue would interact with the materials used for mounting sensors. While the biocompatibility difficulties would be resolved by using natural biomaterials, it would be extremely difficult to replicate the comparable sensing capabilities achieved with conventional materials.

The most widely used production method to date is inkjet printing, which is frequently used to deposit the temperature sensor layer. Because there is reduced material waste and easier manufacture through localized material deposition, these manufacturing procedures are said to be cost-effective. The most challenging elements to successfully develop wearable electronic systems are accuracy, stability, repeatability, long-term reliability, large-scale production at reduced costs, etc. The temperature limitations of biocompatible substrates, which are composed of solution-based nanomaterials, prohibit the printed patterns from undergoing a higher degree of thermal annealing. This is essential to the devices' resilience because insufficient sintering can lead to printed structures that delaminate at lower bending angles and unpredictable sensor responses.

Since most printed materials are amorphous, they are less appealing when it comes to precise reactions to minute instantaneous changes, particularly when caused by temperature variations in the human body. The layers may break due to physiological movements or localized delamination under greater stress conditions. This alters the base resistance and has a major impact on the sensor's repeatability and dependability. The literature has documented that sensors' performance deteriorates following specific bending cycle tests. Therefore, it is crucial to look into materials that function with polymeric substrates and that, even after an endless number of bending cycles, reliably produce signals within the same operating window with very slight deviations. The aging effect of organic compounds is another significant issue that requires careful consideration. In these situations, encapsulation is the best option for protecting the printed sensing layers, which are vulnerable to changing weather. Nevertheless, materials and a thin enough covering are needed so as not to affect the temperature sensor's functionality. It is necessary to ensure that the sensors respond consistently and with negligible variance at close ranges. Additionally, the sensors must be calibrated in accordance with the specified environment and tested in a variety of climatic situations.

It is imperative to do a thorough analysis and comparison of the several types of sensors that can be applied to human skin, including hand bands, body-worn patches, and implantable situations. These sensors should be evaluated for their functionality, user acceptability, data reproducibility and dependability, and most significantly, ease of use. A pragmatic strategy is required to create a wearable sensory system that is robust, user-friendly, clinically viable, and acceptable.

REFERENCES

Bandodkar, A. J., Jia, W., & Wang, J. (2015). Tattoo-based wearable electrochemical devices: a review. *Electroanalysis*, 27(3), 562–572.

Bhelkar, V., & Shedge, D. K. (2017). Different types of wearable sensors and health monitoring systems: a survey. *Proceedings of the 2016 2nd International Conference on Applied and Theoretical Computing and Communication Technology, ICATccT 2016*, 43–48. https://doi.org/10.1109/ICATCCT.2016.7911963.

Chen, W., Kitazawa, M., & Togawa, T. (2009). Estimation of the biphasic property in a female's menstrual cycle from cutaneous temperature measured during sleep. *Annals of Biomedical Engineering*, 37(9), 1827–1838. https://doi.org/10.1007/s10439-009-9746-6.

Dan, L., & Elias, A. L. (2020). Flexible and stretchable temperature sensors fabricated using solution-processable conductive polymer composites. *Advanced Healthcare Materials*, 9(16), 1–13. https://doi.org/10.1002/adhm.202000380.

Davaji, B., Cho, H. D., Malakoutian, M., Lee, J. K., Panin, G., Kang, T. W., & Lee, C. H. (2017). A patterned single layer graphene resistance temperature sensor. *Scientific Reports*, 7(1), 1–10. https://doi.org/10.1038/s41598-017-08967-y.

Fletcher, T., Whittam, A., Simpson, R., & Machin, G. (2018). Comparison of non-contact infrared skin thermometers. *Journal of Medical Engineering and Technology*, 42(2), 65–71. https://doi.org/10.1080/03091902.2017.1409818.

Huang, M., Tamura, T., Chen, W., & Kanaya, S. (2015). Evaluation of structural and thermophysical effects on the measurement accuracy of deep body thermometers based on dual-heat-flux method. *Journal of Thermal Biology*, 47, 26–31.

Itoigawa, K., Ueno, H., Shiozaki, M., Toriyama, T., & Sugiyama, S. (2005). Fabrication of flexible thermopile generator. *Journal of Micromechanics and Microengineering*, *15*(9). https://doi.org/10.1088/0960-1317/15/9/S10.

Jung, M., Jeon, S., & Bae, J. (2018). Scalable and facile synthesis of stretchable thermoelectric fabric for wearable self-powered temperature sensors. *RSC Advances*, *8*(70), 39992–39999. https://doi.org/10.1039/C8RA06664G.

Khan, S., Ali, S., Khan, A., & Bermak, A. (2023). Wearable printed temperature sensors: short review on latest advances for biomedical applications. *IEEE Reviews in Biomedical Engineering*, *16*, 152–170. https://doi.org/10.1109/RBME.2021.3121480

Kim, D. H., Lu, N., Ma, R., Kim, Y. S., Kim, R. H., Wang, S., Wu, J., Won, S. M., Tao, H., Islam, A., Yu, K. J., Kim, T. I., Chowdhury, R., Ying, M., Xu, L., Li, M., Chung, H. J., Keum, H., McCormick, M., Liu, P., Zhang, Y.-W., Omenetto, F. G., Huang, Y., Coleman, T., & Rogers, J. A. (2011). Epidermal electronics. *Science*, *333*(6044), 838–843. https://doi.org/10.1126/science.1206157

Lee, M. H., Guo, R., & Bhalla, A. S. (1998). Pyroelectric sensors. *Journal of Electroceramics*, *2*(4), 229–242. https://doi.org/10.1023/A:1009922522642

Lenhardt, R., & Sessler, D. I. (2006). Estimation of mean body temperature from mean skin and core temperature. *Anesthesiology*, *105*(6), 1117–1121. https://doi.org/10.1097/00000542-200612000-00011

Mah, A. J., Zadeh, L. G., Tehrani, M. K., Askari, S., Gandjbakhche, A. H., & Shadgan, B. (2021). Studying the accuracy and function of different thermometry techniques for measuring body temperature. *Biology*, *10*(12), 1–15. https://doi.org/10.3390/biology10121327

Matsukawa, T., Kashimoto, S., Ozaki, M., Shindo, S., & Kumazawa, T. (1996). Temperatures measured by a deep body thermometer (Coretemp®) compared with tissue temperatures measured at various depths using needles placed into the sole of the foot. *European Journal of Anaesthesiology*, *13*(4), 340–345.

Mehmood, Z., Mansoor, M., Haneef, I., Ali, S. Z., & Udrea, F. (2018). Evaluation of thin film p-type single crystal silicon for use as a CMOS Resistance Temperature Detector (RTD). *Sensors and Actuators, A: Physical*, *283*, 159–168. https://doi.org/10.1016/j.sna.2018.09.062

Memon, S. F., Memon, M., & Bhatti, S. (2020). Wearable technology for infant health monitoring: a survey. *IET Circuits, Devices & Systems*, *14*(2), 115–129. https://doi.org/10.1049/iet-cds.2018.5447

Nag, A., Mukhopadhyay, S. C., & Kosel, J. (2017). Wearable flexible sensors: a review. *IEEE Sensors Journal*, *17*(13), 3949–3960. https://doi.org/10.1109/JSEN.2017.2705700

Parent, B. J., Pachucki, B. M., Ooyama-searls, R. M., Ooyama-searls, R., & Parent, B. (2017). *A Pressure Ulcer Patch Material Study for a Wearable Sensor*. Engineering, Medicine, Materials Science. April 2017.

Perego, P. (2019). Device for mHealth. In G. Andreoni, P. Perego, & E. Frumento (Eds.), M_Health Current and Future Applications, (pp. 87–99). Springer: Cham.

Ring, E. F. J., Thomas, R., Howell, K., & Jones, D. P. (2009). Sensors for medical thermography and infrared radiation measurements. In D. P. Jones (Ed.), *Biomedical Sensors* (pp. 417–441). Momentum: New York.

Tamura, T., Huang, M., & Togawa, T. (2018a). Body temperature, heat flow, and evaporation. In T. Tamura & W. Chen (Eds.), *Seamless Healthcare Monitoring: Advancements in Wearable, Attachable, and Invisible Devices* (pp. 281–307). Springer International Publishing. https://doi.org/10.1007/978-3-319-69362-0_10

Tamura, T., Huang, M., & Togawa, T. (2018b). Current developments in wearable thermometers. *Advanced Biomedical Engineering*, 7, 88–99. https://doi.org/10.14326/abe.7.88

Taylor, N. A. S., Tipton, M. J., & Kenny, G. P. (2014). Considerations for the measurement of core, skin and mean body temperatures. *Journal of Thermal Biology*, 46, 72–101. https://doi.org/10.1016/j.jtherbio.2014.10.006

Vavrinsky, E., Esfahani, N. E., Hausner, M., Kuzma, A., Rezo, V., Donoval, M., & Kosnacova, H. (2022). The current state of optical sensors in medical wearables. *Biosensors*, 12(4), 1–40. https://doi.org/10.3390/bios12040217

Wang, Y. F., Sekine, T., Takeda, Y., Yokosawa, K., Matsui, H., Kumaki, D., Shiba, T., Nishikawa, T., & Tokito, S. (2020). Fully printed PEDOT:PSS-based temperature sensor with high humidity stability for wireless healthcare monitoring. *Scientific Reports*, 10(1), 1–8. https://doi.org/10.1038/s41598-020-59432-2

Yamakage, M., Iwasaki, S., & Namiki, A. (2002). Evaluation of a newly developed monitor of deep body temperature. *Journal of Anesthesia*, 16, 354–357.

Kitamura, K.I., Zhu, X., Chen, W., Nemoto, T. (2010). Development of a new method for the noninvasive measurement of deep body temperature without a heater. Medical Engineering & Physics, 32, 1-6.

Luminescent temperature sensors for environmental monitoring

Shubham Raina and Richa Kothari

15.1 INTRODUCTION

Temperature is an important abiotic factor of the environment that is associated with every aspect of our lives. It significantly affects different ecosystems, biodiversity, health, and more. Its measurement is required for diverse applications such as climate monitoring, green infrastructure, agriculture, renewable energy systems, the healthcare sector, food production, and storage units. Also, global climate patterns can be influenced by even slight temperature variations. Altered climate patterns in turn impact rainfall trends and adversely affect crop yield and agriculture patterns. Also, the increased temperature can decrease dissolved oxygen (DO) in water, impacting aquatic organisms' life. There are different types of temperature sensors available at present time including thermocouples, thermistors, resistance temperature detectors (RTDs), infrared sensors, semiconductor-based sensors, and luminescent temperature sensors. Each of the sensors has different merits and demerits. Among them, an advanced category of instruments used to measure temperature through luminescence phenomena is known as luminescent temperature sensors. Luminescent temperature sensors use the special optical characteristics of luminescent materials to identify temperature changes, as opposed to the more conventional methods that depend on mechanical or electrical principles. One of the main principles behind these sensors is luminescence, which is the ability of certain materials to emit light when excited. One of the major advantages of using luminescent temperature sensors is their ability to deliver non-contact temperature readings. When taking a measurement requires no or very little physical touch with the item under consideration, this becomes even more useful. Using an external light source to illuminate the target material and then analysing its luminescent response allows these sensors to determine the temperature with minimum interference. Applications necessitating real-time temperature monitoring can benefit from luminous temperature sensors due to their quick response times.

DOI: 10.1201/9781032661537-15

15.2 WORKING OF LUMINESCENT TEMPERATURE SENSORS

As luminescence is the emission of light, temperature sensors based on luminescence analyse the intensity of the emitted light to determine the temperature of a substance. Unlike conventional temperature sensors, luminescent sensors obtain temperature measurements by utilising the emission of light from specially designed materials. These substances, known as luminophores, exhibit a fascinating property known as luminescence. The procedure involves luminescence phenomena, such as fluorescence and phosphorescence.

Certain substances can absorb light at a particular wavelength and then emit it at a different wavelength through a process known as fluorescence. The intensity of the emitted light is influenced by the temperature of the substance. Fluorescent materials can be used to create phasic temperature sensors. Phosphorescence is similar to fluorescence, with the only difference being a slight delay in the emission of light. Some materials display a phenomenon where they continue to emit light even after the source of energy that excited them has been turned off. The latency can exhibit variations depending on temperature.

The temperature is determined by analysing the intensity or decay time of the emitted light in both situations. Temperature sensors that utilise luminescence have proven to be highly valuable in a wide range of applications, including both industrial and scientific settings.

In the excitation phase, an external energy source, such as ultraviolet (UV) or near-infrared (NIR) light, is used to illuminate the luminescent material. The light stimulates the electrons in the luminophore molecules, propelling them to higher energy levels.

The phase of emission: In a typical material, these electrons would release their energy as heat. However, the energy transforms and is ultimately manifested as a specific colour of light by a luminophore. This is the level of brightness. After being emitted, the light is captured by a photodetector.

The measurement of temperature relies on the luminescence properties, which vary with temperature changes. As the temperature rises, the interactions between excited electrons and the surrounding elements change. The emitted light's lifetime, colour, and intensity are influenced as a result. Figure 15.1 represents the working of luminescent temperature sensors.

15.3 COMPARISON BETWEEN LUMINESCENT TEMPERATURE SENSORS AND THERMISTOR SENSORS

Luminescent temperature sensors and thermistor sensors are different categories of temperature sensors employed in environmental monitoring applications. Luminescent temperature sensors utilise the phenomenon of

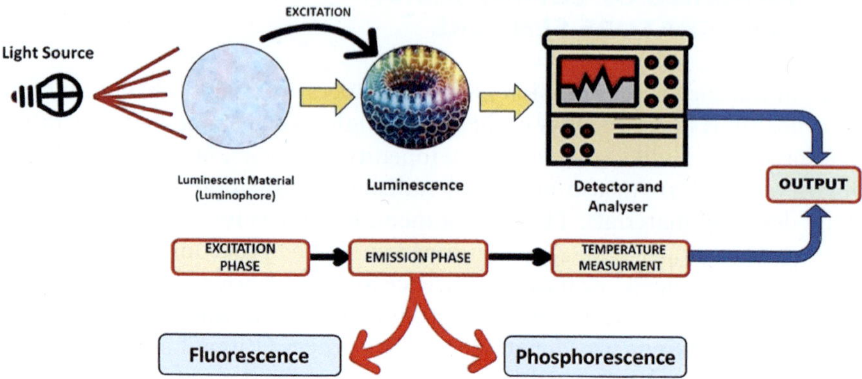

Figure 15.1 Working of luminescent temperature sensors.

luminescence to measure temperature, where the intensity or wavelength of emitted light changes with temperature (Guimaraes et al., 2020). On the other hand, thermistor sensors depend on the change in electrical resistance with temperature (Meijer et al., 2018) to provide accurate temperature readings. Both types of sensors offer unique advantages and limitations that make them suitable for specific environmental monitoring scenarios. Understanding these differences is crucial for selecting the most appropriate sensor for a given application. Luminescent temperature sensors operate by detecting changes in fluorescence emissions that are dependent on temperature. These sensors utilise temperature-sensitive materials such as phosphors or quantum dots, which exhibit varying light properties in response to temperature changes. As the temperature increases, the produced light undergoes a change in intensity or wavelength, which enables more precise temperature measurement.

Luminescent temperature sensors offer notable benefits compared to conventional temperature sensors, such as enhanced capacity for temperature monitoring across various applications. An important benefit of these devices is their ability to perform temperature measurements without physical contact, which is particularly useful in situations where direct touch is neither feasible nor preferred. This characteristic is highly advantageous in fields such as semiconductor manufacturing and medical research. An additional benefit is the ability to use multiplexing, where multiple luminescent sensors with different emission spectra can be used simultaneously, enabling the measurement of various parameters in a single experiment. This feature enhances the efficiency of environmental monitoring systems. Luminescent sensors possess high sensitivity and accuracy, rendering them well-suited for applications necessitating meticulous temperature monitoring, such as climate research and industrial processes.

Thermistor sensors, on the other hand, depend on the resistance of semiconductor materials, which varies with temperature. Thermistors exhibit an exponential change in resistance in response to temperature variations, enabling accurate temperature monitoring. Thermistors are affordable, minuscule, and exhibit rapid responsiveness to fluctuations in temperature. They are extensively utilised in various environmental monitoring applications, such as weather stations and HVAC systems. Although light sensors provide exceptional precision, they may come with disadvantages such as higher expenses and the requirement for specialised equipment. Thermistors, in comparison, are more affordable but may exhibit lower precision than light sensors, particularly in extreme temperature ranges.

The choice between luminous temperature sensors and thermistors ultimately depends on the unique requirements of the environmental monitoring application. The thermistors serve as a cost-effective and practical substitute for general temperature monitoring needs. However, luminous temperature sensors surpass conventional temperature sensors in specific scientific and industrial applications due to their non-contact nature, high sensitivity, accuracy, and multiplexing capabilities (Table 15.1).

Table 15.1 Comparison between luminescent temperature sensors and thermistor sensors

S. No.	Property	Luminescent temperature sensors	Thermistor sensors
1	Sensing mechanism	Based on luminescent materials	Depends on the change in resistance
2	Working principle	Temperature-dependent luminescence	Temperature-dependent resistance
3	Cost	High fabrication cost	Generally cost-effective
4	Sensitivity	High sensitivity	Moderate sensitivity
5	Response time	Generally fast	Fast
6	Durability	Susceptible to photobleaching	Highly durable
7	Size and form factor	Can be miniaturised	Compact, but less miniaturised
8	Intrusiveness	Contact/contactless	Intrusive (requires contact)
9	Cross-sensitivity	Can be designed to sense multiple environmental factors	Generally specific to temperature
10	Application sectors	Environmental monitoring, precision agriculture, climate healthcare, etc.	Temperature measurement in electronic devices, industrial processes, laboratory apparatus, etc.

15.4 APPLICATIONS

Luminescent temperature sensors can be applied over a varied range including industrial processes and quality control, biomedical research and healthcare, climate research and air pollution monitoring, food and pharmaceutical storage, energy efficiency in buildings, aquaculture, and a lot more. Figure 15.2 reflects the different applications of luminescent temperature sensors in environmental monitoring.

15.4.1 Industrial processes and quality control

Luminescent temperature sensors have wide applicability in industrial processes where precise temperature regulation is required. The luminous sensors have great sensitivity and accuracy; therefore, they are suitable for industries such as semiconductor manufacturing units. This is particularly important as even slight fluctuations in temperature can have a significant effect on the quality of the final product. The non-contact design of these sensors offers ideal production conditions and safeguards delicate manufacturing processes.

15.4.2 Biomedical research and healthcare

In biomedical research and healthcare (Rosal et al., 2016), luminescent temperature sensors play a vital role in monitoring and controlling temperature during experiments and medical procedures. For instance, in cellular and

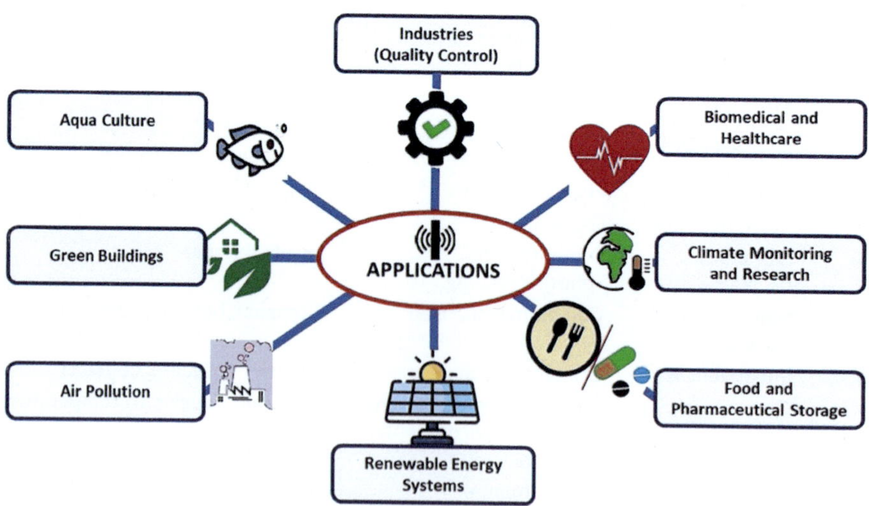

Figure 15.2 Applications of luminescent temperature sensors.

molecular biology research, maintaining specific temperatures is essential for cell cultures and enzymatic reactions. Luminescent sensors provide a non-invasive method for real-time temperature monitoring, contributing to the accuracy of experimental results. For example, a life-threatening disease known as ischemia induces a shortage of different nutrients, oxygen, and glucose, in the body cells. This results in reduced blood flow to different tissues of the body. Further, this condition results in the accumulation of metabolic wastes in the body. Previously, ischemia was detected using infrared thermal cameras, but with progress, now we can detect it using luminescence thermometry. The benefit of using luminescence thermometry over infrared thermal cameras is its potential to detect deeper tissues along with superficial tissues, which was not possible with infrared thermal cameras (Dramićanin, 2018).

Additionally, in medical applications, such as hyperthermia treatments, luminescent sensors enable precise temperature measurements in targeted areas, ensuring the safety and efficacy of therapeutic interventions. In the context of brain activity monitoring, luminescent thermometry presents a minimally invasive approach to measuring brain temperature in freely moving animals, addressing the limitations of traditional methods. Furthermore, the use of luminescent nanothermometers has been demonstrated for reliable and remote monitoring of absolute temperature during liver inflammation, showcasing their potential for biomedical applications (Rodríguez-Sevilla et al., 2022). The non-invasive nature and high sensitivity of these sensors make them promising for various biomedical and healthcare applications, including brain activity monitoring and disease diagnosis (Barbosa et al., 2023).

15.4.3 Climate research and air pollution monitoring

Luminescent temperature sensors can be significant tools in climate research and environmental monitoring. These devices can accurately measure changes in temperature under different environmental conditions and can be easily deployed in such locations. Within the field of climate research, these sensors can be utilised in demanding settings where the utilisation of conventional sensors may be impractical, such as distant areas or harsh landscapes. Luminous sensors are highly effective tools for monitoring wildlife habitats without causing any disruption to the natural environment, as they operate without the need for substantial interaction with the surroundings. The utilisation of these sensors has the potential to offer an enhanced understanding of climatic trends and their effects on various ecosystems. Utilising carbon dots in these sensors can augment their sensing capabilities in various situations, including freshwater ecosystems that are frequently impacted by pollution. Remote aquatic systems and polar ice caps can also be observed and tracked. Techniques such as Multimodal

Non-Contact Luminescence Thermometry rely on temperature-dependent emission qualities to monitor temperature without any physical contact. Chromium-doped oxides have become popular for this purpose (Mykhaylyk et al., 2020).

Utilising luminous temperature sensors in air pollution monitoring can aid in the advancement of efficient air quality control techniques, as temperature fluctuations can impact the dispersion and concentration of pollutants in the atmosphere (Dramićanin, 2020). In addition, the integration of luminous temperature sensors into air quality monitoring systems offers valuable information on temperature changes and enhances the accuracy and dependability of air quality monitoring technologies (Popov et al., 2022). The utilisation of Luminescent Silicon Nanowires (Morganti et al., 2022) also has great potential in the development of efficient sensors that have low energy consumption. Due to their adaptability, they can be utilised for air quality monitoring.

15.4.4 Food and pharmaceutical storage

The food and pharmaceutical sectors rely on maintaining the optimal storage temperature to assure the quality and safety of their products. Luminescent temperature sensors are extremely helpful in environments where accurate and discreet temperature monitoring is required. They guarantee that items with temperature sensitivities are kept under accurate temperature thresholds when incorporated into facilities that house medications (Brites et al., 2023) or perishable goods. This system aids in quality control across several industries by effectively preventing the deterioration or damage of delicate items. These sensors are particularly useful in the food industry, where maintaining the right temperature is crucial to preventing spoilage and ensuring food safety. Additionally, luminescent temperature sensors can also be utilised in laboratories and research facilities to monitor and maintain optimal conditions for experiments and samples. These sensors are also widely used in the food and beverage industry, where maintaining the right temperature is crucial for preserving freshness and preventing spoilage.

15.4.5 Energy efficiency in buildings

Luminescent temperature sensors have diverse applications in renewable energy systems, particularly in building energy efficiency and renewable energy generation. For instance, luminescent solar concentrators (LSCs) based on carbon dots and upconversion nanoparticles have been explored for building integration and renewable energy generation, offering a cost-effective and sustainable approach to energy production.

Additionally, the use of luminescent materials for temperature sensing in LSCs enables multiparametric thermal reading, contributing to the development of autonomous power temperature sensors for building energy applications (Correia et al., 2023). These sensors can provide remote and non-invasive temperature monitoring, enhancing the efficiency and performance of renewable energy systems in buildings. Therefore, the application of luminescent temperature sensors in renewable energy systems, such as LSCs, holds promise for improving energy efficiency and promoting the integration of renewable energy technologies in building and infrastructure applications.

The use of luminescent temperature sensors can significantly improve the energy efficiency of buildings by enhancing heating, ventilation, and air conditioning (HVAC) systems. Luminescent temperature sensors based on LSCs are a promising technology for improving energy efficiency in buildings. LSCs are inexpensive and can be integrated into windows to generate renewable energy (Daigle & O'Brien, 2020). They can also be used for multiparametric thermal reading, as they represent two independent pathways for temperature sensing (Correia et al., 2023). Highly efficient materials, such as carbon dots, can be used to create LSCs with temperature-sensing ability. The emission of these optically active materials shows temperature dependency, which can be used to establish multiple thermometric parameters based on the emission and the electrical (Daigle & O'Brien, 2020). This technology can enable portable optical sensing through multiparametric thermal reading with relative sensitivity, making real-time mobile temperature sensing accessible to all users.

15.5 CHALLENGES

Although luminescent temperature sensors have a wide range of applicability in the environmental field in upcoming times, they suffer many challenges in terms of cost and complexity. Calibration of the sensors, sensitivity to the different temperatures and other environmental factors, as well as material selection and their synthesis and fabrication cost, are also challenging. Figure 15.3 gives an overview of the different challenges associated with the applicability of luminescent temperature sensors in environmental monitoring.

15.5.1 Cost and complexity

One of the major challenges associated with luminescent temperature sensors is the cost and complexity, which can easily limit their implementation to environmental applications. The fabrication cost of the material can

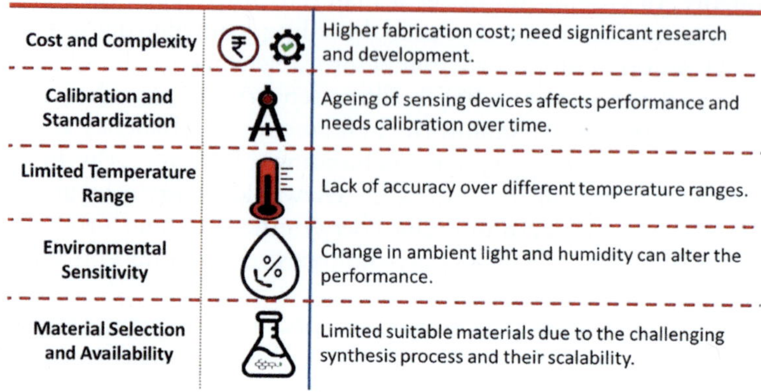

Figure 15.3 Challenges in luminescent temperature sensors for environmental monitoring.

be much higher in comparison to traditional temperature sensors, such as thermistor-based sensors (Dramićanin, 2020). Also, the development of reliable sensing devices needs significant research and development efforts.

15.5.2 Calibration and standardisation

Achieving accurate and consistent measurements with luminescent temperature sensors requires careful calibration and standardisation. The luminescent properties of the sensing materials can be influenced by factors such as environmental conditions and ageing. Therefore, maintaining calibration over time and across different sensors is crucial for ensuring reliable and accurate temperature readings.

Calibration and standardisation present significant challenges to the widespread implementation of luminescent temperature sensors. These sensors require accurate calibration across a wide temperature range, which can be complex and time-consuming. The high relative sensitivities and low-temperature measurement uncertainty of luminescent crossover thermometers highlight the need for precise calibration methods to ensure reliable temperature measurements. Additionally, the development of stable and reliable luminescent materials for temperature sensing, especially at extreme temperatures, requires standardised calibration protocols to validate their performance and accuracy. Furthermore, the integration of luminescent temperature sensors into practical industrial and environmental monitoring systems necessitates standardised calibration procedures to ensure consistent and reliable operation (Brites et al., 2023). For example, small temperature changes in the environment require highly sensitive probes. Highly sensitive materials such as transition metals were previously used for making probes, but currently lanthanide complexes are being in focus for developing probe sensors. But they have stability challenges in aqueous solutions and suspensions. Moreover, certain specimens necessitate UV stimulation, which, particularly in the context of biological matter research, can lead to a substantial presence of fluorescence in the background (Wang et al., 2013). Addressing these calibration and standardisation challenges is crucial for enhancing the accuracy, reliability, and interoperability of luminescent temperature sensors across various applications, including industrial processes, environmental monitoring, and high-temperature measurements.

15.5.3 Limited temperature range

The limited temperature range of luminous temperature sensors poses a significant constraint on the functionality of these devices. Optimal performance of specific bright materials is achieved within specified temperature ranges. If these sensors are used beyond their optimal conditions, they may experience a decrease in accuracy. When it comes to applications that require very high or low temperatures, like cryogenic environments or industrial processes that experience temperature changes, this limitation becomes a significant barrier. Although certain sensors can endure temperatures as high as 1000 K (Runowski et al., 2020), their ability to detect and accurately measure may be diminished in such extreme conditions. Furthermore, several probes are unsuitable for high-temperature environments due to their restricted working ranges. For instance, their effectiveness diminishes beyond 400 K (Getz et al., 2019). The intricate nature of luminous temperature sensors is underscored by their requirement to maintain precision and responsiveness in both exceedingly low and exceedingly high temperatures. It diminishes their efficacy throughout a broad spectrum of temperatures.

15.5.4 Environmental sensitivity

Luminescent temperature sensors can be sensitive to environmental factors such as ambient light and humidity. External light sources may interfere with luminescent signals, affecting the accuracy of temperature measurements. Additionally, changes in humidity levels can impact the properties of luminescent materials (Sekiguchi et al., 2018), requiring additional considerations for applications in diverse environmental conditions.

15.5.5 Material selection and availability

The availability of suitable luminescent materials can be a limitation. Some materials may be rare or challenging to synthesise, limiting the scalability of luminescent sensor production. Researchers face the task of identifying and developing materials that not only exhibit the desired luminescent properties but are also practical for large-scale sensor manufacturing (Stefańska et al., 2023). While luminescent temperature sensors offer numerous advantages, addressing the challenges and limitations is essential for their successful integration into various applications. Ongoing research and technological advancements are necessary to overcome these challenges, improving the reliability, affordability, and versatility of luminescent temperature-sensing technology.

15.6 CONCLUSION

Temperature is an important factor that is needed for environmental monitoring. Luminescent temperature sensors are advanced instruments that have several advantages over conventional sensors such as non-contact applications. They have high sensitivity and accuracy in specific temperature ranges. These sensors have the potential for different environmental applications in industrial processes, quality checks of instruments, agriculture including aquaculture biomedical and climate research, the healthcare sector, air pollution monitoring, and food and pharmaceutical storage. Also, they can be integrated with green buildings for energy efficiency. However, they have several application challenges such as in terms of cost, complexity, limited temperature range, and the availability of suitable luminescent materials. Addressing these challenges is crucial for the successful integration of luminescent temperature-sensing technology into various applications. In the future, there is strong hope for the applicability of luminescent temperature sensors in environmental monitoring sectors.

REFERENCES

Barbosa, I. V., Maia, L. J., Ibanez, A., & Dantelle, G. (2023). Recent developments on BW-II and BW-III ratiometric luminescent nanothermometers for in vivo thermal sensing. *Optical Materials: X, 18*, 100236. https://doi.org/10.1016/j.omx.2023.100236

Brites, C. D. S., Marin, R., Suta, M., Neto, A. N. C., Ximendes, E., Jaque, D., & Carlos, L. D. (2023). Spotlight on luminescence thermometry: basics, challenges, and cutting-edge applications. *Advanced Materials, 35*(36), 2302749.

Correia, S. F., Fu, L., Dias, L. M., Pereira, R. F., de Zea Bermudez, V., André, P. S., & Ferreira, R. A. (2023). An autonomous power temperature sensor based on window-integrated transparent PV using sustainable luminescent carbon dots. *Nanoscale Advances, 5*(13), 3428–3438. https://doi.org/10.1039/D3NA00136A.

Daigle, Q., & O'Brien, P. G. (2020). Heat generated using luminescent solar concentrators for building energy applications. *Energies, 13*(21), 5574. https://www.mdpi.com/1996-1073/13/21/5574.

del Rosal, B., Ximendes, E., Rocha, U., & Jaque, D. (2016). In vivo luminescence nanothermometry: from materials to applications. *Advanced Optical Materials*, 5(1), 1600508. https://doi.org/10.1002/adom.201600508

Dramićanin, M. (2018). Biomedical applications of luminescence thermometry. Miroslav, D. (Ed.) In *Luminescence Thermometry: Methods, Material, and Applications*. Elsevier Science.

Dramićanin, M. D. (2020). Trends in luminescence thermometry. *Journal of Applied Physics*, 128(4), 040902. https://doi.org/10.1063/5.0014825

Getz, M. N., Nilsen, O., & Hansen, P. A. (2019). Sensors for optical thermometry based on luminescence from layered YVO_4: Ln^{3+} (Ln=Nd, Sm, Eu, Dy, Ho, Er, Tm, Yb) thin films made by atomic layer deposition. *Scientific Reports*, 9(1), 10247. https://doi.org/10.1038/s41598-019-46694-8

Guimaraes, L. B., Botas, A. M., Felinto, M. C., Ferreira, R. A., Carlos, L. D., Malta, O. L., & Brito, H. F. (2020). Highly sensitive and precise optical temperature sensors based on new luminescent Tb^{3+}/Eu^{3+} tetrakis complexes with imidazolic counterions. *Materials Advances*, 1(6), 1988–1995. https://doi.org/10.1039/d0ma00201a

Meijer, G. C., Wang, G., & Heidary, A. (2018). Smart temperature sensors and temperature sensor systems. In Nihtianov, S., & Luque, A. (Eds.) *Smart Sensors and MEMs* (pp. 57–85). Woodhead Publishing. https://doi.org/10.1016/B978-0-08-102055-5.00003-6.

Morganti, D., Faro, M. J. L., Leonardi, A. A., Fazio, B., Conoci, S., & Irrera, A. (2022). Luminescent silicon nanowires as novel sensor for environmental air quality control. *Sensors*, 22(22), 8755. https://doi.org/10.3390/s22228755

Mykhaylyk, V., Kraus, H., Zhydachevskyy, Y., Tsiumra, V., Luchechko, A., Wagner, A., & Suchocki, A. (2020). Multimodal non-contact luminescence thermometry with Cr-doped oxides. *Sensors*, 20(18), 5259. https://doi.org/10.3390/s20185259

Popov, A., Timofeyev, M., Bykov, A., & Meglinski, I. (2022). Luminescent upconversion nanoparticles evaluating temperature-induced stress experienced by aquatic organisms owing to environmental variations. *Iscience*, 25(7), 104568. https://doi.org/10.1016/j.isci.2022.104568

Rodríguez-Sevilla, P., Marin, R., Ximendes, E., Del Rosal, B., Benayas, A., & Jaque, D. (2022). Luminescence thermometry for brain activity monitoring: a perspective. *Frontiers in Chemistry*, 10, 941861. https://doi.org/10.3389/fchem.2022.941861

Runowski, M., Woźny, P., Stopikowska, N., Martín, I. R., Lavín, V., & Lis, S. (2020). Luminescent nanothermometer operating at very high temperature—sensing up to 1000 K with upconverting nanoparticles (Yb^{3+}/Tm^{3+}). *ACS Applied Materials & Interfaces*, 12(39), 43933–43941. https://dx.doi.org/10.1021/acsami.0c13011

Sekiguchi, T., Sotoma, S., & Harada, Y. (2018). Fluorescent nanodiamonds as a robust temperature sensor inside a single cell. *Biophysics and Physicobiology*, 15, 229–234. https://doi.org/10.2142/biophysico.15.0_229

Stefańska, D., Kabański, A., Vu, T. H. Q., Adaszyński, M., & Ptak, M. (2023). Structure, luminescence and temperature detection capability of $[C(NH_2)_3]M(HCOO)_3$ (M=Mg^{2+}, Mn^{2+}, Zn^{2+}) hybrid organic–inorganic formate perovskites containing Cr^{3+} ions. *Sensors*, 23(14), 6259. https://doi.org/10.3390/s23146259

Wang, X. D., Wolfbeis, O. S., & Meier, R. J. (2013). Luminescent probes and sensors for temperature. *Chemical Society Reviews*, 42(19), 7834–7869. https://doi.org/10.1039/C3CS60102A

Index